T0215010

Practical Statistics for Astronomers, Second Edition

Astronomy needs statistical methods to interpret data, but statistics is a many-faceted subject that is difficult for non-specialists to access. This handbook helps astronomers analyze the complex data and models of modern astronomy.

This Second Edition has been revised to feature many more examples using Monte Carlo simulations, and now also includes Bayesian inference, Bayes factors and Markov chain Monte Carlo integration. Chapters cover basic probability, correlation analysis, hypothesis testing, Bayesian modelling, time series analysis, luminosity functions and clustering. Exercises at the end of each chapter guide readers through the techniques and tests necessary for most observational investigations. The data tables, solutions to problems, and other resources are available online at www.cambridge.org/9780521732499.

Bringing together the most relevant statistical and probabilistic techniques for use in observational astronomy, this handbook is a practical manual for advanced undergraduate and graduate students and professional astronomers.

JASPER V. WALL is Adjunct Professor in the Department of Physics and Astronomy, University of British Columbia, and Visiting Professor at the University of Oxford, UK.

CHARLES R. JENKINS is a Research Scientist in Earth Sciences and Resource Engineering at the Commonwealth Scientific and Industrial Research Organization (CSIRO), Australia.

Cover insets

Front cover:

A 500-ksec Chandra X-ray image of the Bullet cluster, tracing the bulk of the cluster baryons in the form of the hot intergalactic plasma (Clowe *et al.*, 2006). This gas is centralized and shocked in the collision between the two subclusters, while the dissipationless stars and dark matter (contours) retain the initial shape of their gravitational potentials. Reproduced by kind permission of the authors and the AAS.

Back cover:

(1) An all-sky image from the Planck mission, part of the early release data set. This image of the microwave sky has been synthesized from 12 months of data spanning the full frequency range of Planck, 30–857 GHz. Reproduced with the kind permission of ESA and the Planck Mission Team.

(2) Constraints on the mass densities of dark energy (Ω_Λ) and matter (Ω_m) provided by the supernovae type Ia Hubble diagram, measurements of baryon acoustic oscillations, and the power spectrum of the cosmic microwave background (CMB) (Suzuki *et al.*, 2011, see Chapter 11). The contours show the 68 per cent, 95 per cent and 99.7 per cent confidence regions. Reproduced with the kind permission of the authors, the Supernova Cosmology Project Team and the AAS.

(3) The power spectrum of the fluctuations in the CMB, as determined from the Wilkinson Microwave Anisotropy Probe (WMAP) 7-year data, together with data from other experiments (Komatsu *et al.*, 2010, see Chapter 11). Image reproduced courtesy of the authors, the WMAP team, the NASA GSFC public archive and the AAS.

Cambridge Observing Handbooks for Research Astronomers

Today's professional astronomers must be able to adapt to use telescopes and interpret data at all wavelengths. This series is designed to provide them with a collection of concise, self-contained handbooks that covers the basic principles peculiar to observing in a particular spectra region, or to using a special technique or type of instrument. The book can be used as an introduction to the subject and as a handy reference for use at the telescope, or in the office.

Series editors

Professor Richard Ellis, Department of Astronomy, *California Institute of Technology*

Professor Steve Kahn, Department of Physics, *Stanford University*

Professor George Rieke, Steward Observatory, *University of Arizona*, Tucson

Dr Peter B. Stetson, Herzberg Institute of Astrophysics, *Dominion Astrophysical Observatory*, Victoria, British Columbia

Books currently available in this series:

1. *Handbook of Infrared Astronomy*
   I. S. Glass
4. *Handbook of Pulsar Astronomy*
   D. R. Lorimer, M. Kramer
5. *Handbook of CCD Astronomy*, Second Edition
   Steve B. Howell
6. *Introduction to Astronomical Photometry*, Second Edition
   Edwin Budding, Osman Demircan
7. *Handbook of X-ray Astronomy*
   Edited by Keith Arnaud, Randall Smith and Aneta Siemiginowska
8. *Practical Statistics for Astronomers*, Second Edition
   J. V. Wall, C. R. Jenkins

Practical Statistics for Astronomers

Second Edition

J. V. WALL

University of British Columbia, Vancouver, Canada

C. R. JENKINS

*Commonwealth Scientific and Industrial Research Organization
(CSIRO), Australia*

CAMBRIDGE
UNIVERSITY PRESS

CAMBRIDGE
UNIVERSITY PRESS

University Printing House, Cambridge CB2 8BS, United Kingdom

Cambridge University Press is part of the University of Cambridge.

It furthers the University's mission by disseminating knowledge in the pursuit of
education, learning and research at the highest international levels of excellence.

www.cambridge.org
Information on this title: www.cambridge.org/9780521732499

First published 2012
Reprinted 2014

A catalogue record for this publication is available from the British Library

ISBN 978-0-521-73249-9 Paperback

Additional resources for this publication at www.cambridge.org/9780521732499

Contents

In affectionate memory of Peter Scheuer (1930–2001)
mentor and friend
'$2 + 2 \simeq 5$'

Foreword to first edition

Peter Scheuer started this. In 1977 he walked into JVW's office in the Cavendish Lab and quietly asked for advice on what further material should be taught to the new intake of Radio Astronomy graduate students (that year including the hapless CRJ). JVW, wrestling with simple Chi-square testing at the time, blurted out *'They know nothing about practical statistics...'*. Peter left thoughtfully. A day later he returned. 'Good news! *The Management Board has decided that the students are going to have a course on practical statistics.*' Can I sit in, JVW asked innocently. 'Better news! *The Management Board has decided that you're going to teach it...'*.

So, for us, began the notion of practical statistics. A subject that began with gambling is not an arcane academic pursuit, but it is certainly subtle as well. It is fitting that Peter Scheuer was involved at the beginning of this (lengthy) project; his style of science exemplified both subtlety and pragmatism. We hope that we can convey something of both. If an echo of Peter's booming laugh is sometimes heard in these pages, it is because we both learned from him that a useful answer is often much easier – and certainly much more entertaining – than you at first think.

After the initial course, the material for this book grew out of various further courses, journal articles and the abundant personal experience that results from understanding just a little of any field of knowledge that counts Gauss and Laplace amongst its originators. More recently, the invigorating polemics of Jeffreys and Jaynes have been a great stimulus; although we have tried in this book not to engage too much with 'old, unhappy, far-off things / and battles long ago'.

Amongst today's practitioners of practical statistics, we have had valued discussions with Mark Birkinshaw, Phil Charles, Eric Feigelson, Pedro Ferreira, Paul Francis, Steve Gull, Dave Jauncey, Ofer Lahav, Robert Laing, Tony Lynas-Gray, Donald Lynden-Bell, Louis Lyons, Andrew Murray, John Peacock, Chris

Pritchett, Prasenjit Saha and Adrian Webster. We are very grateful to Chris Blake, whose excellent D.Phil. thesis laid out clearly the interrelation of 2D descriptive statistics; and who has allowed us to borrow extensively from this opus. CRJ particularly acknowledges the Bayesian convictions of the Real Time Decisions group at Schlumberger; Dave Hargreaves, Iain Tuddenham and Tim Jervis. Try betting lives on your interpretation of the Kolomogorov axioms.

JVW is indebted to the Astrophysics Department of the University of Oxford for the enjoyable environment in which much of this was pulled together. The hospitality of the Department Heads – Phil Charles and then Joe Silk – is greatly appreciated; the stimulation, kindness, technical support and advice of colleagues there has been invaluable. Jenny Wall gave total support and encouragement throughout; the writing benefited greatly from the warmth and happiness of her companionship.

CRJ wishes to acknowledge the support of Schlumberger Cambridge Research for the writing of this book, as part of its 'Personal Research Time' initiative. The encouragement of the lab's director, Mike Sheppard, catalysed its completion. Programme manager Ashley Johnson created the necessary space in a busy research group. Fiona Hall listened, helped with laughter through the long period of gestation and took time out from many pressing matters to support that final burst of writing.

Foreword to second edition

Teaching is highly educational for teachers. Teaching from the first edition revealed to us how much students enjoyed Monte Carlo methods, and the ability with such methods to test and to check every derivation, test, procedure or result in the book. Thus, a change in the second edition is to introduce Monte Carlo as early as possible (Chapter 2). Teaching also revealed to us areas in which we assumed too much (and too little). We have therefore aimed for some smoothing of learning gradients where slope changes have appeared to be too sudden. Chapters 6 and 7 substantially amplify our previous treatments of Bayesian hypothesis testing/modelling, and include much more on model choice and Markov chain Monte Carlo (MCMC) analysis. Our previous chapter on 2D (sky distribution) analysis has been significantly revised. We have added a final chapter sketching the application of statistics to some current areas of astrophysics and cosmology, including galaxy formation and large-scale structure, weak gravitational lensing, and the cosmological microwave background (CMB) radiation.

We received very helpful comments from anonymous referees whom CUP consulted about our proposals for the second edition. These reviewers requested that we keep the book (a) practical and (b) concise and – *small*, or 'backpackable', as one of them put it. We have additional colleagues to thank either for further discussions, finding errata or because we just plain missed them from our first edition list: Matthew Colless, Jim Condon, Mike Disney, Alan Heavens, Martin Hendry, Jim Moran, Douglas Scott, Robert Smith and Malte Tewes. Jonathan Benjamin, Chris Blake, Adam Moss, John Peacock and Sanaz Vafaei provided valuable input on additional material in this second edition.

During preparation of the second edition, JVW is grateful to the National Science and Engineering Council of Canada for support, and to the Physics and Astronomy Department of the University of British Columbia for its hospitality. Matt Wall provided JVW with welcome and expert computer-system assistance.

As before, our greatest debt of gratitude is to our partners Jenny Wall and Fiona Hall for their unflagging understanding and support.

Note on notation

Here are some of the symbols used in the mathematical parts of this book. The list is not complete, but does include notation of more than localized interest. Some symbols are used with different meanings in different parts of the book, but in context there should be no possibility of confusion.

a_{lm}: coefficients of a spherical harmonic expansion

$\mathcal{B}$: Bayes factor

C: usually the covariance (or error) matrix, characterizing a multivariate Gaussian

c_l: coefficients of the angular power spectrum

$\mathrm{cov}[x, y]$: covariance of two random variables x and y.

D: Kolmogorov–Smirnov test statistic

$E[X]$: expectation or ensemble average. Also denoted $< X >$

f, F: probability density distributions and cumulative probability density distributions, respectively; in Chapter 9, Fourier pairs

$\mathcal{F}$: variable distributed according to the F distribution

$\mathcal{H}$: Hessian matrix

H_0, H_1: null hypothesis and alternative hypothesis

$\mathcal{K}$: Kaplan–Meier estimator

L: intrinsic luminosity

$\mathcal{L}$: likelihood

ML: maximum likelihood, maximum likelihood method

MLE: maximum likelihood estimator

$N(S)$: flux density distribution, or source count

$\mathcal{P}$:   posterior odds

$P(N)$:   counts-in-cells probability of finding N objects in a cell

P_l: Legendre polynomials

prob(...):   probability of the indicated event. In the case of a continuous variable, the probability density

prob($A \mid B$):   probability of A, given B

R: distance

r: product–moment coefficient

$\mathcal{R}$:   Rayleigh test statistic

S: mean square deviation of a set of data; in Chapter 8, flux density

$\mathcal{S}$:   test statistic for a particular orientation of the principal axis of the orientation matrix

S_e: sample cumulative distribution, as used in the Kolmogorov–Smirnov test

t: variable distributed according to the t distribution

U: Wilcoxon–Mann–Whitney test statistic

V, $V_{\max}$:   volume contained within R; the maximum volume, corresponding to the greatest distance consistent with an object still appearing in a catalogue

var[x]:   variance of a random variable x

$w(\theta)$:   two-point angular correlation function

$\overline{X}$:   sample average of a set of data

$X_1, X_2, \ldots$:   usually a specific set of data; instances of possible data, denoted x. We try to keep to this distinction by using upper case for particular values and lower case for algebraic variables (although not with Greek letters, or statistics like t where lower case is standard)

y, z: excess variance and skewness of clustered counts-in-cells

Y_{lm}:   spherical harmonics

$\vec{\alpha}$:   vector, usually a vector of parameters

Γ: Gehan test statistic

η: luminosity distribution

κ: Kendall test statistic

μ, σ: usually the mean and standard deviation of a Gaussian distribution; μ may also be the parameter of a Poisson distribution

μ_n: nth central moment of a distribution

ρ: covariance coefficient of a bivariate Gaussian; in Chapter 8, the luminosity function

$\varsigma(\theta, \phi)$: surface density of objects on the sky

σ_s: sample standard deviation

ϕ: space distribution

χ^2: variable distributed according to the χ^2 distribution

1

Decision

Statistics, the most important science in the whole world: for upon it depends the practical application of every other science and of every art.

(Florence Nightingale)

If your experiment needs statistics, you ought to have done a better experiment.

(Ernest Rutherford)

Science is about decision. Building instruments, collecting data, reducing data, compiling catalogues, classifying, doing theory – all of these are tools, techniques or aspects which are necessary. But we are not doing science unless we are deciding something; *only decision counts*. Is this hypothesis or theory correct? If not, why not? Are these data self-consistent or consistent with other data? Adequate to answer the question posed? What further experiments do they suggest?

We decide by comparing. We compare by describing properties of an object or sample, because lists of numbers or images do not present us with immediate results enabling us to decide anything. Is the faint smudge on an image a star or a galaxy? We characterize its shape, crudely perhaps, by a property, say the full-width half-maximum, the FWHM, which we compare with the FWHM of the point-spread function. We have represented a data set, the image of the object, by a *statistic*, and in so doing we reach a decision.

Statistics are there for decision and because we know a background against which to take a decision. To this end, every measurement we make, and every parameter or value we derive requires an *error estimate*, a measure of range (expressed in terms of probability) that encompasses our belief of the true value of the parameter. We are taught this by our masters in the course of interminable undergrad lab experiments. Why? It is because no measured quantity or property is of the slightest use in decision and therefore in science, unless it has a 'range quantity' attached to it.

A *statistic* is a quantity that summarizes data; it is the ultimate data-reduction. It is a property of the data and nothing else. It may be a number, a mean for example, but it does not have to be. It is a basis for using the data or experimental result to make a decision. We need to know how to treat data with a view to decision, to obtain the right *statistics* to use in drawing *statistical inference*. (It is the latter which is the branch of science; at times the term *statistics* is loosely used to describe both the descriptive values and the science.)

The opening quotes indicate a mixed press. Nightingale was a pioneer of applied statistics and graphical presentation. Her message is clear, but suggests the age-old confusion between statistics and data. Rutherford's message also appears clear and uncompromising, but it can only hold in some specialized circumstances. For a start, astronomers are not always free to do better experiments. The laboratory is the big stage; the Universe is an experiment we cannot re-run. Attempting to understand astrophysics and cosmology from one freeze-frame in the spacetime continuum requires some reconsideration of the classical scientific method. This scientific method of *repetition* of experimentally reproduced results does not apply. Thus, the first issue for astronomers: we cannot always re-roll the dice, and anyway, repetition implies similar conditions. We are never at the same spacetime coordinates.

There is thus need for a certain rigour in our methodology. The inability to re-roll dice has led and still leads astronomers into some of the greatest errors of inference. It becomes tempting to the point of irresistibility *to use the data on which a hypothesis was proposed to verify that hypothesis.*

Example *The Black Cloud* (Hoyle, 1958). The Black Cloud appears to be heading for the Earth. The scientific team suggests that this proves the cloud has intelligence. Not so, says the dissenting team member. Why? A golf ball lands on a golf course which contains 10^7 blades of grass; it stops on one blade; the chances are 1 in 10^7 of this event occurring by chance. This is not so amazing – the ball had to land somewhere. It would only be amazing if the experiment were re-run to test the newly formulated hypothesis (e.g. the blade being of special attractive character; the golfer of unusual skill) and the event was repeated. However, the importance of deciding if the Black Cloud knew about the Earth cannot await the next event or the sequence of events, and tempts the rush to judgement in which initial data, hypothesis and test data are combined; so in many instances in astronomy and cosmology.

The most obvious area in which this offence is committed is in claims of physical association of objects of small angular separation on the sky; or similarly, claims of alignment of objects in close proximity on the sky. Most such claims are bogus because they use the object grouping in which the association or the alignment was originally noted in subsequent tests of significance. The original data may be used *to formulate a hypothesis only*; testing must await examination of fresh and unbiased data which do not include the original data. It is essential to divorce hypothesis–formulation data from hypothesis–test data. There is no set of tests which can cope with a-posteriori statistics, or will ever be able to do so.

A second difference for astronomers stems from the first – the remoteness of our objects and the inability to re-run our experiments precisely means that we do not necessarily know the underlying distributions of the variables measured. The essence of classical statistical analysis is (i) the formulation of hypothesis, (ii) the gathering of hypothesis–test data via experiment, and (iii) the construction of a test-statistic. But making a decision on the basis of the test-statistic may demand that the sampling distribution of the statistic be known before a decision can be made. How else could we decide if the value we got was normal or abnormal? It may well be the case that no one, physicist, sociologist, botanist, ever does know these underlying distributions exactly; but astronomers are worse off than most because of our necessarily small samples and our inability to control experiments, leading to poor definitions of the underlying distributions.

Astronomers cannot avoid statistics and there are at least the following reasons for this unfortunate situation.

(i) Error (range) assignment – ours, and the errors assigned by others: what do they mean?
(ii) How can data be used best? Or at all?
(iii) Correlation, testing the hypothesis, model fitting; how do we proceed?
(iv) Incomplete samples, samples from an experiment which cannot be re-run, upper limits; how can we use these to best advantage?
(v) Others describe their data and conclusions in statistical terms. We need some self-defence.
(vi) But above all, we must decide. The decision process cannot be done without some methodology, no matter how good the experiment. Rutherford may not have known when he was using statistics.

This is not a book about statistics, the values or the science. It is about how to get results in astronomy, using statistics, data analysis and statistical inference.

Consider first how we do science in order to see at what point 'statistics' enter(s) the process.

1.1 How is science done?

In simplest terms, each experiment goes round a loop which can be characterized by six stages:

1. Observe: with an observing or data-gathering programme, record or collect the data.
2. Reduce: clean up the data to remove experimental effects, i.e. flat field it, calibrate it.
3. Analyse: obtain the numbers from the clean data – intensities, positions. Produce from these summary descriptors of the data which enable comparison or modelling – descriptors that lead to reaching the decision which governed the design of the experiment; and which are *statistics*.
4. Conclude: carry through a process to reach a decision. Test the hypothesis; correlate; model, etc.
5. Reflect: what has been learnt? Is the decision plausible? Is it unexpected? At which experimental stage must re-entry be made to check? What is required to confirm this unexpected result? Or – what was inadequate in the experimental design? How should the next version be defined? Is an extended or new hypothesis suggested? Far too little time is spent here; perhaps the pressure of observing application deadlines and/or the perceived need to publish get the better of us.
6. Experiment design: if the hypothesis is important enough; if the data warrant it; if previous experimental experience suggests it is possible; if technical advances make it feasible – then the next experiment needs to be designed. This may (and usually does) take the form of thinking out an observing proposal, writing and submitting it. It may take the form of re-design of an instrument on a current telescope. It may take the form of a proposal to build a new instrument. It may take the form of designing a new telescope or space mission, a process which, in itself, may occupy much of a research career. The latest such projects involve multi-nation collaborations on scales of billions of dollars. The timescales from initial plans to realization may range to 40 years (e.g. the James Webb Space telescope; the Square Kilometre Array).

And so back to stage 1.

This process is a loop and 'experiments' may begin at different points. For instance, we disbelieve someone else's conclusions based on their published data set. We enter at point (3) or even (4); and we may then go around the data-gathering cycle ourselves as a result. Or we enter at (5), looking at an old result in the light of new and complementary ones from other fields – and proceed to (6) and back to (1) ...

Table 1.1　*Stages in astronomy experimentation*

Stage	How	Examples	Considerations
Observe	In person? Remotely? Depends on facility	Experiment design: calibration integration time *Stats*	What is wanted? Number of objects *Stats*
Reduce	Algorithms	Flat field Flux calibration	Data integrity Signal-to-noise *T Stats*
Analyse	Parameter estimation, hypothesis testing *T Stats*	Intensity measurements Positions *T Stats*	Frequentist, Bayesian? *T Stats*
Conclude	Hypothesis testing *T Stats*	Correlation tests Distribution tests *T Stats*	Believable, repeatable, understandable? *T Stats*
Reflect	Carefully; far too little time is invested here	Mission achieved? A better way? 'We need more data'? *T Stats*	The next observations *T Stats*
Design	Hone the mission; build science case *Stats*	New observations/ instrument/ telescope/space mission	Feasibility – cost, team design, experience, human resources; simulations, predictions *Stats*

Of course it could be argued that (6) should start the process, but we need some knowledge base before we start designing.

All too often we use (3) to set up the tests at (4). This carries the charge of mingling hypothesis and data, as in the Black Cloud example.

Table 1.1 summarizes the process. Points in Table 1.1 at which recourse to statistics or to statistical inference is important have been indicated by *Stats*; a *T* appears when the issue applies to theorists as well as to experimentalists. Few are the regions in which we can ignore statistics and statistical inference. *Experiment design needs to consider from the start* what statistic or summarized data form is required to achieve the desired outcome. There are then checks throughout the experiment, and finally there is analysis in which the measured statistics are used in inference. Applied statistics in the guise of forecasting is increasingly used in astronomy instrument/survey/experiment design.

1.2 Probability; probability distributions

The concept of *probability* is crucial in decision processes, and there is a commonly accepted relationship between probability and statistics. In a world in which our statistics are derived from finite amounts of data, we need probabilities as a basis for inference. For example, limited data yields us only a partial idea of the point-spread function, such as the FWHM; we can only assign probabilities to the range of point-spread functions roughly matching this parameter.

We all have an inbuilt sense of probability. We know, for example, that the height of adults is anything from, say, 1.5 to 2.5 m. We know this from the totality of the population, all adults. But we know what a tall person is – and it is not necessarily somebody who is 2.5 m tall. The *distribution* is not flat; it peaks at around 1.7 m. The distribution of the heights of all adults, normalized to have an area of 1.0, is the measured *probability density function*, often called the probability distribution. (We meet them in a more rigorous context in Chapter 2.) The tails contain little area; and it is the tails that give us the decision: we probably call somebody tall when they are taller than 75 per cent of us.

We have made a decision based on a statistic, by relating that statistic to a probability distribution; we have decided that the person in question was tall. Note also what we did – observe, reduce, analyse, conclude, probably all in one glance. We did not do this rigorously in making a quantitative assessment of just how tall, which would have required a detailed knowledge of the distribution of height and a quantitative measurement. And reflect? Context of our observation? Why did we wish to register/decide that the person was tall? What next as a result? How was this person selected from the population? The brain has not only done the five steps but has also set the result into an extensive context; and this in processing the single glance.

The probability distribution in our minds – the heights of adults – is unlikely to have a mathematical description; it is one determined by counting enough of the population (probably subconsciously) so that it is well defined. There are distributions for which mathematical description is very precise, such as the Poisson and Gaussian (Normal) distributions, and there are many cases in which we have good reason to believe that these must represent the underlying probability distributions well.

This is also an example of a 'ruling-out'; here we ruled out the hypothesis that the person is of 'ordinary' height. There is a different type of statistical inference, the 'ruling-in' process, in which we compute the probability of getting a given result, and if it is 'probable', we accept the original hypothesis.

It is also an example of 'counting' to find the probabilities, the frequency distribution. There are other ways of assigning probabilities, including opinion and states of knowledge; and, in fact, there are instances in which we are moderately comfortable with the paradoxical notion of assigning probabilities to unique events. It is essential that our view of statistics and statistical inference be broad enough to take such probability concepts on board.

1.3 Bolt-on statistics?

With regard to statistics and probability, in many of the conversations we have had with users of the first edition of this book, we found that 'statistics' is often seen as a bolt-on addition to scientific analysis, a technological feature rather like dentistry; necessary, somewhat unpleasant, but *a solved piece of technology*. In the aftermath of the global financial crisis beginning in 2008, the role of quantitative finance was widely discussed. One of the failures that was identified was the failure of statistical models of *risk* – failures that had consequences costing trillions of dollars. Why the contradiction? Surely if statistics and probability were that routine, things could not have gone wrong quite so badly?

The answer is that there is a very wide range of degree of certainty associated with the application of statistics. An early distinction was drawn by Knight (1921), who was curious about why some businesses made huge profits and some only modest ones. His suggestion was that the run-of-the-mill firms dealt in *risk*, whereas the very successful ones (with an obvious selection effect in operation) dealt in *uncertainty*. What did Knight mean by these terms, especially 'uncertainty', which we often use interchangeably with words like random, stochastic or probability?

Take a concept, implicit in our usual undergraduate lab statistics training, in which we think we know the mean and standard deviation of our normally distributed observable. To put the implication at its starkest, this means that we know every single observation that we will ever make; only the order is unknown. This is melodramatic phrasing, but it expresses the extraordinary power of the assumption that we know a probability distribution. Often, when we start out in statistics, we have an uneasy feeling that we are getting something for nothing. In fact, there is a high price to pay in the scope of the assumptions we make, either openly or unknowingly. This illustrates what Knight meant by mere risk, in a business context: *risk involves only known probabilities*. A casino is an example. Unless the roulette wheels are improperly engineered, the management of a casino can predict its profits, as long as customers keep coming through the doors with the same amount of money in their pockets. The probabilities are known, and the casino management knows exactly how to set its margins to attain a given return.

Of course, not even a casino operates in this ideal environment. Taleb (2010) gives the example of the single biggest loss experienced by a casino of which he had apparently intimate knowledge: in a show put on to entertain idle patrons, a performing tiger ate its trainer, with consequent eye-watering claims for trauma, loss of earnings to the bereaved family, and so on. This is what Knight meant by *uncertainty* and Taleb by his term 'black swans' – not only may the probabilities not be known, *they may not even have been considered.*

Returning to the more familiar ground of astronomy, what do we learn for the application of statistics to our subject? There is exactly the same continuum between risk and uncertainty, reflected in the robustness of the assumptions we make in order to pursue our statistical analyses. Do we know the parameters of the distributions we assume? Probably not, but we can estimate them from the data. How well we do this depends on how much data we have. If we have a lot, we wonder if it is 'all the same', or whether the underlying parameters are actually varying within the data set. Indeed, the very form of the distributions we assume is an issue. Gaussian? To the extent that the central limit theorem (Section 2.4.2.3) holds, perhaps. More realistically, we need a range of distributions, each with its own prior probability and parameters . . . and so on, up the hierarchy of complexity towards greater uncertainty.

As you embark on this little handbook, remember that the statistics you will encounter represent a model of the world, in the same messy, complicated, intuition-needing sense as the astronomy to which you may think you can 'bolt it on'. Making the measurement is the easy part, understanding the error is the hard part; but as you will see if you persist with us, there is a framework (formally, the framework of Bayesian inference, aided by the concept of hyperparameters) that allows us to bound our ignorance and control its consequences – if we are fortunate. If we are not, of course, we may be eaten by a tiger.

1.4 Probability and statistics in inference: an overview of this book

Statistics are combinations of the data that do not depend on any unknown parameters. The average is a common example. When we calculate the average of a set of data, we expect that it will bear some relation to the true, underlying mean of the distribution from which our data were drawn. In the classical tradition, we calculate the sampling distribution of the average, the probabilities of the various values it may assume as we (hypothetically) repeat our experiment many times. We then know the *probability* that some range around our single measurement will contain the true mean. This is information that we can use to take decisions.

This is precisely the utility of statistics – they are laboriously discovered combinations of observations which converge, for large sample sizes, to some underlying parameter we want to know (say, the mean). Useful statistics are actually rather few in number.

We meet the issues of probability distributions, statistics, the relation between these, and the role of random-number analyses in Chapters 2 and 3. The long development of these concepts is outlined in Table 1.2, a sketch of the timeline of the development of probability and statistics. Origins of statistical inference can be traced back to Aristotle (384–322 BC) who developed a logic framework and stated a version of Occam's Razor. For a fascinating historical study of statistics and probability, see the erudite books by Anders Hald (1990, 1998).

In Chapter 2 we also meet a radically different way of making inferences – the *Bayesian approach*, totally distinct in its logic from the 'classical' or 'frequentist' approach just discussed. The Bayesian approach focuses on the probabilities right away, without the intermediate step of statistics. In the Bayesian tradition, we invert the reasoning just described. The data, we say, are unique and known; it is the mean that is unknown, that should have probability attached to it. Without using statistics, we instead calculate the probability of various values of the mean, given the data we have. This also allows us to make decisions. In fact, as we shall see, this approach comes a great deal closer to answering the questions that scientists actually ask. This drastic change in approach came painfully and relatively recently – see Table 1.2. From Chapter 2 on, we invoke both methodologies to greater or lesser extent; we explain why in context.

Chapter 4 *Correlation and association* provides our first look at a practical area of statistics, namely correlations, searches for them in data sets as well as tests of their significance. This area of statistics might well be the one which most readily refutes the charge that statistics as a science has not discovered anything.[1] The original regression lines of Francis Galton ('regression to mediocrity') played a major role in genetics, while subsequently the germ theory of disease (John Snow) and the expansion of the Universe (Edwin Hubble) both emerged from correlation analyses.

---

[1] ... but serves only as the lamp-post serves the drunken man: for support rather than for illumination (Andrew Lang, nineteenth-century poet and philosopher).

Table 1.2 *A brief history of probability and statistics*

Year	Individual(s)	Key words	Events
~1340	William of Ockham, or Occam	**Occam's Razor**	'It is useless to do with more what can be done with less.' Ockham, an ordained Franciscan, was excommunicated for his views on separation of church and state, amongst other things. In addition to the application of the principle in statistics and data modelling, Hawking (1988) attributes the discovery of quantum mechanics to it.
1654	Pascal, Fermat	**odds, probability theory**	Gombaud, Chevalier de Mere & Mitton pose questions on gambling odds to Pascal in ~1654. Seven letters exchanged between Blaise Pascal & Pierre de Fermat are the genesis of probability theory.
1657	Huygens	**probability**	First publication on probability, 14 problems (+ solutions) in gambling, based on the Pascal–Fermat correspondence; the only publication on the subject for 50 years.
1662	Graunt	**descriptive statistics, life tables, survival analysis**	Publication of Graunt's *Observations on the Bills of Mortality*; first known collection and analysis of data for statistical purposes; start of actuarial risk analysis.
1692	Huygens, Arbuthnot	**probability**	*Of the Laws of Chance, or, a method of Calculation of the Hazards of Game . . .* ; Arbuthnot's translation of Huygens' work becomes the first English publication on probability.
1665–1676	Newton, Leibniz	**calculus**	Newton & Leibniz independently discover calculus; their dispute runs for decades. Probability theory can proceed.
1687	Newton	**binomial distribution**	In the monumental *Principia*, Newton changes the direction of physics and mathematics forever; the book includes the binomial probability distribution.

Table 1.2 (*cont.*)

Year	Individual(s)	Key words	Events
1693	Halley	**risk analysis, actuarial analysis**	Halley publishes *An Estimate of the Degrees of Mortality of Mankind.*
1711	de Moivre	**statistical independence**	Concept of statistical independence introduced, expressed in ratios of products of numbers of wins and losses.
1713	J. Bernoulli	**binomial distribution, law of large numbers**	N. Bernoulli publishes posthumously his uncle's *Ars Conjectandi* proving the binomial distribution of Newton, developing issues of probability, and introducing the (weak) law of large numbers.
1730	de Moivre	**central limit theorem**	The central limit theorem in the special case of the binomial distribution.
1733	de Moivre	**Normal distribution**	The Normal distribution shown to be an approximation of the binomial distribution.
1749, 1791	Achenwall, Sinclair	**'Statistik' = statistics**	The study of social data (Graunt, Halley and many others) suggests to Achenwall that dealing with natural 'states' of society should be referred to as **Statistik**. Sinclair's 21-volume *Statistical Account of Scotland* (1791) establishes the term.
1749	Mayer	**combining observations**	Prior to 1750, with the exception of Tycho Brahe, most observers believed that combining observations led to divergence from the best estimate. Mayer shows that the reverse is the case.
1756	Bayes	**systematic errors**	In a letter from Bayes: 'The more observations you make with an imperfect instrument the more it seems to be that the error in your conclusion will be proportional to the imperfection of the instrument . . .'

(*cont.*)

Table 1.2   *(cont.)*

Year	Individual(s)	Key words	Events
1757, 1760	Boscovich, Maire	**combining observations**	In 1757 they publish a synopsis of ideas on combining observations; a full description of the method appears in 1760.
1763	Bayes	**Bayes' theorem**	Bayes' theorem is presented to the Royal Society posthumously by Richard Price: 'I now send you an essay which I have found among the papers of our deceased friend Mr Bayes, and which, in my opinion, has great merit...' The theorem is finally accepted by the great Laplace in 1781.
1787	Laplace	**combining observations**	While publishing proof of the stability of the solar system, Laplace improves on the method of Mayer to combine observations.
1805	Legendre	**least squares**	In his treatise on comet orbits, Legendre develops the method of least squares.
1809–1810	Gauss, Laplace	**Normal distribution**	77 years after de Moivre, Gauss shows that observational errors are expected to have a Normal (Gaussian) distribution. Laplace provides a much better derivation.
1812	Laplace	**probability theory**	Laplace publishes his landmark *Théorie Analytique des Probabilités*.
1835	Quetelet	**Normal distribution**	Quetelet, astronomer, statistician, social scientist, publishes a statistical study of human properties showing the ubiquitous nature of the Normal distribution; he is a major influence in promoting the use of statistical studies in social and astronomical contexts.
1837	Poisson	**Poisson distribution, law of large numbers**	Poisson publishes *Recherchés sur la Probabilité des Jugements ...* in which he introduces the distribution now bearing his name, and coins the phrase 'Law of Large Numbers'.

Table 1.2   (*cont.*)

Year	Individual(s)	Key words	Events
1885–9	Galton	**regression, correlation, bivariate Gaussian**	The amazing Galton, cousin of Darwin, African explorer, statistician, psychologist, biologist, criminologist (fingerprints), meteorologist, the original behavioural geneticist, and publisher of 350 books and papers: he introduces regression plots (the heights of sons against their fathers), correlation plots and explores the bivariate Gaussian to describe these.
1893	Pearson	**standard deviation**	Pearson introduces the term, already known as error of mean square or mean error, and it becomes the accepted way of describing distribution spread.
1897	Pearson	**product moment coefficient**	Pearson introduces the standard (frequentist) method of describing the strength of a correlation.
1900	Pearson	**chi-square test**	Pearson's polemical paper developing chi-square demonstrates inter alia that a long run of bad luck on the roulette wheel at Monte Carlo could not have been due to chance.
1904	Spearman	**rank correlation coefficient**	Spearman develops a non-parametric way of testing for correlation even when data are not on ordinal scales.
1908	Gosset	**'Student's' t distribution**	Gosset derives the t distribution to test for differences between means of observed distributions.
1913	Eddington	**Eddington bias**	Distortion of number counts from measurement error, introduced in a two-page paper in *Monthly Notices of the Royal Astronomical Society*.
1920	Malmquist	**Malquist bias**	Malmquist discovers the luminosity–distance correlation from a survey of fixed sensitivity, the bias that will plague astronomers for the next century.

(*cont.*)

Table 1.2   (*cont.*)

Year	Individual(s)	Key words	Events
1922	Fisher	**maximum likelihood**	Fisher develops the concept and consequences of maximum likelihood.
1925	Fisher	**ANOVA and much more**	Fisher publishes the massively influential *Statistical Methods for Research Workers*, which includes the new analysis of variance and describes for the first time the full panoply of *t* test, chi square, the *F* test, randomized design, and significance testing. What we now call Fisher Information was also introduced in this year.
1933	Neyman and Pearson	**the null hypothesis**	Neyman and Pearson introduce hypothesis testing, the null hypothesis, and Type I and Type II errors.
1933	Hoteling	**PCA**	Principal component analysis, the right way to search for dependencies in multivariate data, is developed and published.
1933	Kolmogorov	**Kolmogorov axioms, Kolmogorov–Smirnov test**	Kolmogorov presents his probability axioms, a basis for a self-consistent theory of probability; and the Kolmogorov–Smirnov non-parametric test to search for significant difference between distributions.
1935	Fisher	**'The Design of Experiments'**	The standard for decades for the integrated design and statistical analysis of experiments.
1939	Jeffreys	**theory of probability**	The first attempt to develop a fundamental theory of scientific inference based on Bayesian statistics. His ideas were well ahead of their time.
1953	Neyman, Scott, Shane	**clustering of galaxies**	A key paper identifying 'contagion' in the clustering of galaxies; Neyman had been working on the statistics of epidemics and so noticed the mathematical similarities between clusters of disease outbreaks and the clustering of galaxies.

Table 1.2 (*cont.*)

Year	Individual(s)	Key words	Events
1953	Metropolis *et al.*	**the Metropolis algorithm**	*Equations of State Calculations by Fast Computing Machines* introduces an idea which will come of age several decades later and make Bayesian inference possible for real problems.
~1955	Cooley, Tukey	**FFT**	Cooley and Tookey derive the Fast Fourier Transform algorithm, all-pervasive in image compression, CMB cosmology, aperture synthesis, etc.
~1960	Fisher–Pearson vs. Jaynes–Jeffreys	**Bayes' revival**	The powerful and productive rivals Pearson and Fisher, while disagreeing violently and publicly on more than one issue, were frequentist in approach and so influential in the first half of the 20th century that Bayesian methods were all but forgotten. The rise of computing power and more importantly of mathematicians such as Jaynes and Jeffreys, who thought deeply about methodology, began a sea change. 1965–80 was a period of intense debate amongst statisticians who divided themselves into frequentists and Bayesians. The outcome is the current ascendancy of Bayesian methods, with frequentist methods still in use in many traditional areas of sample testing such as polling and medical drug research where simple yes-no answers are requested; and in areas where no model exists such as non-parametric testing. Modern approaches combine the methodologies.
1957	Scheuer	$p(D)$	The 'probability of Deflection', confusion limit analysis, developed for interferometric radio surveys, has application in all domains in which more than one signal contributes within the resolution of the instrument.

(*cont.*)

Table 1.2   (*cont.*)

Year	Individual(s)	Key words	Events
1965	Schmidt, Rowan-Robinson	V/V_{max} **test**	A method for systematic examination of spatial distribution of extragalactic objects.
1968	Jaynes	**'On Prior Probabilities'**	Jaynes' powerful article is as good a point as any to mark the start of the new Bayesian offensive.
1977	Peebles *et al.*	**correlation functions**	The correlation functions for galaxies, powerful tools in examining galaxy formation, evolution and cosmology.
1978	Gull, Daniel	**maximum entropy**	*Image reconstruction from incomplete and noisy data* is published in *Nature*.
1979	Efron	**bootstrap**	Efron publishes the bootstrap test, referred to in *numerical recipes* as 'quick and dirty Monte Carlo'.
1985	Feigelson, Nelson	**survival analysis**	The foundations of non-parametric univariate survival analysis and its application to astronomical data.

Chapter 5, *Hypothesis testing* is basically frequentist; here we describe the classical tests based on statistics. Chapter 5 describes both parametric and non-parametric (or 'distribution-free') tests. The latter is a methodology of doing frequentist-type hypothesis testing *without knowing what the underlying probability distribution actually is*, i.e. without having such a distribution which is characterized by parameters. This is of particular importance for astronomers for a number of reasons, as we have mentioned.

So why this plethora of approaches: parametric versus non-parametric/ frequentist versus Bayesian? Part of this is the very different ways in which we encounter data, ours or other people's, or perhaps statistics – data descriptors – presented to us in lieu of data. For instance, it is important to recognize that data may be in a number of forms depending on what *measurement scale* is used; see Table 1.3. When it does come to us in forms other than those immediately recognizable, i.e. on numerical scales (magnitudes, redshifts, etc.), there are still formally valid ways of carrying out statistical inference, as we shall see. However, our options may be limited and, in particular, Bayesian techniques may not be applicable.

Table 1.3 *Measurement scales*

Scale type	Also called	Description: example
Nominal/categorical	Binned	**gender:** male / female
Ordinal/ranking	Ordered	**army ranks:** private, corporal, sergeant, . . .
Ratio	Numerical/ measures	uniformly calibrated scale with zero point: **temperature in kelvin**
Interval	Numerical/ measures	uniformly calibrated scale: **time**, whose beginning (and end) we do not know, but which we arbitrarily 'zero-point' in many ways to give it ratio scales, e.g. Gregorian calendar, Julian Day, UTC, GMT

Chapter 6 describes data modelling and parameter estimation, and is a mixture of frequentist and Bayesian techniques. Chapter 7 describes more advanced Bayesian techniques, including model selection, Markov chain Monte Carlo analysis and hyperparameters.

Chapters 8, 9, 10 and 11 have a strong astronomical orientation. Chapter 8 is about detection, catalogues, surveying, luminosity functions, incomplete samples and confusion (object-blending). Chapter 9 deals with 1D statistics, i.e. spectral scans, or time-sequenced observations, the sampling and analysis of such data. Chapter 10 is termed *Statistics of large-scale structure*, as it deals, in the main, with the 2D distribution of objects on the sky. Chapter 11 is a descriptive epilogue, considering the integral role of statistics in current astrophysics and cosmology by way of examples in galaxy distribution, weak gravitational lensing and cosmic microwave background (CMB) studies.

1.5 How to use this book

This is not a textbook of statistical theory, a guide to numerical analysis, or a review of published work. It is a practical manual, which assumes that proofs, numerical methods and citation lists can easily be found elsewhere. This book sets out to tell it from an astronomer's perspective, and our main objective is to help in gaining familiarity with the broad concepts of statistics and probability, to understand their usefulness and to feel confident in applying them. Work through the examples and exercises; they are drawn from our experience and have been chosen to clarify the text. They vary in difficulty, from one-page calculations to mini-projects. Some need data; these may be simulated. If preferred, example data sets are available on the book's website – as are the solutions to the exercises. Aim to become confident in the use of Monte Carlo simulations to check any calculations, and to try out ideas. Remember, in this subject we can do useful and revealing experiments – in the computer. Do not be ashamed to let simulations guide your mathematical intuition!

For further details on statistical methods and justification of theory, there is no substitute for a full textbook. None of our topics is arcane and they will be found in the index of many books on probability and statistics. We have found several particularly helpful and these are described in Appendix A. We need to mention one indispensable reference work here, though: *Numerical Recipes* (Press *et al.*, 2007), which points the way on numerous statistical issues, provides the means for numerically solving most problems, and contains comprehensive, humorous and wise advice.

There is little algebra in this book; it would have greatly lengthened and cluttered the presentation to have worked through details. Likewise, we have not explained how various integrals were done or eigenvalues found. These things can be done by computers; packages such as the superb MATHEMATICA, used for many of the calculations in this book, can deal swiftly with more mathematical technology than any of us know. Using these packages frees us all up to think about the problem to hand, rather than searching in vain for missing minus signs or delving into handbooks for integrals which never seem to be there in quite the needed form.

We have not attempted exhaustive referencing. Rather, we have given enough key references to provide entry-points to the literature. Online bibliographic databases provide excellent cross-referencing, showing who has cited a paper and whom it cites; it is the work of minutes to collect a comprehensive reading-list on any topic. The lecture notes for many excellent university courses are now on the Web; a well-phrased search may well yield useful material to help with whatever is puzzling you.

Finally, use this book as you need it. It can be read from front to back, or dipped into. Of course, no interesting topic is self-contained, but we hope that the cross-referencing will connect all the technology needed to explore a particular topic.

Exercises

1.1 Discovery. At first sight, discovery of a new phenomenon may not read as an experiment as described in Section 1.1. But it is. Describe the discovery of pulsars (Hewish *et al.*, 1968) in terms of the six experimental stages.

1.2 Significance. The significance of a certain conclusion depends very strongly on whether the most luminous known quasar is included in the data set. The object is legitimately in the data set in terms of pre-stated selection criteria. Is the conclusion robust? Believable?

2

Probability

God does not play dice with the Universe.
(Albert Einstein[1])

Whether He does or not, the concepts of probability are important in astronomy for two reasons.

1. Astronomical measurements are subject to random measurement error, perhaps more so than most physical sciences because of our inability to re-run experiments and our perpetual wish to observe at the extreme limit of instrumental capability. We have to express these errors as precisely and usefully as we can. Thus, when we say 'an interval of 10^{-6} units, centred on the measured mass of the Moon, has a 95 per cent chance of containing the true value', it is a much more quantitative statement than 'the mass of the Moon is 1 ± 10^{-6} units'. The second statement really only means anything because of some unspoken assumption about the distribution of errors. Knowing the error distribution allows us to assign a probability, or measure of confidence, to the answer.

2. The inability to do experiments on our subject matter leads us to draw conclusions by contrasting properties of controlled samples. These samples are often small and subject to uncertainty in the same way that a Gallup poll is subject to 'sampling error'. In astronomy we draw conclusions such as: 'the distributions of luminosity in X-ray-selected Type I and Type II objects differ at the 95 per cent level of significance.' Very often the strength of this conclusion is dominated by the number of objects in the sample and is virtually unaffected by observational error.

---

[1] The first documented statement to this effect, rather more elegantly phrased, appears in a 1926 letter to his friend Max Born (AEA 8-180). Einstein made several public and private comments of a similar nature.

This chapter begins with a discussion of what *probability* is, and proceeds to introduce the concepts of *conditionality* and *independence*, providing a basis for the consequent discussion of *Bayes' theorem*, with *prior* and *posterior probabilities*. Only at this point is it safe to consider the concept of *probability distributions*; some common probability distributions are compared and contrasted. We then introduce *Monte Carlo generators*, thereby providing tools with which to examine probabilities, probability distributions and relations between sample size and uncertainty, *inter alia*. This all sets the stage for the following chapter, dealing with *statistics* themselves, the penultimate product of data reduction – if conclusions/discoveries are considered as the ultimate product. The issues of *expectation* and *errors*, dependent on the distributions and statistics, are discussed in the final section of the following chapter.

2.1 What is probability?

The development in this chapter drew heavily on the writings of Jaynes (Jaynes, 2003, 1986, 1983, 1976). Another fundamental reference, rather heavy going, is Jeffreys (1961).

The study of probability began with the analysis of games of chance involving cards or dice. Because of this background we often think of probabilities as a limiting case of a frequency. Many textbook problems are still about dice, hands of cards, or coloured balls drawn from urns; in these cases it seems obvious to take the probabilities of certain events according to the ratio

number of favourable events/total number of events

and the probability of throwing a six with one roll of the dice is 'obviously' 1/6.

This probability derives from what Laplace called the 'principle of indifference', which in effect tells us to assign equal probabilities to events unless we have any information distinguishing them. In effect we have done the following calculation:

probability of one spot $= x$

probability of two spots $= x$

probability of three spots $= x$

and so on; this is the principle of indifference step. Further, we believe that we have identified all the cases; with the convention that the probability of a certain event (anything between one and six spots) is unity, we have

$$6x = 1.$$

This calculation, apparently trivial as it is, shows a vitally important feature: we cannot usefully define probability by this kind of ratio. We have had to assume that each face of the die is equally probable to start with – thus, the definition of probability becomes circular.

If we can *identify* equally likely cases, then *calculating* probabilities amounts simply to enumerating cases – not always easy, but straightforward in principle. However, identifying equally likely cases requires much more thought.

Many interesting and useful calculations can be done using the principle of indifference, either directly or by exploiting its applicability to aspects of the problem. For example, we may know that a die is biased, the faces are not 'equally likely'. However, given some details of, say, the mass distribution of the die, we may be able to calculate the probabilities of the faces using an assumption that the initial direction of the throw is isotropic – in which case the principle of indifference applies to throw-directions.

Sometimes we estimate probabilities from data. The probability of our precious observing run being clouded out is estimated by

number of cloudy nights last year / 365

but two issues arise. One is the limited data – we suspect that 10 years' worth of data would give a different, more accurate result. The second issue is simply the identification of the 'equally likely' cases. Not all nights are equally likely to be cloudy, some student of these matters tells us; it is much more likely to be cloudy in winter. What is 'winter', then? A set of nights equally likely to be cloudy?

We can only estimate the probabilities correctly once we have identified the equally likely cases, and this identification is the subjective, intuitive step that is built into our reasoning about data from apparently malevolent instrumentation in an uncertain world.

It is common to define probabilities as empirical statements about frequencies, in the limit of large numbers of cases – our 10 years' worth of data; but, as we have seen, this definition must be circular because selecting the data depends on knowing which cases are equally likely. Defining probabilities in this way is sometimes called 'frequentist'.

So, what is probability? The notion we adopt for the present is that probability is *a numerical formalization of our degree or intensity of belief*. In everyday speech we often refer to the probability of unique events, showers of rain or election results. In the desiccated example of throwing dice, *x* measures the strength of our belief that any face will turn up. Provided that the die is not loaded, this belief is 1/6, the same for each face.

Ascribing an apparently subjective meaning to probability in this way needs careful justification. After all, one person's degree of belief is another person's certainty, depending on what is known. We can only reason as best we can with the information we have; if our probabilities turn out to be wrong, the deficiency is in what we know, not the definition of probability. We just need to be sure that two people with the same information will arrive at the same probabilities. It turns out that this constraint, properly expressed, is enough to develop a theory of probability which is mathematically identical to the one often interpreted in frequentist terms.

A useful set of properties of probability can be deduced by formalizing the 'measure of belief' idea. The argument is originally due to Cox (1946) and goes as follows: if *A*, *B* and *C* are three events and we wish to have some measure of how strongly we think each is likely to happen, then, for consistent reasoning, we should at least apply the rule *if A is more likely than B, and B is more likely than C, then A is more likely than C*. Remarkably, this is sufficient to put constraints on the probability function which are identical to the Kolmogorov axioms of probability, proposed some years before Cox's paper:

- any random event *A* has a probability prob(*A*) between 0 and 1;
- the sure event has prob(*A*) = 1;
- If *A* and *B* are exclusive events, then prob(*A* or *B*) = prob(*A*) + prob(*B*).

The Kolmogorov axioms are a sufficient foundation for the entire development of mathematical probability theory, by which we mean the apparatus for manipulating probabilities once we have assigned them.

Example Before 1987, four naked-eye supernovae had been recorded in 10 centuries. What, before 1987, was the probability of a bright supernova happening in the twentieth century?

There are three possible answers.

(1) Probability is meaningless in this context. Supernovae are physically determined events and, when they are going to happen, can, in principle, be accurately calculated. They are not random events.

From this God's-eye viewpoint, probability is indeed meaningless; events are either certain or forbidden. 'God does not play dice...'.

(2) From a frequentist point of view our best estimate of the probability is 4/10, although it is obviously not very well determined.

This assumes that supernovae were equally likely to be reported through-out 10 centuries, which may well not be true. Eventually some degree of belief about detection efficiency will have to be made explicit in this kind of assignment.

(3) We could try an a-priori assignment. In principle we might know the stellar mass function, the fate and lifetime as a function of mass, and the stellar birth rate. We would also need a detection efficiency. From this we could calculate the mean number of supernovae expected in 1987, and we would put some error bars around this number to reflect the fact that there will be variation caused by factors we do not know about – metallicity, perhaps, or location behind a dust cloud, and so on.

The belief-measure structure is more complicated in this detailed model but it is still there. The model deals in populations, not individual stars, and assumes that certain groups of stars can be identified which are equally likely to explode at a certain time.

Suppose now that we sight supernova 1987A. Is the probability of there being a supernova later in the twentieth century affected by this event?

Approach (1) would say no – one supernova does not affect another. Approach (2), in which the probability simply reflects what we know, would revise the probability upward to 5/10. Approach (3) might need to adjust some aspects of its models in the light of fresh data; predicted probabilities would change.

Probabilities reflect what we know – they are not things with an existence all of their own. Even if we could define 'random events' (approach 1), we should not regard the probabilities as being properties of supernovae.

2.2 Conditionality and independence

Two events A and B are said to be *independent* if the probability of one is unaffected by what we may know about the other. In this case, it follows (not trivially!) from the Kolmogorov axioms that

$$\text{prob}(A \text{ and } B) = \text{prob}(A)\text{prob}(B). \tag{2.1}$$

Sometimes independence does not hold, so that we would also like to know the *conditional probability*: the probability of A, given that we know B. The definition is

$$\text{prob}(A \mid B) = \frac{\text{prob}(A \text{ and } B)}{\text{prob}(B)}. \tag{2.2}$$

If A and B are independent, knowing that B has happened should not affect our beliefs about the probability of A. Hence, $\text{prob}(A \mid B) = \text{prob(A)}$ and the definition reduces to $\text{prob}(A \text{ and } B) = \text{prob}(A)\text{prob}(B)$ again.

If there are several possibilities for event B (label them $B_1, B_2, \ldots$) then we have that

$$\text{prob}(A) = \sum_i \text{prob}(A \mid B_i)\text{prob}(B_i). \tag{2.3}$$

A might be a cosmological parameter of interest, while the Bs are not of interest. They might be instrumental parameters, for example. Knowing the probabilities $\text{prob}(B_i)$ we can get rid of these 'nuisance parameters' by a summation (or integration). This is called *marginalization.*

Example Take the familiar case in astronomy where some 'remarkable' event is observed, for example two quasars of very different redshifts close together on the sky. The temptation is to calculate an a-priori probability, based on surface densities, of two specified objects being so close. However, the probability of the two quasars being close together is conditional on having noticed this fact in the first place. Thus, the probability of the full event is simply $\text{prob}(A \mid A) = 1$, consistent with how we should expect to measure our belief in something that we already know. We can say nothing further, although we might be able to formulate a hypothesis to carry out an experiment.

Consider now the very different case in which we wish to know the probability of finding two objects of different types, say galaxy and quasar, within a specified angular distance r of each other. To be specific, we plan to search some fixed solid angle Ω. The surface densities in question are ς_G and ς_Q. On finding a galaxy, we will search around it for a quasar. We need

$$\text{prob}(\text{G in field and Q within } r)$$
$$= \text{prob}(\text{Q within } r \mid \text{G in field})\text{prob}(\text{G in field}).$$

This assumes that the probabilities are independent, obviously what we would like to test. A suitable model for the probabilities is the Poisson distribution, and in the interesting case where the probabilities are small we have

$$\text{prob}(\text{G in field}) = \varsigma_G \Omega$$

and

$$\text{prob}(\text{Q within } r) = \pi r^2 \varsigma_Q.$$

The answer we require is therefore

$$\text{prob(G in field and Q within } r) = \varsigma_G \varsigma_Q \Omega \pi r^2.$$

This is symmetrical in the quasar and galaxy surface densities as we would expect; it should not matter whether we searched first for a galaxy or for a quasar. Note the strong dependence on the search area that is specified *before the experiment*; if there is obscurity about this, then the probabilities are not well determined.

As an extension of this example, it is possible to calculate the probability of finding triples of objects aligned to some small tolerance (Edmunds & George, 1985). If the objects are all the same, the probability of a linear triple depends on the cube of the surface density and search area.

2.3 ... and Bayes' theorem

Bayes'[2] theorem is a simple equality, derived by equating prob(A and B) with prob(B and A). This gives the 'theorem':

$$\text{prob}(B \mid A) = \text{prob}(A \mid B)\text{prob}(B)/\text{prob}(A). \qquad (2.4)$$

In this, the denominator is a normalizing factor. The theorem is particularly useful when interpreted as a rule for induction; the data, the event A, are regarded as succeeding B, the state of belief preceding the experiment. Thus prob(B) is the *prior probability* which will be modified by experience. This experience is expressed by the *likelihood* prob($A \mid B$). Finally prob($B \mid A$) is the *posterior probability*, the state of belief after the data have been analysed.

Bayes' theorem by itself is a perfectly innocent identity, a mathematical truism. It acquires its force from its interpretation. To see what this force is, we return to the familiar and simple problem of drawing those coloured balls from urns. It is clear, even automatic, what to calculate; if there are M red balls and N white balls, the probability of drawing three red balls and two white ones is ...

---

[2] Who was Bayes? Thomas Bayes (1702–1761), English vicar, mathematician, statistician. His bibliography consisted of three works: one (by the vicar) on divine providence, the second (by the mathematician) a defence of the logical bases of Newton's calculus against the attacks of Bishop Berkeley, and the third (by the statistician and published posthumously) the famous *Essay towards Solving a Problem in the Doctrine of Chances*. There is speculation that it was published posthumously because of the controversy which Bayes believed would ensue. This must be an a-posteriori judgement. Surely Bayes could never have imagined the extent of this controversy without envisaging the nature of modern scientific data.

As a series of brilliant scientists realized, and as a series of brilliant scientists did not, this is generally not the problem we face. As scientists, we more often have a datum (three red balls, two white ones) and we are trying to infer something about the contents of the urn. This is sometimes called the problem of 'inverse probability'. (An exact analogy is the inverse problem of a sample of objects 'drawn' or obtained in a sky survey to a fixed sensitivity, and attempting to infer from the sample how these types of object are distributed throughout the Universe.) How does Bayes' theorem help? We interpret it to be saying

prob(**contents of urn** | **data**) $\propto$ prob(**data** | **the contents of the urn**)

and of course we can calculate the right-hand side, given some assumptions.

The urn example illustrates the principles involved; these are far more interesting than coloured balls.

Example There are N red balls and M white balls in an urn; we know the total $N + M = 10$, say. We draw $T = 3$ times (putting the balls back after drawing them) and get $R = 2$ red balls. How many red balls are there in the urn?

Our model (hypothesis) is that the probability of a red ball is

$$\frac{N}{N + M}.$$

We assume that the balls are not stratified, arranged in pairs, or anything else 'peculiar'. The probability of getting R red balls, the likelihood, is

$$\binom{T}{R} \left(\frac{N}{N + M} \right)^R \left(\frac{M}{N + M} \right)^{T-R}.$$

This is the number of permutations of the R red balls amongst the T draws, multiplied by the probability that R balls will be red and $T - R$ will not be red. (This is a *binomial* distribution; see Section 2.4.2.1.)

Thus we have the *probability (data, given the model)* part of the right-hand side of Bayes' theorem. We also need *probability (model)*, or the prior. We assume that the only uncertain bit of the model is N, which to start with we take as being uniformly likely between zero and $N + M$. Without bothering with the details at the moment, we plot up the left-hand side of Bayes' theorem (the posterior probability) as a function of N – see Figure 2.1. For a draw of, say, three red balls in five tries, the posterior probability peaks at

6; for 30 out of 50, the peak is still at 6 but other possibilities are much less likely.

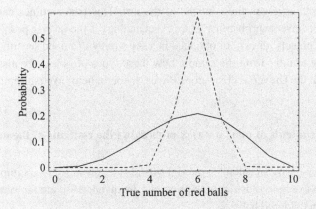

Figure 2.1   The probability distribution of the number of red balls in the urn, for five drawings (solid curve) and 50 drawings (dashed curve).

This seems unsurprising and in accord with common sense – but notice that we are speaking now of the probability of there being 1, 2, 3, ... red balls in a unique urn that is the subject of our experiment. We are describing our state of belief about the contents of the urn, given what we know (the data, and our prior information).

The key point of this example is that we have succeeded in answering our scientific question: we have made an inference about the contents of the urn, and can make probabilistic statements about this inference. For example, the probability of the urn containing three or fewer red balls is 11 per cent. We are assigning probabilities to these statements to N because we are using probability to reflect our degree of certainty. Our concern, as experimental scientists, is with what we can infer about the world from what we know.

Bayes' theorem allows us to make inferences from data, rather than compute the data we would get if we happened to know all the relevant information about our problem.

This may seem academic; but suppose we had data from two populations and wanted to know if the means were different. Many chapters of statistics textbooks answer the opposite question for us: given populations with two different means, what data would you get? The combination of interpreting probability as a consistent measure of belief, plus Bayes' theorem, allows us to answer the question we wish to pose: given the data, what are the probabilities of the parameters contained in our statistical model?

Another very significant point about this example is the use of prior information; again, we assigned probabilities to N to reflect what we know. Notice that although the word 'prior' suggests 'before the experiment', it really means 'what we know apart from the data'. Sometimes this can have a dramatic, even disconcerting effect on our inferences; see Section 2.5. Sometimes we even need a 'probability of a probability':

Example Return to the question of supernova rate per century and consider how to estimate this; call this ρ. Our data are four supernova in 10 centuries. Our prior on ρ, expressing our total ignorance, is uniform between 0 and 1; we have no preconceptions or information about ρ. A suitable model for prob(data $\mid \rho$) is the binomial distribution (Section 2.4.2.1), because in any century we either get a supernova or we do not (neglecting here the possibility of two supernovae in a century). Our posterior probability is then

$$\text{prob}(\rho \mid \text{data}) \propto \binom{10}{4} \rho^4 (1 - \rho)^6 \times \text{prior on } \rho.$$

We follow Bayes and Laplace in taking the prior to be uniform in the range 0 to 1. Then, to normalize the posterior probability properly, we need

$$\int_0^1 \text{prob}(\rho \mid \text{data}) d\rho = 1,$$

resulting in the normalizing constant

$$\int_0^1 \binom{10}{4} \rho^4 (1 - \rho)^6 d\rho,$$

which happens to be

$$\frac{\Gamma(10)\Gamma(4)}{\Gamma(14)} = B[5, 7],$$

where Γ is the Gamma function and B is the (tabulated) beta function. In general, for n supernovae in m centuries, the distribution is

$$\text{prob}(\rho \mid \text{data}) = \frac{\rho^n (1 - \rho)^{m-n}}{B[n + 1, m - n + 1]}.$$

Our distribution ($n = 4$, $m = 10$) peaks – unsurprisingly – at $4/10$, as shown in Figure 2.2.

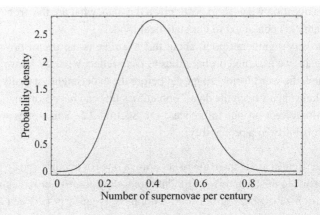

Figure 2.2   The posterior probability distribution for ρ, given that we have four supernovae in 10 centuries.

As the sample size increases, the distribution becomes narrower so that the peak posterior probability is more and more closely defined by the ratio of successes (supernovae, in our example) to sample size. This result is sometimes called the law of large numbers, expressing, as it does, the frequentist idea of a large number of repetitions resulting in a converging estimate of probability.

The key step in this example is ascribing a probability distribution to ρ, *in itself a probability*. This makes no sense in a frequentist approach, nor indeed in any interpretation of probabilities as objective. Even if we are prepared to leap this metaphysical hurdle, in very many cases the assignment of a prior probability is much more difficult than in this example. Indeed, the assignment of priors in the current example is very simple. For a long time the objection to Bayesian methods focused on the Bayes/Laplace uniform prior.

Both Jeffreys (1961) and Jaynes (1968) discuss the prior on ρ, arguing that in many cases a uniform prior is far too agnostic. By intricate arguments, they arrive at other possibilities:

$$\text{prob}(\rho) = \frac{1}{\rho(1-\rho)}$$

and the 'Haldane prior'

$$\text{prob}(\rho) = \frac{1}{\sqrt{\rho(1-\rho)}}.$$

These are intended to reflect the fact that in most experiments, we are expecting, with good reason, a yes or no answer.

Assigning priors when our knowledge is rather vague can be quite difficult, and there has been a long debate about this. Some 'obvious' priors (like the one we might use for location, simply uniform from $-\infty$ to ∞) are not normalizable and can sometimes get us into trouble. Out of the enormous literature on this subject, try Lee (2004) for an introduction, and Jaynes' writings for some fascinating arguments. One of the ways of determining a prior is the maximum entropy principle; we will see an example of such a prior later (Section 7.9). A common prior for a scale factor σ is Jeffreys' prior, uniform in $\log \sigma$.

Example The use of Bayes' theorem as a method of induction can be neatly illustrated by our supernova example. For simplicity, imagine that we establish our posterior distribution at the end of the nineteenth century, so that it is $\rho^4(1 - \rho)^6/B[5, 7]$, as shown earlier. At this stage, our data are four supernovae in 10 centuries. Reviewing the situation at the end of the twentieth century, we take this as our prior. The available new data consist of one supernova, so that the likelihood is simply the probability of observing exactly one event of probability ρ, namely ρ. The updated posterior distribution is

$$\text{prob}(\rho \mid \text{data}) = \frac{\rho^5(1 - \rho)^6}{B[6, 7]}$$

which peaks at $\rho = 5/11$, as we might expect.

In these examples, we have focused on the peak of the posterior probability distribution. This is one way amongst many of attempting to characterize the distribution by a single number. Another choice is the posterior mean, defined by

$$< \rho > = \int_0^1 \rho \, \text{prob}(\rho \mid \text{data}) \mathrm{d}\rho. \tag{2.5}$$

If we have had N successes and M failures, the posterior mean is given by a famous result called Laplace's rule of succession:

$$< \rho > = \frac{(N + 1)}{(N + M + 2)}.$$

In our example, at the end of the nineteenth century Laplace's rule would give $5/12$ as an estimate of the probability of a supernova during the twentieth century. This differs from the $4/10$ derived from the peak of the posterior probability, and it will do so in general.

Unless posterior distributions are very narrow, attempting to characterize them by a single number is frequently misleading. How best to characterize

Table 2.1 *The common probability density functions*

Distribution	Density function	Mean	Variance	Raison d'etre
Uniform	$f(x; a, b)$ $= 1/(b-a)\, a < x < b$ $= 0,\, x < a, x > b$	$(a+b)/2$	$(b-a)/12$	In the study of rounding errors; as a tool in studies of other continuous distributions.
Binomial	$f(x; p, q)$ $= \frac{n!}{x!(n-x)!} p^x q^{n-x}$	np	npq	x is the number of 'successes' in an experiment with two possible outcomes, one ('success') of probability p, and the other ('failure') of probability $q = 1 - p$. Becomes a Normal distribution as $n \to \infty$.
Poisson	$f(x; \mu) = e^{-\mu} \mu^x / x!$	μ	μ	The limit for the binomial distribution as $p \ll 1$, setting $\mu \equiv np$. It is the 'count-rate' distribution, e.g. take a star from which an average of μ photons are received per Δt (out of a total of n emitted; hence $p \ll 1$); the probability of receiving x photons in Δt is $f(x; \mu)$. Tends to the Normal distribution as $\mu \to \infty$.
Normal (Gaussian)	$f(x; \mu, \sigma)$ $= \frac{1}{\sigma\sqrt{2\pi}}$ $\times \exp[-(x-\mu)^2/2\sigma^2]$	μ	σ^2	The essential distribution: see text. The central limit theorem ensures that the majority of 'scattered things' are dispersed according to $f(x; \mu, \sigma)$
Chi-square	$f(\chi^2; \nu)$ $= \frac{\chi^{2(\nu/2-1)}}{2^{\nu/2}\Gamma(\nu/2)} \exp(-\chi^2/2)$	ν	2ν	Vital in the comparison of samples, model testing; characterizes the dispersion of observed samples from the expected dispersion, because if x_i is a sample of ν variables normally and independently distributed with means μ_i and variances σ_i^2, then $\chi^2 = \sum_{i=1}^{N}(x_i - \mu_i)^2/\sigma_i^2$ obeys $f(\chi^2; \nu)$. Invariably tabulated and used in integral form. Tends to Normal distribution as $\nu \to \infty$.
Student t	$f(t; \nu) = \Gamma[(\nu+1)/2]$ $\times \frac{(1+t^2/\nu)^{-(\nu+1)/2]}}{\sqrt{\pi\nu}\Gamma(\nu/2)}$	0	$\nu/(\nu - 2)$ (for $\nu > 2$)	For comparison of means, Normally distributed populations; if $n x_i$'s are taken from a Normal population (μ, σ), and if x_s and σ_s are determined, then $t = \sqrt{n}(\bar{x}_s - \mu)/\sigma_s$ is distributed as $f(t, \nu)$ where 'degrees of freedom' $\nu = n - 1$. Statistic t can also be formulated to compare means for samples from Normal populations with the same σ, different μ. Tends to Normal as $\nu \to \infty$.

the distribution depends on what is to be done with the answer, which in turn depends on having a carefully posed question in the first place.

2.4 Probability distributions

2.4.1 Concept

We have referred several times to *probability distributions*. The basic idea is intuitive; here is a little more detail.

Consider the boring experiment in which we toss four 'fair' coins. The probability of no heads is $(1/2)^4$; of one head $4 \times (1/2)^4$; of two heads $6 \times (1/2)^4$, etc. The sum of the possibilities for getting no heads to four heads is readily seen to be 1.0. If x is the number of heads $(0, 1, 2, 3, 4)$, we have a set of probabilities $\text{prob}(x) = (1/16, 1/4, 3/8, 1/4, 1/16)$; we have a *probability distribution*, describing the expectation of occurrence of event x. This probability distribution is discrete; there is a discrete set of outcomes and so a discrete set of probabilities for those outcomes.

In this sort of case we have a mapping between the outcomes of the experiment and a set of integers. Sometimes the set of outcomes maps onto real numbers instead, the set of outcomes no longer containing discrete elements. We deal with this by the contrivance of 'discretizing' the range of real numbers into little ranges within which we assume that the probability does not change. Thus, if x is the real number that indexes outcomes, we associate with it a probability *density* $f(x)$; the probability that we will get a number 'near' x, say within a tiny range δx, is $\text{prob}(x)\,\delta x$. We loosely refer to probability 'distributions', whether or not we are dealing with discrete outcomes.

Formally: if x is a continuous random variable, then $f(x)$ is its *probability density function*, commonly termed *probability distribution*, when

(i) $\text{prob}[a < x < b] = \int_a^b f(x)\mathrm{d}x$;
(ii) $\int_{-\infty}^{\infty} f(x)\mathrm{d}x = 1$, and
(iii) $f(x)$ is a single-valued non-negative number for all real x.

The corresponding *cumulative probability distribution function* is $F(x) = \int_{-\infty}^{x} f(y)\mathrm{d}y$. Probability distributions and distribution functions may be similarly defined for sets of discrete values of x; and distributions may be *multivariate*, functions of more than one variable.

2.4.2 Some common distributions

The better-known probability density functions appear in Table 2.1 together with location (where is the 'centre'?) and dispersion (what is the 'spread'?)

quantifiers. These quantifiers can be given by the first two *moments of the distributions* (Section 3.1):

$$\mu_1(\text{mean}) = \mu = \int_{-\infty}^{\infty} x f(x)\, dx \tag{2.6}$$

$$\mu_2(\text{variance}) = \sigma^2 = \int_{-\infty}^{\infty} (x - \mu_1)^2 f(x)\, dx \tag{2.7}$$

The square root of the variance, σ, is known as the standard deviation. Table 2.1 gives some indication of how or where each distribution arises. Three of them are of great importance, the binomial, Poisson, and Gaussian or Normal.

2.4.2.1 Binomial distribution

There are two outcomes – 'success' or 'failure'. This common distribution gives the chance of n successes in N trials, where the probability of a success at each trial is the same, namely ρ, and successive trials are independent. This probability is then

$$\text{prob}(n) = \binom{N}{n} \rho^n (1 - \rho)^{N-n}. \tag{2.8}$$

The leading term, the combinatorial coefficient, gives the number of distinct ways of choosing n items out of N:

$$\binom{N}{n} = \frac{N!}{n!(N-n)!}. \tag{2.9}$$

This coefficient can be derived in the following way. There are $N!$ equivalent ways of arranging the N trials. However, there are $n!$ permutations of the successes, and $(N - n)!$ permutations of the failures, which correspond to the same result – namely, exactly n successes, arrangement unspecified. Since we require not just n successes (probability p^n) but exactly n successes, we need exactly $N - n$ failures, probability $(1 - p)^{(N-n)}$ as well. The binomial distribution follows from this argument. The binomial distribution has a mean value given by

$$\sum_{n=0}^{N} n\, \text{prob}(n) = Np$$

and a variance or mean square value of

$$\sum_{n=0}^{N} (n - Np)^2 \text{prob}(n) = Np(1 - p).$$

Example Suppose we know, from a sample of 100 galaxy clusters selected by automatic pattern-recognition techniques, that 10 contain a dominant central galaxy. We plan to check a different sample of 30 clusters, now selected by X-ray emission. How many of these clusters do we expect to have a dominant central galaxy? If we assume that the 10 per cent probability holds for the X-ray sample, then the chance of getting n dominant central galaxies is

$$\text{prob}(n) = \binom{30}{n} 0.1^n 0.9^{30-n}.$$

For example, the chance of getting 10 is about 1 per cent; if we found this many, we would be suspicious that the X-ray cluster population differed from the general population.

Suppose we made these observations and did find 10 centrally dominated clusters. What can we do with this information?

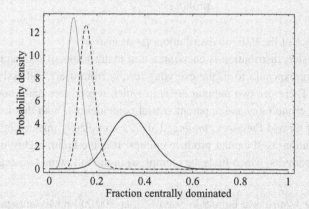

Figure 2.3 The posterior probability distribution for the observation that 10/30 X-ray-selected clusters are centrally dominated. The dark black line uses a uniform prior distribution for this fraction; the dashed line uses the prior derived from an assumed previous sample in which 10 out of 100 clusters had dominant central members. The light curve shows the distribution for this earlier sample.

The Bayesian thing to do is a calculation that parallels the supernova example. Assuming that the X-ray galaxies are a homogeneous set, we can deduce the probability distribution for the fraction of these galaxies that have a dominant central galaxy. A relevant prior would be the results for the original larger survey. Figure 2.3 shows the results, making clear that the data are not really sufficient to alter our prior very much. For example, there is only a 10 per cent chance that the centrally dominant fraction exceeds even 0.2; and, indeed, Figure 2.3 shows that the possibility of it being as high as 33 per cent is completely negligible. Our X-ray clusters differ markedly from the general population.

The binomial distribution is the parent of two other famous distributions, the Poisson and the Gaussian.

2.4.2.2 Poisson distribution

The Poisson distribution derives from the binomial in the limiting case of very rare (independent) events and a large number of trials, so that although $p \to 0, Np \to$ a finite value. Calling the finite mean value $\mu_1 = \mu$, the Poisson distribution is

$$\text{prob}(n) = \frac{\mu^n}{n!} e^{-\mu}. \tag{2.10}$$

The variance of the Poisson distribution, μ_2, is also μ.

The Poisson distribution is encountered in many walks of life, and despite our lifelong exposure to it, the everyday results from it are endlessly misinterpreted. There are two popular areas in which it produces daily newspaper headlines: crime rates and apparent spatial coincidences, both due to what is commonly termed *Poisson clumping*. This is a completely misleading term as the distribution itself cannot produce clumps. It is the brain, highly geared to detecting patterns, which finds the clumps, as in the following two examples.

Example Consider a homicide rate of 7 per 100 000 inhabitants per year (from Wikipedia, as for New York 2006; note that 48 cities of population 250 000 or greater in the USA had a higher rate). The low probability implies that the Poisson distribution applies. A population of 8 165 000 yields 572 homicides in the year, an average of 11 per week. What is the probability according to Poisson of 44 homicides in a week? The answer is about 1.5 per cent, i.e. roughly every two years there will be as many as 44 homicides in a week. You can be sure that the newspapers will headline 'Crime Wave' on such occasions – followed by political posturing and subsequent polarization of 'we need more police and tougher laws' versus 'we need to improve education/opportunities/social conditions'. There will be irresistible demands for action. There is no doubt merit in taking action;

but the fact remains that random numbers drawn from Poisson distributions do not provide a good basis for decision.

The second example considers 2D Poisson distributions, as in the scattering of seeds at random in a field. Divide the field into little square cells: the probability of a seed falling into any one is small so that Poisson applies.

Example Suppose we place dots in two dimensions as in Figure 2.4:

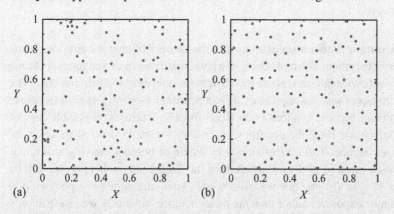

Figure 2.4   Poisson in two dimensions. In which figure are the dots randomly placed?

which diagram of the two in Figure 2.4 has the dots randomly distributed? Both have about 100 dots in a square of unity in size. Figure 2.4(b) looks the better bet, but in fact is not random at all. The dots were first placed on an $X - Y$ grid and then shifted at random slightly away from the grid on a scale of about 0.1 of the grid interval. Figure 2.4(a) is truly random, and paradoxically this is the one in which we can 'clearly' see clumps ('Poisson clumping'), voids, and even a semi-circle.

Here is trouble. The real-life encounter is the situation in which an interested party is trying to claim that, e.g., the occurrence of a disease is associated with the positions of power stations. The effect is the examination of two distributions (disease cases and power stations) overlaid, both of which will certainly exhibit some randomness with a dose of Poisson clumping, and some non-randomness. It is difficult to establish true probability or true evidence in the face of these effects, especially when we add in the effect of a-posteriori reasoning (Chapter 1). Moreover, a correlation does not imply a causal connection; see Chapter 4.

We mentioned sky distributions in Chapter 1, and, in particular, examining whether one type of extragalactic object is associated with another type. The difficulty of rigour is again severe, for the same reasons – two distributions, both with Poisson clumping, both probably non-random to start with, and a-posteriori statistics once again.

Finally, the Poisson distribution may play its biggest role in the lives of astronomers via the photons with which we measure emission from our chosen objects.

Example Poisson statistics govern the number of photons arriving during an integration. The probability of a photon arriving in a fixed interval of time is (often) small, at least at wavelengths shorter than the infrared (IR). The arrivals of successive photons are independent (apart from small correlations arising because photons obey Bose–Einstein statistics, negligible for our purposes). Thus, the conditions necessary for the Poisson distribution are met. Hence, if the integration over time t of photons arriving at a rate λ has a mean of $\mu = \lambda t$ photons, then the fluctuation on this number will be $\sigma = \sqrt{\mu}$. (In practice we usually only know the number of photons in a single exposure, rather than the mean number; obviously we can then only estimate the μ. This is the subject of an exercise in the next chapter.) There are the following limiting cases:

1. Suppose we are detecting our objects with no effective background either from the sky or from our instrumentation. The photons we receive are solely from the objects measured. This idealized situation means that with $\mu = \lambda t$, the scatter on μ is (Poisson) $\sigma = \sqrt{\lambda t}$. If we 'integrate' more by simply waiting for more photons, the *photon-limited* case,

$$\sigma \propto \sqrt{t}, \text{ while signal} \propto t.$$

Thus, *signal/noise* $\propto \sqrt{t}$.

2. Now suppose our object is barely visible against the sky background, as in charge-coupled device (CCD) imaging of very faint objects in the optical regime. Our signal is still $\mu = \lambda t$, but our noise is $\sigma = \sqrt{\lambda_{\text{sky}} t}$. The net result is

$$S/N \propto \frac{\lambda t}{\sqrt{\lambda_{\text{sky}} t}}.$$

We again get

$$S/N \propto \sqrt{t},$$

but note how much harder we have to work! The sky emission $\lambda_{sky} t$ will be much higher than the λt (the *sky-limited case*) so that to achieve similar S/N to the simple photon-limited case, we shall have to integrate much longer. Moreover, the fact that integration goes as $\sqrt{t}$ implies that there is an effective limit to how good an observation can be. For example, if an S/N of 2 is achieved in 2 hours, to get this S/N to 4 will require four times as long, 8 hours, a whole night of precious telescope time.

3. Now suppose we have extremely bright photon-limited objects, for which we require very short exposures only, because there is vast signal, e.g. CCD observations of bright stars. Then, with huge signal and short exposures, the *readout noise* of the device, fixed, time-independent and placed on top of the signal, may dominate the uncertainty. Calling this fixed error $\sigma_{readout}$,

$$S/N \propto \frac{\lambda t}{\sigma_{readout}}, \ or \ \propto t$$

for CCD of readout noise $\sigma_{readout}$.

4. Lastly, at the long-wavelength end of the spectrum, sub-millimetre and radio wavelengths, the flood of (relatively feeble) photons from objects is so enormous that the Poisson situation of rare events no longer applies. We are dealing with the *receiver-limited* case; S/N is governed by receiver sensitivity. The physical situation is quite different, but curiously, the result is a familiar one. We now have such a quantity of photons from the object – call this flux of photons S – that it is the receiver noise (assuming this to be thermal or roughly equivalent to thermal) which requires the integration:

$$S/N \propto \frac{S}{\sigma_{rec}/\sqrt{t}}, \ or \ \propto \sqrt{t}$$

for a receiver of thermal noise σ_{rec}.

2.4.2.3 Gaussian (Normal) distribution

Both the binomial and the Poisson distributions tend to the Gaussian distribution (Figure 2.5), large N in the case of the binomial, large μ in the case of the Poisson. The (univariate) Gaussian (Normal) distribution is

$$\mathrm{prob}(x) = \frac{1}{\sigma\sqrt{2\pi}} \exp\left[-\frac{(x-\mu)^2}{2\sigma^2} \right] \tag{2.11}$$

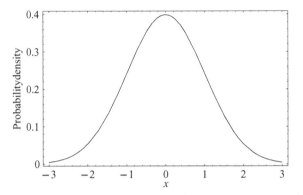

Figure 2.5   The Normal (Gaussian) distribution. The area under the curve is 1.00; the area between $\pm 1\sigma$ is 0.68; between $\pm 2\sigma$ is 0.95; and between $\pm 3\sigma$ is 0.997.

from which it is easy to show that the mean is μ and the variance is σ^2 (Section 3.1). For the binomial when the sample size is very large, the discrete distribution tends to a continuous probability density

$$\text{prob}(n) = \frac{1}{\sigma\sqrt{2\pi}} \exp\left[-\frac{(n-\mu)^2}{2\sigma^2}\right]$$

in which the mean $\mu = Np$ and variance $\sigma^2 = Np(1-p)$ are still given by the parent formulae for the binomial distribution. Here is an instance of the discrete changing to the continuous distribution: in this approximation we can treat n as a continuous variable (because n changes by one unit at a time, being an integer, and so the fractional change $1/n$ is small). The Poisson distribution performs similarly – as μ is increased, the Poisson distribution becomes more symmetrical and when $\mu \geq 20$, the distribution is virtually indistinguishable from a Gaussian with mean = μ and $\sigma = \sqrt{\mu}$.

Example Opinion-sampling, and '... expected to be correct within 2.1 per cent 19 times out of 20'. The phrase has become commonplace when presenting results from opinion polls. Where does it come from? *Sampling theory* is a subject in its own right – there are countless books. In opinion-polling, a representative sample of the population is asked the relevant question. (This statement is doubly loaded: how to choose the relevant sample and how to choose the question?) Suppose that total population opinion favours option A over option B by a ratio of 3:1, i.e. 75 per cent of the population favour option A, and $p = 0.75$. Then suppose that 4000 representative people were asked for their opinion; $N = 4000$. How accurate would our result be? The result is given by the binomial distribution, but as $pN = 3000.0$, far above the value of ~ 20 required for the binomial distribution to approximate the Gaussian distribution, we can use the Gaussian distribution. For this we get $\mu = 3000.0$ and $\sigma = 27.4$. We know that 95 per cent of the area of this distribution lies between $\pm 1.96\sigma = \pm 53.7$, and as a percentage of the 4000, this is 2.1 per cent. 0.95 is 19/20, i.e. if we repeated the survey, 19 times out of 20 we should get a result within 2.1 per cent of the true proportion.

Pollsters rarely sample more than 5000 people; thus, they generally quote accuracies of this order. The reason for this limited sampling (for any total size of population) is (a) sampling is costly, and (b) the result is unlikely to improve in practice. The '2.1 per cent' is only the *sampling error*, and at this level, other uncontrollable errors come to be of similar or even greater magnitude. These factors include (a) sampling bias, the difficulty of choosing a representative sample when you cannot ask all members of the population (the whole art of *sampling design*), and (b) voter fickleness, if, e.g., opinion polls are to do with elections. This comes in (at least) two forms – (a) sample members may not give their true opinions (particularly if those opinions are at extreme ends of the political spectrum), and (b) late factors, e.g. a telling debate exchange or a world event, may change voter opinions at the last moment.

The true importance of the Gaussian distribution and its dominant position in experimental science stems from the *central limit theorem*. A non-rigorous statement of this is as follows.

Form averages M_n from repeatedly drawing n samples from a population x_i with finite mean μ, variance σ^2. Then the distribution of

$$\left[\frac{(M_n - \mu)}{\sigma/\sqrt{n}} \right] \rightarrow \text{Gaussian distribution}$$

with mean 0, variance 1, as n $\rightarrow \infty$.

This is a remarkable theorem. What it says is that provided certain conditions are met – and they are in many situations – a little bit of averaging

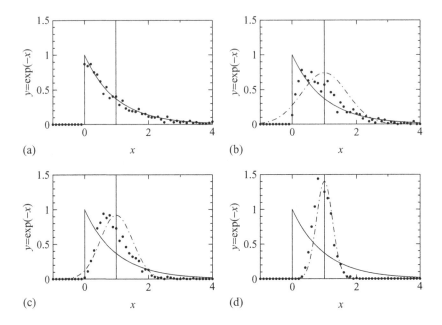

Figure 2.6   An indication of the power of the central limit theorem. The panels show successive amounts of 'integration': in (a), a single value has been drawn; in (b), 200 values have been taken from an average of two values; (c), 200 values from an average of four; (d), 200 values from an average of 16.

will produce a Gaussian distribution of results *no matter what the shape of the distribution from which the sample is drawn*. Even eyeball integration counts. It means that errors on averaged samples will always look 'Gaussian'. The reliance on Gaussian distributions, made valid by the unsung hero of statistical theory and, indeed, experimentation, the central limit theorem, shapes our entire view of experimentation. It is this theorem which leads us to describe our errors in the universal language of sigmas, and, indeed, to argue our results in terms of sigmas as well, which we explicitly or implicitly recognize as describing our place within or at the extremities of the Gaussian distribution.

Figure 2.6 demonstrates the compelling power of the central limit theorem. Here we have brutally truncated an exponential, clearly an extremely non-Gaussian distribution. The histogram obtained in drawing 200 random samples from the distribution (see Figure 2.6) follows it closely. When 200 values resulting from averaging just four values have been drawn, the distribution is already becoming symmetrical; by the time 200 values of 16 averages have been taken, it is virtually Gaussian.

Before leaving the central limit miracle and Gaussian distributions, it is important to emphasize how tight the tails of the Gaussian distribution are (Table B.2). The range $\pm 2\sigma$ encompasses 95.45 per cent of the area. Thus, the infamous 2σ result has a less than 5 per cent chance of occurring by chance. But we scoff – because the error estimates are difficult to make, and observers are optimistic. Things upset the distribution; there are outlying points. Thus, astronomers feel it necessary to quote results in the range 3σ to even 10σ, casting inevitable doubt on the belief in their own error estimates. In fact, experimentalists are aware of another key feature of the central limit theorem; the convergence to a Gaussian happens fastest at the centre of the distribution, but the wings may converge much more slowly to a Gaussian form. Interesting results (the 10σ ones) of course acquire their probabilistic interpretation from knowing the shape of the tails to high accuracy. In practice it is virtually impossible to be certain that an error distribution is truly Gaussian because the amount of data required to check the tails is, well, astronomical.

2.4.2.4 Power-law distribution

A very different type of distribution figures prominently in the life of astronomers – the power-law distribution. Take N as the number of objects or events that have a measured property (say luminosity) either greater than a value L (integral form) or within the bin dL centred on L. With γ as the power-law exponent,

$$N(> L) = K\,L^{\gamma+1}, \text{ integral form, } N > L, \text{ or} \tag{2.12}$$

$$dN = (\gamma + 1)\,K\,L^{\gamma}\,dL, \text{ differential form, } dN \text{ objects in } dL. \tag{2.13}$$

This is a *scale-free* or *scale-independent* distribution, because if $f(x) = x^{\gamma}$, then $f(ax) = a^{\gamma} \cdot x^{\gamma} = \text{const} \cdot x^{\gamma} = \text{const} \cdot f(x)$, the definition of scale independence.

This distribution is so different that intuition gained from our experience with Gaussians and progenitors binomial and Poisson becomes dangerous. As it stands, the mean and variance are both infinite, but in real life, this is not the case; something always physically limits each end. Nevertheless, steep power laws can appear in astronomy extending over decades, and when confronted with such distributions, astronomers have made serious mistakes. Trying to describe results for objects selected from power-law distributions in terms of means and sigmas becomes hugely misleading at best, meaningless at worst. The slopes of the power laws are generally steep and inverse, that is to say there are many more objects of small L than large. This leads to very strong biases when objects are drawn from such a parent population, particularly if relatively large measurement errors are involved.

Power laws arise in many walks of life (see, e.g. Ball, 2004): they describe the distribution of fluctuations in the economic market, growth rates of firms, the distribution of salaries, and the size distributions of, e.g., avalanches, earthquakes and forest fires. The important thing is *criticality*. For instance, like the onset of an avalanche to adding sand at the apex of a sand-pile to a point where it suddenly becomes unstable. At criticality, there is no prescription as to whether a small region will slide and stop; or whether the entire side of the pile will be collectively triggered to break away and collapse. (These experiments have been done.) Earthquakes, stock-market fluctuations and forest-fires can be seen to follow similar patterns. So do many phenomena in the physical sciences, critical exponents for fluids, interconnectivity sub-networks on the Internet (the system is scale free), etc. The main feature of the power law in each case is that *the exponent is negative*; there are many more small things than large – many little sandslips go nowhere while very rare are the catastrophic collapses of the pile.

We meet these probability distributions in astronomy daily – the Salpeter Mass Function, magnitude or source counts (surface densities of objects on the sky), luminosity functions, the primordial fluctuation spectrum and more. There are always more faint objects than bright, more low-mass or low-luminosity objects than high; so that the exponent γ is invariably negative. The power law in pure form does not, of course, obey the formal definition of a probability distribution (Section 2.4.1). However, there are those physical limits which generally set upper and lower bounds.

Example Consider the counts of objects from one 15-arcmin-square region of the UK Schmidt Telescope survey, R-band, centre selected at random (but at high Galactic latitude), and with magnitudes measured off the plates by the CCD system of the SuperCosmos Sky Survey. Figure 2.7 shows the image and the results.

Power laws (fitted via maximum likelihood; see Section 6.1) describe the data well. Most objects in the total count are stars. The objects classified as galaxies are seen to have a significantly steeper slope, as expected for a distant population distributed approximately uniformly about us.

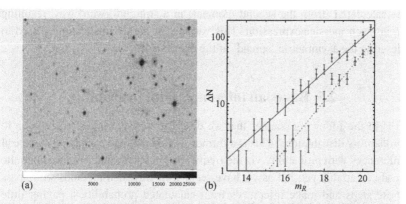

Figure 2.7 (a) A 15-arcmin square of sky from the R-band UKSTU sky survey at RA 22h, Dec −18°. The scanning process recognizes about 750 images in the area. (b) The number–magnitude count (a 'source count' at other wavelengths) for all objects (dots) and for objects classified as galaxies (triangles). The data are plotted in numbers of objects in the area in equal bins of 0.4 mag, a 'differential count'. Power laws have been fitted to the data. As magnitude is an inverse logarithmic scale, $m_1 - m_2 = -2.5\log(L_1/L_2)$ where L is luminosity, the power-law index is positive; of course, it would be negative if the plot were in terms of apparent luminosities.

Be aware of the many pitfalls of the power law. This distribution has no saving grace via approximations to familiar well-bounded distributions. We have mentioned selection bias. Add these issues: (a) is this power law an integral or differential distribution? This is one common way of getting the index wrong by unity. (b) Is the binning on a uniform or a log scale? If a differential distribution is binned via a uniform $\Delta \log L$ scale, instead of via ΔL, the slope is reduced by unity (Figure 2.7). (c) There is no characteristic scale or spread for such a distribution, although in practice the physical limits always manage to provide high and low end-stops. Do not rely on these to make power laws tractable in terms of our normal usage of means and standard deviations. See the final example in Section 2.6.

For example, given a fixed range of a power law between, say, a and b, the mean (from the first moment) can be calculated:

$$\mu = \left(\frac{\gamma + 1}{\gamma + 2}\right) \left[\frac{b^{\gamma+2} - a^{\gamma+2}}{b^{\gamma+1} - a^{\gamma+1}}\right]. \tag{2.14}$$

The expression breaks down for $\gamma = -1$ or -2; results can readily be derived for these special cases. This mean in the interval a to b is useful in showing how skewed the distribution over the interval is, and also in providing suitable abscissa points for plotting a power law. The variance about this mean can also be calculated (from the second moment) in a straightforward way, resulting in an even messier expression. This variance is highly misleading if used to describe the asymmetric 'spread' of the distribution.

2.5 Bayesian inferences with probability

It is in the following chapter that we describe *statistics* and their relation to probability distributions. This is a further step on the road to making statistical inferences with probability via the frequentist/classical route. We are not quite ready to do this yet; but as we have seen, with Bayesian methods we bypass these steps and make inferences from calculated probabilities, paying little regard to the name of the distribution we have calculated (if, indeed, it has a name). So what, then, is the relation between Bayesian inference and the probability distributions we have been describing?

One such connection is in estimating the parameters of assumed probability distributions, i.e. we are assuming a model for our data and wish to find out how this model is characterized. In this we are essentially *data modelling*; see Chapters 6 and 7. We have a probability distribution $f(\text{data} \mid \vec{\alpha})$ in mind and we wish to know the parameter vector $\vec{\alpha}$. The Bayesian route is clear: compute the posterior distribution of $\vec{\alpha}$, as we have shown in several examples in this chapter.

Example Suppose we have N data X_i, drawn from a Gaussian of known variance σ but unknown μ. The parameter we want is μ. To proceed, we need a prior on μ; we take the so-called 'diffuse' prior, where

$$\text{prob}(\mu) = \text{constant}$$

over some wide range of μ, the range defined by our knowledge of the problem. Of course we might have more precise information available. From Bayes, the posterior distribution follows at once:

$$f(\mu \mid \text{data}) \propto \exp\left[-\frac{\sum_{i=1}^{N}(X_i - \mu)^2}{2\sigma^2}\right]$$

and with some simplification (including the absorbing of terms not depending on μ into the '$\propto$')

$$f(\mu \mid \text{data}) \propto \exp\left[-\frac{\frac{1}{N}\sum_{i=1}^{N}(X_i - \mu)^2}{2\frac{\sigma^2}{N}}\right]$$

so that the average of the data is distributed around μ, with variance σ^2/N.

This method is related to the classical technique of maximum likelihood. If the prior is 'diffuse', as in the example, then the posterior probability is proportional to the likelihood term $f(\text{data} \mid \vec{\alpha})$. Maximum likelihood picks out the mode of the posterior, the value of $\vec{\alpha}$ which maximizes the likelihood. This amounts to characterizing the posterior by one number, an approach which is often useful because of powerful theorems on maximum likelihood. We consider this in more detail in Section 6.1; some exercises at the end of this chapter illustrate the procedure.

Following the discussion of probability distributions, we are now able to consider some more detailed Bayesian problems, with surprising outcomes.

Example Suppose that we make an observation with a telescope at a randomly selected position in the sky. Our model of the data (an event labelled D, consisting of the single measured flux density f) is that it is distributed in a Gaussian way (Section 2.4.2.3) about the true flux density S with a variance (Section 2.4.2) σ^2. The extensive body of source counts also tells us the a-priori distribution of S; for the purposes of this example, we approximate this information by the simple prior for a static Euclidean universe

$$\text{prob}(S) = K S^{-5/2}$$

describing our prior state of knowledge. K normalizes the counts to unity; there is presumed to be one source in the beam at some flux-density level. The probability of observing f when the true value is S we take to be

$$\exp\left[-\frac{1}{2\sigma^2}(f - S)^2\right].$$

Bayes' theorem then tells us that

$$\text{prob}(S \mid D) = K' \exp\left[-\frac{1}{2\sigma^2}(f - S)^2\right] S^{-5/2},$$

with the normalizations condensed into the single parameter K'. If we were able to obtain n independent flux measurements f_i, then the result would be

$$\text{prob}(S \mid D) = K'' \exp\left[-\frac{1}{2\sigma^2}\sum_{i=1}^{n}(f_i - S)^2\right] S^{-5/2}.$$

Suppose, for specific example, that the source counts were known to extend from 1 to 100 units, the noise level was $\sigma = 1$, and the data were 2, 1.3, 3, 1.5, 2 and 1.8. In Figure 2.8 are the posterior probabilities for the first two, then four, then six measurements. The increase in data gradually overwhelms the prior but the prior affects conclusions markedly (as it should) when there are few measurements.

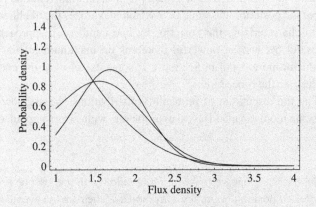

Figure 2.8   Measurement of flux density given a power-law prior (source count) and a Gaussian error distribution. The posterior probability distribution for flux density is plotted for two, four and then six of the measurements listed in the text; the form of the curve approaches Gaussian as numbers increase.

If, subsequently, we looked at a survey plate of the region we had observed, and found that the emission was from some category of object (say, a quasar) with different source counts, our prior would change and so would the posterior probability. In turn, our idea of the most probable flux density would also change.

In this example, the prior seems to be well determined. However, in some cases we wish to estimate quantities where the argument is not so straightforward. What would we take as the prior in the previous example if we were making the first ever measurements at a new wavelength?

2.6 Monte Carlo generators

We need to come clean about random numbers at this point – we have already used random-number (Monte Carlo) generation (Figure 2.6). In fact there are frequent occasions in probability calculations, hypothesis testing and model-fitting (see Chapters 5 and 7) when it is essential to have recourse to a set of numbers distributed perhaps how we *guess* the data might be. We may wish to test a test to see if it works as advertised; we might need to test efficiency of tests; we might wish to determine how many iterations we require; or we might even want to test that our code is working. We need random numbers, either uniformly distributed, or drawn randomly from a parent population of known frequency distribution. If we introduce them here, then nothing is sacred – much of what we say from here on, and of what we have said to this point, is open to direct simulation by the reader through random-number generation – commonly known as 'Monte Carlo'.

It is important not to compromise any such analyses with bad random data. *Numerical Recipes* (Press *et al.*, 2007) presents a number of methods for random-number generation, from single expressions to powerful routines. A key issue is *cycle length*; how long is it before the pseudo-random cycle is repeated? (Or, how many random numbers do you need?) At some level, therefore, it is necessary to understand the characteristics of the generator. Moreover, it is essential to follow the prescribed implementation precisely. It may be tempting to try some 'extra randomizing', for example by combining routines or by modifying seeds. Be very scared of any such process. Finally it is easy to forget that the routines generate *pseudo-random numbers*. Run them again from the same starting point and you will get the same set of numbers.

How do we draw a set of random numbers following a *given* frequency distribution? Suppose, following the cautions above, that we have a way of producing random deviates that are uniformly distributed over the range 0.0–1.0, in, say, the variable α; and we have a functional form for our frequency distribution $dn/dx = f(x)$. We need a transformation $x = x(\alpha)$ to distort the uniformity of α to follow $f(x)$. But we know that

$$\frac{dn}{dx} = \frac{dn}{d\alpha}\frac{d\alpha}{dx} = \frac{d\alpha}{dx} \tag{2.15}$$

as $dn/d\alpha$ is uniform by assumption; thus

$$\alpha(x) = \int^x f(x)dx, \tag{2.16}$$

from whence the required transformation $x = x(\alpha)$, the *inverse* of Equation (2.16), i.e. solve (2.16) for x.

Example Thus, e.g., the example in Section 6.1: the source-count random distribution is $f(x)dx = -1.5x^{-2.5}dx$, a 'Euclidean' differential source count. Here $d\alpha = -1.5x^{-2.5}dx$, $\alpha = x^{-1.5}$, and the transformation is the inverse, $x = \alpha^{-1/1.5}$.

The very same procedure works if we do *not* have a functional form for $f(x)dx$. If this is a histogram, we need simply to calculate the integral version, and perform the inverse function operation as above. Simple interpolation of a curve is generally necessary.

Example Figure 2.9 shows an example of choosing uniformly distributed random numbers and transforming them to follow the frequency distribution prescribed by a given histogram.

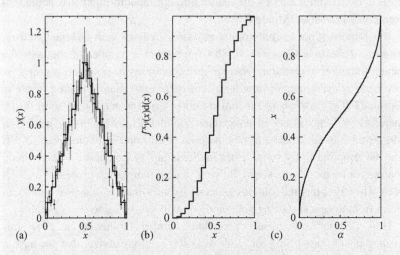

(a)   (b)   (c)

Figure 2.9   An example of generating a Monte Carlo distribution following a known histogram. (a) The step-ladder histogram, with points from 2000 trials, produced by (i) integrating the function (b) and (ii) transforming the axes to produce the inverse function of the integrated distribution (c). The points with $\sqrt{N}$ error bars in the left diagram are from drawing 2000 uniformly distributed random numbers and transforming them according to the right diagram.

How do we draw numbers obeying some arbitrary distribution? The prescription above is all very well, and works when integration of the function can be done; it cannot in many cases, the Gaussian being an obvious one. There is another method, the *rejection method*, of generating random numbers to a prescription, and it can be coded in just a few lines. Details are in Lyons (1986) and Press *et al.* (2007).

The algorithm assumes that we need random numbers distributed according to some awkward function f, but which is 'covered' by a function g from which we can get random numbers. This means that $f \leq Mg$ always, for some positive constant M. The steps of the algorithm are simple.

  (i) Draw a random number X_i from g.
 (ii) Draw a random number U_i, uniformly distributed between 0 and 1.
(iii) If $f(X_i)/Mg(X_i) \geq U_i$, accept X_i.
(iv) Otherwise reject X_i and repeat.

A nice proof of why this works is in MacKay (2003). While simple, the method will become very inefficient in high dimensions – but it is at least a simple route to getting samples from a multivariate distribution.

As a Monte Carlo product to which we shall turn a number of times during the course of this, let us invent a very simple universe. We will add complexity to it as we go; and it will serve to illustrate a number of specific issues.

Example Figure 2.10 shows the opening stages of creating our own little universe. It is in Euclidean space; its extent is from 0 (the centre) to 1.0 (the edge; we told you it was a model). We fill it uniformly with 10^6 objects; to do so, we simply need 10^6 values of (r, θ, ϕ), with $0 < r < 1, 0 < \phi < 2\pi$ and $0 < \theta < \pi$ where these are the conventional spherical coordinates. These must be chosen such that each volume element $r^2 \sin\theta \, dr \, d\theta \, d\phi$ is uniformly populated. As volume is $\propto r^3$, we need an inverse function to convert our uniform random numbers (0.0 to 1.0) to random distances R via the inverse function, i.e. $R = \mathcal{R}^{1/3}$ where $\mathcal{R}$ is the random number selected uniformly over the range 0 to 1. We assign every object the same luminosity of $L = 10$ units. We suppose our universe is clean and non-relativistic, so that the flux received from each object is L/r^2.

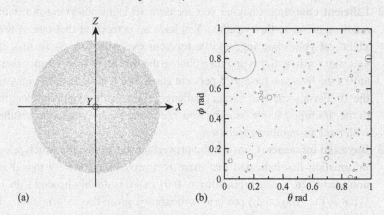

(a)   (b)

Figure 2.10   (a) Looking straight down the Y-axis of our 3D toy universe with $R_{max} = 1.0$ and uniformly distributed objects plotted in (x, y, z); this is a thin slice $\pm\Delta y$ about $y = 0$; (b) assigning all objects the same luminosity L, pretending that we are at the centre and observing with a telescope of sensitivity $0.1L/R_{max}^2$, here is a typical patch of sky. The sizes of the circles are proportional to the flux measured by our telescope. Note the similar sorts of pattern to those seen in Figure 2.4, the 'Poisson clumping'.

We shall return to random numbers – in Chapters 6 and 7 in particular – where we discover how to use them to do difficult integration via Monte Carlo techniques (Section 7.6), and how to do some important tests such as bootstrap and jackknife (Section 6.6). In the interim, we urge you to gain familiarity with random number generation: it is mandatory for examining what data can tell you via probabilities and statistical inference. Besides, you may find it quite entertaining; see the following exercises.

Exercises

2.1 **A warm-up on coin-tossing.** This is not an astronomical problem but does provide a warm-up exercise on probability and random numbers. Every computer has a way of producing a random number between 0 and 1. Use this to simulate a simple coin-tossing game where player A gets a point for heads, player B a point for tails. Guess how often in a game of N tosses the lead will change; if A is in the lead at toss N, when was the previous change of lead most likely to be? And by how much is a player typically in the lead? Try to back up these guesses with calculations. For many more game-based illustrations of probability, see Haigh (1999).

2.2 **Efficient choosing.** Imagine you are on a 10-night observing run with a colleague, in settled weather. You have an agreement that one of the nights, of your choosing, will be for your exclusive use. Show that, if you wait for five nights and then choose the first night that is better than any of the five, you have a 25 per cent chance of getting the best night of the 10. For a somewhat harder challenge, show that the optimum length of the 'training sample' is a fraction $1/e$ of the total. Check your results with random-number generation.

2.3 **Bayesian inference.** Consider the proverbial bad penny, for which prior information has indicated that there is a probability of 0.99 that it is unbiased ('OK'); or a probability of 0.01 that it is double-headed ('dh'). What is the (Bayesian) posterior probability, given this information, of obtaining seven heads in a row? In such a circumstance, how might we consider the fairness of the coin? Or of the experimenter who provided us with the prior information? What are the odds on the penny being fair?

2.4 **Laplace's rule and priors.** Laplace's rule (Section 2.3) $\overline{\rho} = (N+1)/(N+M+2)$ depends on our prior for ρ. If we have one success and no failures, consider what the rule implies, and discuss why this is odd. How is the rule changed for alternative priors, for example Haldane's?

2.5 Bayesian reasoning in an everyday situation. The probability of a certain medical test being positive is 90 per cent, if the patient has disease D. If your doctor tells you the test is positive, what are your chances of having the disease? If your doctor also tells you that 1 per cent of the population have the disease, and that the test will record a false positive 10 per cent of the time, use Bayes' theorem to calculate the chance of having D if the test is positive. Simulate the experiment via Monte Carlo.

2.6 Inverse χ^2 statistic. For a Gaussian of known mean (say, zero), show that the posterior distribution for the variance is inverse χ^2. Use the 'Jeffreys' prior' for the variance: $\text{prob}(\sigma) = 1/\sigma$. Comment on the differences between this result and the one obtained by using a uniform prior on σ.

2.7 Maximum likelihood and the Poisson distribution. Suppose we have data which obey a Poisson distribution with parameter μ, and in successive identical intervals we observe $n_1, n_2 \ldots$ events. Form the likelihood function by taking the product of the distributions for each n_i, and differentiate to find the maximum likelihood estimate of μ. Is it what you expect?

2.8 Maximum likelihood and the exponential distribution. Suppose we have data $X_1, X_2 \ldots$ from the distribution $1/2a \exp(- \mid x \mid /a)$. Compute the posterior distribution of a for a uniform prior, and Jeffreys' prior $\text{prob}(a) \propto 1/a$. Do the differences seem reasonable? Which prior would you choose? If a were known, but the location μ was to be found, what would be the maximum likelihood estimate?

2.9 Birth control. Imagine a society where boys and girls were (biologically) equally likely to be born, but families cease producing children after the birth of the first boy. Are there more males than females in the population? Attack the problem in three ways: pure thought, by a Monte Carlo simulation, and by an analytic calculation.

2.10 Univariate random numbers. Work out the inverses of the integral functions required to generate (a) $f(x) = 2x^3$, (b) a power law, representative of luminosity functions, $f(x) = x^{-\gamma}$. Use these results to produce random experiments following these probabilities by drawing 1000 random samples uniformly distributed between 0 and 1; verify by comparison with the given functions.

2.11 Make your own toy universe. Set up a toy universe as in the example. With a single luminosity of, say, $L = 10$ units and the telescope sensitive to fluxes down to $0.1L/r_{\max}^2$, calculate the number density per unit area as a function of received flux, and plot this source count. Now assign luminosities at random to the objects, following a power law in luminosity with slope of -3. 'Resurvey' the sky and calculate the source count again. Understand the results. This example is a prelude to Chapter 8.

3
Statistics and expectations

Lies, damned lies and statistics.

(Anon)

In embarking on statistics we are entering a vast area, enormously developed for the Gaussian distribution in particular. This is classical territory; historically, statistics were developed because the approach now called Bayesian had fallen out of favour. Hence, direct probabilistic inferences were superseded by the indirect and conceptually different route, going through statistics and intimately linked to hypothesis testing. The use of statistics is not particularly easy. The alternatives to Bayes' methods are subtle and not very obvious; they are also associated with some fairly formidable mathematical machinery. We will avoid this, presenting only results and showing the use of statistics, while trying to make clear the conceptual foundations.

3.1 Statistics

Statistics are designed to summarize, reduce or describe data. The formal definition of a *statistic* is that it is some function of the data alone. For a set of data $X_1, X_2, \ldots$, some examples of statistics might be the average, the maximum value or the average of the cosines. Statistics are therefore combinations of finite amounts of data. In the following discussion, and indeed throughout, we try to distinguish particular fixed values of the data, and functions of the data alone, by upper case (except for Greek letters). Possible values, being variables, we will denote in the usual algebraic spirit by lower case.

The summarizing aspect of statistics is exemplified by those describing (a) *location* and (b) *spread* or *scatter*.

(a) The location of the data can be indicated by various combinations:

 Average, denoted by overlining: $\overline{X} = 1/N \sum_{i=1}^{N} X_i$

 Median: arrange X_i according to size; renumber. Then $X_{\text{med}} = X_j$ where $j = N/2 + 0.5$, N odd, $X_{\text{med}} = 0.5(X_j + X_{j+1})$ where $j = N/2$, N even.

 Mode: X_{mode} is the value of X_i occurring most frequently; it is the location of the peak in the histogram of X_i.

(b) Statistics indicating the scale or amount of scatter in the data are, for example,

 Mean deviation: $\overline{\Delta X} = \frac{1}{N} \sum_{i=1}^{N} | X_i - X_{\text{med}} |$.

 Mean square deviation: $S^2 = \frac{1}{N} \sum_{i=1}^{N} (X_i - \overline{X})^2$.

 Root-mean-square deviation: rms $= S$.

We are so familiar with statistics like these that a result such as '$D = 8.3 \pm 0.1$ Mpc' provokes no questions. But what does it mean? It does not tell us the probability that the true value of D is between 8.2 and 8.4. We usually *assume* that a Gaussian distribution applies, placing our faith in the central limit theorem. Knowing the distribution of the errors allows us to make probabilistic statements, which are what we need. After all, if there were only a 1 per cent chance that the interval [8.2, 8.4] contained the true value of D, we might not regard the stated error as being very useful.

So this is one key aspect of statistics; they are associated with distributions. In fact they are most useful when they are estimators of the parameters of distributions. In quoting our measurement of D, we are hoping that 8.3 is an estimate of the parameter μ of some Gaussian, while 0.1 is an estimate of σ.

The other key aspect of statistics is that they are to be interpreted in a classical, not Bayesian framework. We need to look carefully at this distinction; it parallels our discussion of those coloured balls in the urn. Assuming a true distance D_0, a classical analysis tells us that D is (say) Normally distributed around D_0, with a standard deviation of 0.1. So we are to imagine many repetitions of our experiment, each yielding a value of the estimate D which dances around D_0. We might form a *confidence interval* (such as [8.2, 8.4]) which will also dance around randomly, but will contain D_0 with a probability we can calculate. Just as in the case of the coloured balls, this approach assumes the thing we want to know, and tells us how the data will behave.

A Bayesian approach circumvents all this; it deduces directly the probability distribution of D_0 from the data. It assumes the data, and tells us the thing we want to know. There are no imagined repetitions of the experiment. Conceptually it is clearer than classical methods, but these are so well developed and established (particularly for the Gaussian) that we will give some explanation of classical statistics now, and, indeed, use classical results in many places in this book.

It is worth remembering, however, that statistics of known usefulness are quite rare; the intensive development of statistics based on the Gaussian should not blind us to this fact. In many cases of astronomical interest we may need to derive useful statistics for ourselves. By far the easiest method for doing this is maximum likelihood (Section 6.1) and this is so close to a Bayesian method that we may expect to be doing Bayesian, not classical, inference in any new case where we cannot draw immediately on classical results.

To repeat, statistics are properties of the data and only of the data; they summarize, reduce or describe the data. Parameters such as the μ and σ of the Poisson and Gaussian distributions define these distribution functions and are *not* statistics. But we may anticipate that our data do follow these or other distributions and we may therefore wish to relate statistics from the data to parameters describing the distributions.

This is done through *expectations* or *expectation values*, long-run average properties depending on distribution functions. The expectation $E[f(x)]$ of some function f of a random variable x, with distribution function g, is defined as

$$E[f(x)] = \int f(x)g(x)\,dx, \qquad (3.1)$$

i.e. the sum of all possible values of f, weighted by the probability of their occurrence. We can think of the expectation as being the result of repeating an experiment many times, and averaging the results. We might, for example, compute a mean value of a fixed number of data, $\overline{X}$; if we repeat the experiment many times, we will find that the average of $\overline{X}$ will converge to the true mean value, the expectation of the function $f(x) = x$:

$$E[x] = \int xg(x)\,dx. \qquad (3.2)$$

Note that the expectation is not necessarily to be understood as referring to a very large sample; we can ask for the expectation value of a combination of a finite number of data – like, for example, an average.

The statistic S^2 we introduced before should likewise converge to the variance, defined by

$$\text{var}[(x - \mu)^2] = E[(x - \mu)^2] \qquad (3.3)$$
$$= \int (x - \mu)^2 g(x)\,dx. \qquad (3.4)$$

However, as we shall see, we do occasionally have to take some care that the integrals actually exist.

Example Take our favourite distribution, the Gaussian. The probability of getting a datum x near μ is

$$g(x \mid \mu, \sigma) = \frac{1}{\sqrt{2\pi}\sigma} \exp \frac{-(x - \mu)^2}{2\sigma^2}$$

but what are these parameters μ and σ? It is not difficult to show (changing variables and using standard identities) that

$$E[x] = \int xg(x \mid \mu, \sigma) \, dx = \mu, \tag{3.5}$$

and

$$E[(x - \mu)^2] = \int (x - \mu)^2 g(x \mid \mu, \sigma) \, dx = \sigma^2. \tag{3.6}$$

We would therefore expect that the average $\overline{X}$ and mean-square deviation S^2 would be related to μ and σ^2. As any statistics text will show, indeed $\overline{X}$ and S^2, although they are functions only of the data and therefore show random variation, will converge to μ and σ^2 when we have a lot of data.

Other distributions give different results. For the exponential distribution

$$f(x) = \frac{1}{2a} \exp\left(-\frac{|x|}{a}\right),$$

the expectation of $|x|$ is the width parameter a. The pathological Cauchy distribution

$$f(x) = \frac{1}{\pi(1 + x^2)}$$

has the alarming property that the expectation of the average of N data is, again, exactly the same Cauchy distribution; the location can just as well be estimated with just one datum. The difficulty arises because the distribution has such wide wings. In astronomy, power-law distributions (Section 2.4.2.4) are common. It is worth checking any piece of remembered statistics, as it is almost certain to be based on the Gaussian distribution.

Other expectations of theoretical importance are known as the *nth central moments*:

$$\mu_n = \int (x - \mu)^n g(x) \, dx \tag{3.7}$$

where g is some probability distribution. They are estimated analogously by suitable averages to the way in which mean and variance were estimated in the previous example. They are sometimes useful for characterizing the shape of distributions, although they are very sensitive to outliers. Two descriptors using moments are common: skewness, $\beta_1 = \mu_3^2$, indicates deviation from symmetry (= 0 for symmetry about μ); and kurtosis, $\beta_2 = \mu_4/\mu_2^3$, indicates the degree of peakiness (= 3 for the Gaussian distribution).

As an estimate of the degree of concentration of an unknown distribution near its mean, the Chebyshev inequality is sometimes useful: for any positive integer n, and data X drawn from a distribution of mean μ and variance σ^2,

$$\text{prob}(|\, X - \mu \,| > n\sigma) \leq \frac{1}{n}. \tag{3.8}$$

This is very conservative but is sometimes better than nothing as an estimate.

3.2 What should we expect of our statistics?

We have but a few of the data X_i but we want to know how all of them are organized; we want an approximation to their *probability* or *frequency* distribution and we want it for as little effort as we can get away with (efficiently) and as accurately as possible (robustly). Suppose, for instance, that we are drawing samples from a population obeying a Gaussian defined by $\mu = 0$, $\sigma = 1$. Figure 3.1 conveys some indication of how size of sample would affect estimates of these parameters.

There are, then, at least four requirements for statistics.

1. They should be *unbiased*, meaning that the expectation value of the statistic turns out to be the true value. For the Gaussian distribution (Section 2.4.2.3), for data X_i, $\overline{X}$ is indeed an unbiased estimate of the mean μ, but the unbiased estimate of the variance σ^2 is

$$\sigma_s^2 = \frac{1}{N-1} \sum_{i=1}^{N} (X_i - \overline{X})^2$$

which differs from the expectation value of S^2 by the factor $N/(N-1)$. The factor is confusing: σ_s^2, sometimes referred to as the *sample variance*, is the estimator for the *population variance* σ^2. (The difference is understandable as follows. The X_i of our sample are first used to get $\overline{X}$, an estimate of μ, and although this is an unbiased estimate of μ it is the estimate which yields a *minimum* value from the sum of the squares of the deviations of the

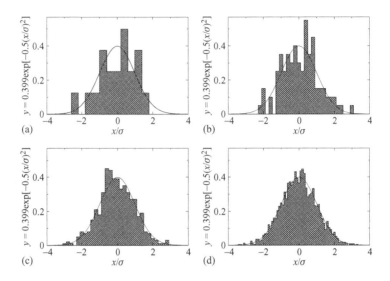

Figure 3.1 x_i drawn at random from a Gaussian distribution of $\sigma = 1$: (a) 20 values, (b) 100 values, (c) 500 values, (d) 2500 values. The average values of x_i are 0.003, 0.080, −0.032 and −0.005; the median values 0.121, 0.058, −0.069 and −0.003; and the rms values 0.968, 1.017, 0.986 and 1.001. Solid curves represent Gaussians of unit area and standard deviation.

sample, and thus a low estimate of the variance. A full analysis provides the appropriate correction factor $N/(N − 1)$; of course the difference disappears as $N \to \infty$.)

2. They should be *consistent*, the case if the descriptor for arbitrarily large sample size gives the true answer. As we have seen, the rms is a consistent measure of the standard deviation of a Gaussian distribution in that it gives the right answer for large N; but it is a biased estimator for small N unless modified by the factors just discussed.

3. The statistic should obey *closeness*, yielding smallest possible deviation from the truth. The Cauchy distribution (Section 3.1) looks innocent enough, somewhat similar to a Gaussian, even. But with infinite variance, trying to estimate dispersion via the standard deviation would yield massive scatter and little information.

4. The statistic should be *robust*. For example, if we have a fundamentally symmetric distribution of data but a few experimental errors creep in, *outliers* appearing at the ends of the distribution, then, as a measure of central location, the *median* is far more robust than the *average* – it is less affected by the outliers.

3.3 Simple error analysis

3.3.1 Random or systematic?

The average is a very common statistic; it is what we are doing all the time, for example, in 'integrating' on a faint object. The variance on the average is

$$S_m^2 = E\left[\left(\frac{1}{N}\sum_{i=1}^{N} X_i - \mu\right)^2\right]$$

which, after some manipulation, is

$$S_m^2 = \frac{\sigma^2}{N} + \frac{1}{N^2}\sum_{i \neq j} E[(X_i - \mu)(X_j - \mu)]. \tag{3.9}$$

Neglecting the last term for the moment, the first term expresses the generally held belief – the error on the average, or mean, of some data diminishes, like $\sqrt{N}$, as the amount of data is increased. This is one of the most important tenets of observational astronomy, and many other sciences too.

Now for the last term: apart from infinite variances (e.g. the Cauchy distribution, the familiar and comforting $\sqrt{N}$ result holds only when this last term is zero. The term contains the *covariance*, defined as

$$\text{cov}[X_i, X_j] = E[(X_i - \mu_i)(X_j - \mu_j)]; \tag{3.10}$$

it is closely related to the correlation coefficient between x_i and x_j (Section 4.2). We are keeping the subscripts now because of the possibility that the data from the i-pixel, spectral channel or time slot, are not independent of the data from the jth position. In the simplest cases, the data are independent and identically distributed (probability of x_i and x_j = probability of x_i × probability of x_j) and then the covariance is zero. This is a condition for the familiar $\sqrt{N}$ averaging away of noise; our assumption is that noise from one datum to the next or one pixel to the next is independent.

Example Suppose we had a time series, say of photometric measurements X_i. Here the i's index time of observation. It might be a reasonable assumption that the measurements were identically distributed and independent of each other. In this case, the probability distribution would be the same for each time, and so can just be written $g(x \mid \text{parameters})$. The covariance term is then just

$$\operatorname{cov}[X_i, X_j] = E[(X_i - \mu_i)(X_j - \mu_j)]$$
$$= \int (x_i - \mu_i)g(x_i \mid \ldots)\, \mathrm{d}x_i \int (x_j - \mu_j)g(x_j \mid \ldots)\, \mathrm{d}x_j$$
$$= 0 \tag{3.11}$$

because, by definition of μ, each integral must separately be zero.

Often this simple situation does not apply. One possibility is that $\operatorname{cov}[x_i, x_j]$ depends only on a 'distance' $(i - j)$. If the data are indexed in some meaningful way, for example as a time series, the data are then called *stationary*. As a second possibility, in photometric work it is quite likely that if one measurement is low, because of cloud, then the next few will be low too. (We speak of the dreaded $1/f$ noise; more of this in Section 9.7.) Then the probability distribution becomes multivariate and the simple factorizations do not apply:

$$E[(X_i - \mu_i)(X_j - \mu_j)] = \int \int (x_i - \mu_i)(x_j - \mu_j)g(x_i, x_j \mid \ldots)\, \mathrm{d}x_i \mathrm{d}x_j$$

so that we need to know more about the observational errors – in other words, how to write down the joint distribution $g(x_i, x_j \mid \ldots)$ – before we can assess how the average of data will behave. In these more complicated cases, the averaging away is almost certain to be slower than $\sqrt{N}$.

A common distinction is made in experimental subjects between *random* and *systematic* errors, random errors being considered as those showing the $\sqrt{N}$ diminution. In reality there is a continuum, with the covariance frequently non-zero. At the other far extreme, systematic errors persist no matter how much data are collected. If you are observing Arcturus when you should be observing Vega, the errors will never average away, no matter how persistent you are. Systematic errors can only be reduced by thorough understanding of the experimental equipment and circumstances; 'random' errors may be more or less random, depending on how correlated they are with each other.

3.3.2 Error propagation

Often, the thing we need to know is some more or less complicated function of the measured data. Knowing data error, how do we estimate error in the desired quantity?

If the errors are small, by far the easiest way is to use a Taylor expansion. Suppose we measure variables $x, y, z, \ldots$ with independent errors $\delta X, \delta Y, \delta Z, \ldots$ and we are interested in some function $f(x, y, z, \ldots)$. The change in f caused by the errors is, to first order

$$\delta F = \frac{\partial f}{\partial x}\bigg|_{x=X}\,\delta X + \frac{\partial f}{\partial y}\bigg|_{y=Y}\,\delta Y + \frac{\partial f}{\partial z}\bigg|_{z=Z}\,\delta Z + \cdots$$

The variance of a sum is the total of the variances of the individual terms (because the errors are assumed to be independent), so we get

$$\mathrm{var}[f] = \left(\frac{\partial f}{\partial x}\right)^2\bigg|_{x=X}\sigma_x^2 + \left(\frac{\partial f}{\partial y}\right)^2\bigg|_{y=Y}\sigma_y^2 + \left(\frac{\partial f}{\partial z}\right)^2\bigg|_{z=Z}\sigma_z^2 + \cdots,$$

$$(3.12)$$

where the σ represent the variances in each of the variables.

These considerations lead to a well-known result for combining measurements: if we have n independent estimates, say X_j, each having an associated error σ_j, the best combined estimate is the weighted mean,

$$\overline{X}_w = \sum_{j=1}^{n} w_j \overline{X}_j \bigg/ \sum_{j=1}^{n} w_j,$$

where the weights are given by $w_j = 1/\sigma_j^2$, the reciprocals of the sample variances. The variance of $\overline{X}_w$ is

$$\sigma_w^2 = 1 \bigg/ \sum_{j=1}^{n} 1/\sigma_j^2.$$

Example Suppose (i) $f(x, y) = x/y$. Then the Taylor series rule gives us immediately

$$\frac{\mathrm{var}[f]}{f^2} = \left(\frac{\sigma_x}{x}\right)^2 + \left(\frac{\sigma_y}{y}\right)^2;$$

we simply add up the relative errors. If (ii), $f(x) = \log x$, then the rule gives

$$\mathrm{var}[f] = \left(\frac{\sigma_x}{x}\right)^2$$

and the error in the log is just the relative error in the quantity we have measured.

3.3.3 Combining distributions

Often this method is not good enough – we may need to know details of the probability distribution of a derived quantity. The simplest case is a transformation from the measured x, with probability distribution g, to some derived quantity $f(x)$ with probability distribution h. Since probability is conserved, and remembering that what we call a probability distribution is strictly a probability *density* distribution, we have the requirement that

$$h(f)\,\mathrm{d}f = g(x)\,\mathrm{d}x \tag{3.13}$$

so that h involves the derivative $\mathrm{d}f/\mathrm{d}x$. Some care may be needed in applying this simple rule if the function f is not monotonic.

Example Suppose we are taking the logarithm of some exponentially distributed data. Here, $g(x) = \exp(-x)$ for positive x, and $f(x) = \log(x)$. Applying our rule gives

$$h(f) = \exp(-\exp(f))\exp(f)$$

which, as we might expect (Figure 3.2) has a pronounced tail to negative values and is correctly normalized to unity. Our simpler methods would give us $\delta h = \delta x/x$, which evidently cannot give a good representation of the asymmetry of h. Quoting '$h \pm \delta h$' is clearly not very informative.

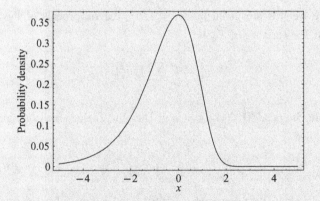

Figure 3.2   The probability distribution of logarithm of data drawn from an exponential distribution.

This technique rapidly becomes difficult to apply for more than one variable, but results for some useful cases are as follows.

1. Suppose we have two identically distributed independent variables x and y, both with distribution function g. What is the distribution of their sum $z = x + y$? For each x, we have to add up the probabilities of all the numbers $y = z - x$ that yield the z in which we are interested. The probability distribution $h(z)$ is therefore

$$h(z) = \int g(z - x)g(x)\,\mathrm{d}x, \tag{3.14}$$

where the probabilities are simply multiplied because of the assumption of independence. h is therefore the *autocorrelation* (Section 9.2) of g. The result generalizes to the sum of many variables, and is often best calculated with the aid of the Fourier transform of the distribution g. This transform is called the *characteristic function*.

2. Quite often we need the distribution of the product or quotient of two variables. Without details, the results are as follows:
For $z = xy$, the distribution of z is

$$h(z) = \int \frac{1}{|x|} g(x) g(z/x) \, dx \qquad (3.15)$$

and of $z = x/y$ is

$$h(z) = \int |x| g(x) g(zx) \, dx. \qquad (3.16)$$

In almost any case of interest, these integrals are too hard to do analytically.

Example One exception of interest is the product of two Gaussian variables of zero mean; this has applicability for a radio-astronomical correlator, for instance. Leaving out the mathematical details, the result emerges in the form of a standard modified Bessel function. The input Gaussians are of zero mean and variance σ^2. The distribution of the product is

$$h(z) = \frac{2}{\pi \sigma^2} K_0 \left[\frac{|z|}{\sigma^2} \right]$$

which, as Figure 3.3 shows, is quite unlike a Gaussian. It has a logarithmic singularity at zero but is normalized to unity.

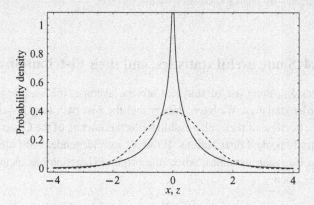

Figure 3.3 The probability distribution of the product of two identical Gaussians – the original Gaussian is the dashed curve.

The case of the ratio is equally instructive. Here we get

$$h(x) = \frac{1}{\pi} \frac{1}{1 + z^2},$$

a Cauchy distribution. It has infinite variance and, as we see in Figure 3.4, the variance of the original Gaussian surprisingly does not affect the form of the answer. This is a somewhat unrealistic case – it corresponds to forming the ratio of data of zero signal-to-noise ratio – but illustrates that ratios involving low signal-to-noise are likely to have very broad wings. The Bessel function distribution will, on averaging, succumb to the central limit theorem; this is not the case for the Cauchy distribution. In general, deviations from Normality will occur in the tails of distributions, the outliers that are so well known to all experimentalists.

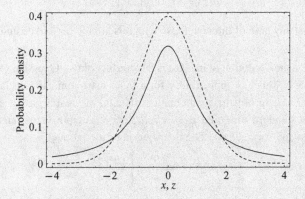

Figure 3.4 The probability distribution of the ratio of two identical Gaussians – the original Gaussian is the dashed curve.

3.4 Some useful statistics, and their distributions

For N data X_i, some useful statistics are the average, the sample variance and the order statistics. We have already met the first two; they acquire their importance because of their relationship to the parameters of the Gaussian, and other centrally peaked distributions. If the X_i are independent and identically distributed Gaussian variables, where the original Gaussian has mean μ and variance σ^2, then:

(i) the average or mean value $\overline{X}$ follows a Gaussian distribution around μ, with variance σ^2/N. We have met this result before (Section 2.5).

(ii) the sample variance σ_s^2 is distributed like $\sigma^2 \chi^2/(N-1)$, where the chi-square variable has $N-1$ degrees of freedom (Table B.3).

(iii) the ratio

$$\frac{\sqrt{N}(\overline{X} - \mu)}{\sigma_s^2}$$

is distributed like the t statistic, with $N-1$ degrees of freedom. This ratio has an obvious usefulness, telling us how far our average might be from the true mean (Section 5.2, Table B.3).

(iv) if we have two independent samples (size N and M) drawn from the same Gaussian distribution, then the ratio of the sample variances $\sigma_{s_1}^2$ and $\sigma_{s_2}^2$ follows an F distribution. This allows us to check if the data were indeed drawn from Gaussians of the same width (Section 5.2, Table B.4).

The order statistics are simply the result of arranging the data X_i in order of size, relabelled as $Y_1, Y_2, \ldots$ So Y_1 is the smallest value of the X's, and Y_N the largest. Maximum values are often of interest, and the median $Y_{N/2}$ (N even) is a useful robust indicator of location. We might also form robust estimates of widths by using order statistics to find the range containing, say, 50 per cent of the data. Both the density and the cumulative distribution are therefore of interest.

Suppose that the distribution of x is $f(x)$, with cumulative distribution $F(x)$. Then the distribution g_n of the nth order statistic is

$$g_n(y) = \frac{N!}{(n-1)!(N-n)!}[F(y)]^{n-1}[1-F(y)]^{N-n} f(y) \qquad (3.17)$$

and the cumulative distribution is

$$G_n(y) = \sum_{j=n}^{N} \binom{N}{j} [F(y)]^j [1-F(y)]^{N-j}. \qquad (3.18)$$

Example The Schechter luminosity function (Section 8.3) $x^\gamma \exp(-x/x^*)$ is a useful model of the luminosity function for field galaxies. The observed value of γ is close to unity, but we will take $\gamma = 1/2$ for convenience in ensuring that the distribution can be normalized over the range zero to infinity; we also take $x^* = 1$. If we select 10 galaxies from this distribution, the maximum of the 10 will follow the distribution shown in the figure. We see that the distribution is quite different to the

Schechter function, with a peak quite close to x^*. If we choose 100 galaxies, then, of course, the distribution moves to brighter values.

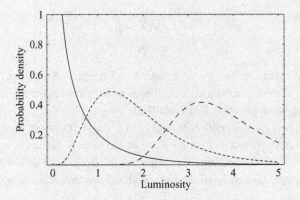

Figure 3.5   The Schechter luminosity function (solid curve) and the distribution of the maximum of 10 and 100 samples from the distribution, plotted as short- and long-dash curves, respectively.

3.5 Uses of statistics

So far, we have concentrated on defining statistics and noticing that they (a) may estimate parameters of distributions and (b) will be distributed in some more or less complicated way themselves. Their use then parallels the Bayesian method.

Firstly, we may use them to estimate parameters; but the way in which they do this is more subtle than the Bayesian case. We do not get a probability distribution for the parameter of interest, but a distribution of the statistic, *given the parameter*. As noted in the introductory section of this chapter, the confidence interval is the usual way of making use of a statistic as an estimator.

Secondly, we may test hypotheses. This again parallels the Bayesian case, but the methods are much further apart conceptually. Recall the case discussed in Section 2.5, where we have estimated parameters α and β. Using statistics, we would have two data combinations A (estimating α) and B (estimating β). How would we answer a question like 'is $\alpha > \beta$' in this approach? The classical method entails finding some new combination – say $t = A - B$ – and then computing its distribution on the hypothesis that $\alpha - \beta = 0$. We then find the probability of the observed value of t, or bigger, occurring given this hypothesis; and if the probability is small, we would conclude that the data were unlikely to have occurred by chance. The hint, of course, is that indeed $\alpha > \beta$, (or $\alpha < \beta$) but we do not know the probability of this. We could only calculate it if we had some specific values of α and β in mind; more of this in Section 5.1.

This classical approach is the basis of numerous useful tests, and we discuss some of them in detail in Chapters 4 and 5. However, there is no doubt that the method does not quite seem to answer the question we had in mind, although often its results are indistinguishable from the more intelligible Bayesian approach. The same decisions get taken.

Perhaps the most difficult part of this testing procedure is the implicit use of data corresponding to events that did not occur – the 'observed value of t, or bigger' referred to. Jeffreys (1961) wrote: '...a hypothesis that may be true may be rejected because it has predicted observable results that have not occurred. This seems a remarkable procedure.'

However, using large but unobserved values of the test statistic usually does not matter much; in cases of interest, our statistic will be unlikely anyway, and larger values will be even less likely. The conceptual puzzle remains and can lead to paradoxes.

Exercises

3.1 Means and variances. Find the mean and variance of a Poisson distribution and of a power law; find the variance ($= \infty$) of a Cauchy distribution.

3.2 Simple error analysis. Derive the well-known results for error combining, for two products, and the the sum and difference of two quantities, from the Taylor expansion of Section 3.3.2.

3.3 Combining Gaussian variables. Use the result of Section 3.3.2 for errors on z when $z = x + y$ to find the distribution of the sum of two Gaussian variables. Test this with a Monte Carlo experiment.

3.4 Average of Cauchy variables. Show that the average value of Cauchy-distributed variables has the same distribution as the original data. Use characteristic functions and the convolution theorem.

3.5 Poisson statistics. Draw random numbers from Poisson distributions (Section 2.6) with $\mu = 10$ and $\mu = 100$. Taking 10 or 100 samples, find the average and the rms scatter. How close is the scatter to $\sqrt{\mu}$?

3.6 Robust statistics. Make a Gaussian with outliers by combining two Gaussians, one of unit variance, one three times wider. Leave the relative weight of the wide Gaussian as a parameter. Compare the mean deviation with the rms, for various relative weights. How sensitive are the two measures of scatter to outliers? Repeat the exercise, with a width derived from order statistics. Check your results with a Monte Carlo experiment.

3.7 Change of variable. Suppose that ϕ is uniformly distributed between zero and 2π. Find the distribution of $\sin \phi$. How could you find the distribution of a sum of sines of independent random angles?

3.8 Order statistics. We record a burst of N neutrinos from a supernova, and the probability of recording a neutrino at time t is, in suitable units, $\exp(t - t_0)$ where t_0 is the time of emission. The maximum likelihood estimate of t_0 is just T_1, the time of arrival of the first neutrino. Use order statistics (Section 3.4) to show that the average value of T_1 is just $t_0 + \frac{1}{N}$. Is this maximum likelihood estimator biased, but consistent (i.e. the correct answer as $N \to \infty$)?

4

Correlation and association

It is difficult to understand why statisticians commonly limit their inquiries to Averages, and do not revel in more comprehensive views.

(Francis Galton, 1889)

When we make a set of measurements, it is instinct to try to correlate the observations with other results. One or more motives may be involved in this instinct. For instance we might wish (a) to check that other observers' measurements are reasonable, (b) to check that our measurements are reasonable, (c) to test a hypothesis, perhaps one for which the observations were explicitly made, or (d) in the absence of any hypothesis, any knowledge or anything better to do with the data, to find if they are correlated with other results in the hope of discovering some new and universal truth.

4.1 The fishing trip

Take the last point first. Suppose that we have plotted something against something, on a fishing expedition of this type. There are grave dangers on this expedition, and we must ask ourselves the following questions.

1. Does the eye see much correlation? If not, calculation of a formal correlation statistic is probably a waste of time.
2. Could the apparent correlation be due to selection effects? Consider, for instance, the beautiful correlation in Figure 4.1, in which Sandage (1972) plotted radio luminosities of sources in the 3CR catalogue as a function of distance modulus. At first sight, it *proves* luminosity evolution for radio sources. Are the more distant objects (at earlier epochs) clearly not the more powerful? In fact, as Sandage recognized, it proves nothing of the

71

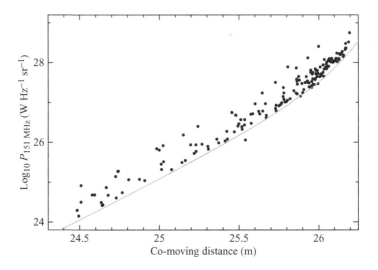

Figure 4.1   Radio luminosities of 3CR radio sources versus distance modulus. The curved line represents the survey limit, the limit imposed by forming a catalogue from a flux-limited sample (Section 8.2).

kind. The sample is flux- (or apparent intensity) limited; the solid line shows the flux-density limit of the 3CR catalogue. The lower right-hand region can never be populated; such objects are too faint to show above the limit of the 3CR catalogue. But what about the upper left? Provided that the luminosity function (the true space density in objects per megaparsec^3) slopes downward with increasing luminosity, the objects are bound to crowd towards the line. This is about all that can be gleaned immediately from the diagram – the space density of powerful radio sources is less than the space density of their weaker brethren. The plot shows *Malmquist bias* (Section 8.2) hard at work.

Astronomers produce many plots of this type, and will describe purported correlations in terms such as 'The lower right-hand region of the diagram is unpopulated because of the detection limit, but there is no reason why objects in the upper left-hand region should have escaped detection . . . ' True, but nor can they escape probability; the upper left of Sandage's diagram is not filled with quasars and radio galaxies because we need to sample large spheres about us to have a hope of encountering a powerful radio source. Small spheres, corresponding to small redshifts and distance moduli, will yield us only low-luminosity radio sources because their space density is so much higher. The lesson applies to any proposed correlation for variables

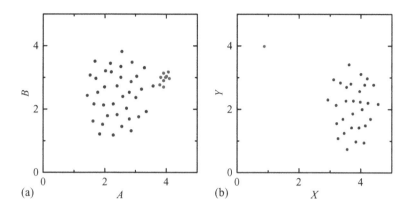

Figure 4.2   Dodgy correlations: in each case formal calculation will indicate that a correlation exists to a high degree of significance.

with steep probability density functions dependent upon one of the variables plotted.

3. If we are happy about (2), we can try formal calculation of the significance of the correlation as described in Section 4.2. Further, if there is a correlation, does the regression line (Section 6.2) make sense?

4. If we are still happy, we must return to the plot to ask if the formal result is realistic. A rule of thumb – if 10 per cent of the points are grouped by themselves so that covering them with the thumb destroys the correlation to the eye, then we should doubt it, no matter what significance level we have found. Beware, in particular, of plots which look like those of Figure 4.2, plots which strongly suggest selection effects, data errors or some other form of statistical conspiracy.

5. If we are *still* confident, we must remember that a correlation does not prove a causal connection. The essential point is that correlation may simply indicate a dependence of both variables on a third variable. Cigarette manufacturers said so for years; but finding the physical attribute which caused heart/lung disease *and* the desire to smoke proved difficult. But there are famous instances, many, e.g. the correlation between quality of children's handwriting and their height, and between the size of feet in China and the price of fish in Billingsgate Market. For the former, the hidden variable is *age* (are tall children cleverer? No, but older), while for the latter, it is *time*.

There are, in fact, ways of searching for intrinsic correlation between variables when they are known to depend mutually upon a third variable. The

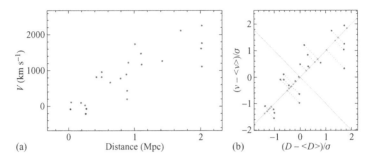

Figure 4.3   (a) An early Hubble, diagram (Hubble, 1936); recession velocities of a sample of 24 galaxies versus distance measure. (b) The same plot but with data normalized by standard deviation; the lines represent principal components, as described in Section 4.5.

problem, however, when on the fishing trip, is how to know about a third variable, how to identify it when we might suspect that it is lurking. We consider it further in Sections 4.3 and 4.5.

Finally we must not get too discouraged by all the foregoing. Consider Figure 4.3, a ragged correlation if ever there was one, although there are no nasty groupings of the type rejected by the rule of thumb. It is, in fact, one of the earliest 'Hubble diagrams' – the discovery of the recession of the nebulae, and the expanding universe (Hubble, 1936).

4.2 Testing for correlation

In dealing with correlations, we encounter in detail many important aspects of the use of probability and statistics. The foregoing problem appears simple: we have a set of measurements (X_i, Y_i) and we ask (formally) if they are related to each other.

To make progress, we have to make 'related' more precise. The best-developed way of doing this – although not necessarily relevant – is to model our data as a bivariate or joint Gaussian of *correlation coefficient* ρ:

$$
\text{prob}(x, y \mid \sigma_x, \sigma_y, \rho) = \frac{1}{2\pi \sigma_x \sigma_y \sqrt{1 - \rho^2}}
$$
$$
\times \exp\left(\frac{-1}{2(1 - \rho^2)} \left(\frac{(x - \mu_x)^2}{\sigma_x^2} + \frac{(y - \mu_y)^2}{\sigma_y^2} - \frac{2\rho(x - \mu_x)(y - \mu_y)}{\sigma_x \sigma_y} \right) \right).
$$
$$(4.1)$$

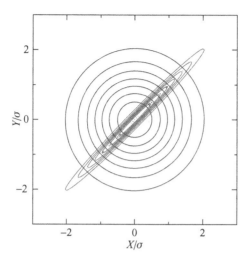

Figure 4.4 Linear contours of the bivariate Gaussian probability distribution assuming zero means; the near-circular contours represent $\rho = 0.01$, a bivariate distribution with little connection between x and y, while the highly elliptical contours represent $\rho = 0.99$, indicative of a strong correlation between x and y. Negative values of ρ reverse the tilt, and indicate what is loosely referred to as anticorrelation.

This model is so well developed that 'correlation' and '$\rho \neq 0$' are nearly synonymous; if $\rho \to 0$, there is little correlation, while if $\rho \to 1$, the correlation is perfect; see Figure 4.4.

The parameter ρ is the *correlation coefficient*, and in the above formulation, it is given by

$$\rho = \frac{\text{cov}[x, y]}{\sigma_x \sigma_y}, \tag{4.2}$$

where cov is the covariance (Section 3.3.1) of x and y, and σ_x^2 and σ_y^2 are the variances. The correlation coefficient can be estimated by

$$r = \frac{\sum_{i=1}^{n}(X_i - \overline{X})(Y_i - \overline{Y})}{\sqrt{\sum_{i=1}^{n}(X_i - \overline{X})^2 \sum_{i=1}^{n}(Y_i - \overline{Y})^2}}. \tag{4.3}$$

r is known as the Pearson product–moment correlation coefficient (Fisher, 1944).

Assuming here and in Figure 4.4 (with no loss of generality) that means are zero, the contours in the figure will have dropped by $1/e$ from the maximum at the origin when

$$\frac{1}{1-\rho^2}\left(\frac{x^2}{\sigma_x^2} + \frac{y^2}{\sigma_y^2} - \frac{2\rho xy}{\sigma_x\sigma_y}\right) = 1, \tag{4.4}$$

or, in matrix notation, when

$$(x \ y)\frac{1}{1-\rho^2}\begin{pmatrix} \frac{1}{\sigma_x^2} & -\frac{\rho}{\sigma_x\sigma_y} \\ -\frac{\rho}{\sigma_x\sigma_y} & \frac{1}{\sigma_y^2} \end{pmatrix}\begin{pmatrix} x \\ y \end{pmatrix} = 1. \tag{4.5}$$

The inverse of the central matrix is known as the *covariance matrix* or *error matrix*

$$C = \begin{pmatrix} \sigma_x^2 & \mathrm{cov}(x, y) \\ \mathrm{cov}(x, y) & \sigma_y^2 \end{pmatrix}. \tag{4.6}$$

The off-diagonal elements of the covariance matrix can be estimated by

$$\frac{1}{N-1}\overline{(X_i - \overline{X_i})(X_j - \overline{X_j})}.$$

The matrix is particularly valuable in calculating propagation of errors, but there are numerous applications, for example in principal component analysis (Section 4.5) and in maximum-likelihood modelling (Section 6.1).

The multivariate Gaussian is one example of a class of multivariate distribution functions that depend only on the data vector $\vec{x}$ via a so-called quadratic form

$$\vec{x}^T C \vec{x}.$$

The multivariate Gaussian is the most familiar of these. The quadratic form defines an ellipse so that all of this class of functions have ellipsoidal equiprobable contours.

An important special case of random-number generation involves the multivariate Gaussian, with given σ_i and ρ_i. This is crucial for checking many tests or model-fitting routines and, thanks to this discussion of error matrices, quite simple to formulate.

- Set up the error matrix and determine the covariance matrix from it. (For the bivariate case, the error matrix is $e_{1,1} = \sigma_x^2$, $e_{2,1} = e_{1,2} = \mathrm{cov}[x, y] = \rho\sigma_x\sigma_y$, $e_{2,2} = \sigma_y^2$, as we have seen.)
- Find the eigenvalues and eigenvectors of the covariance matrix.
- Combine the eigenvectors, the column vectors, into the transformation matrix T, the matrix that diagonalizes the covariance matrix.

- Then draw (x',y') Gaussian pairs, uncorrelated, with variances equal to the two eigenvalues. Compute the (x, y) pairs according to

$$\begin{pmatrix} x \\ y \end{pmatrix} = [T] \cdot \begin{pmatrix} x' \\ y' \end{pmatrix} \tag{4.7}$$

The points in Figure 6.2 were obtained in this manner.

To return to the point at issue: for bivariate data, what we really want to know is whether or not $\rho = 0$; it is this condition for which we are testing. Using the bivariate Gaussian is a very specific model; a Gaussian is assumed, it allows only two variances, and assumes that both x and y are random variables. Thus, σ_x and σ_y include both the errors in the data, and their intrinsic scatter – all presumed Gaussian. The model does not apply, for example, to data where the x-values are well defined and there are 'errors' only in y, perhaps different at different x. In such cases we would use model-fitting, perhaps of a straight line (Sections 6.1, 6.2). This is a different problem. These effects mean that we have to approach the correlation coefficient with caution, as the way we set up our experiment may result in graphs like those of Figures 4.1 or 4.2.

As always, there are two quite different ways of proceeding from this point, Bayesian and non-Bayesian.

4.2.1 Bayesian correlation-testing

The Bayesian approach is to use Bayes' theorem to extract the probability distribution for ρ from the likelihood of the data and suitable priors. Since we want to know about ρ independently of any inference about the means and variances, we have to integrate these 'nuisance variables' out of the full posterior probability $\text{prob}(\rho, \sigma_x, \sigma_y, \mu_x, \mu_y \mid \text{data})$. For the bivariate Gaussian model, the result is given by Jeffreys (1961) as

$$\text{prob}(\rho \mid \text{data}) \propto \frac{(1 - \rho^2)^{\frac{N-1}{2}}}{(1 - \rho r)^{N-\frac{3}{2}}} \left(1 + \frac{1}{N - 1/2} \frac{1 + r\rho}{8} + \cdots \right). \tag{4.8}$$

The Bayesian test for correlation is thus simple: compute r from the (X_i, Y_i), and calculate $\text{prob}(\rho)$ for the range of ρ of interest.

Example We generated 50 samples from a bivariate t distribution (Section 2.4.2) with three degrees of freedom. The true correlation coefficient was 0.5. The large tails of the distribution produce outliers, not accounted for by the assumed Gaussian used in interpreting the r statistic. Figure 4.5 shows what Equation (4.7) gives: the distribution of ρ peaks at around 0.2. If we now remove the samples outside 4σ, the distribution peaks at around 0.5 and is appreciably narrower. The method is thus fairly robust, although obviously affected by being used with the 'wrong' distribution.

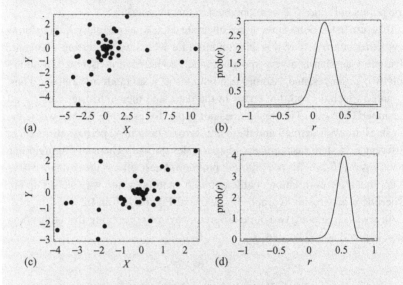

Figure 4.5   50 X_i, Y_i chosen at random from a bivariate t distribution with $\rho = 0.5$. The Jeffreys' probability distribution of correlation coefficient ρ is shown, peaking at around 0.2 for the upper panel. The data have been restricted to $\pm 4\sigma$ in the lower panel; the distribution now peaks at 0.54.

Given this probability distribution for ρ, we can answer questions such as 'what is the probability that $\rho > 0.5$?' or (perhaps more usefully) 'what is the probability that ρ from data set A is bigger than ρ from data set B?' (see Section 5.2.3). As is often the case, the utility of the Bayesian approach is not that prior information is accurately incorporated, but rather that we get an answer to the question we really want to ask.

Jeffreys used a uniform prior for ρ – not obviously justifiable, and certainly not correct if ρ is close to 1 or -1, as he points out, but in these cases a statistical test is a waste of time anyway.

Example An interesting use of Jeffreys' distribution is to calculate the probability that ρ is positive, as a function of sample size (Figure 4.6). This tells us how much data we need to be confident of detecting correlations.

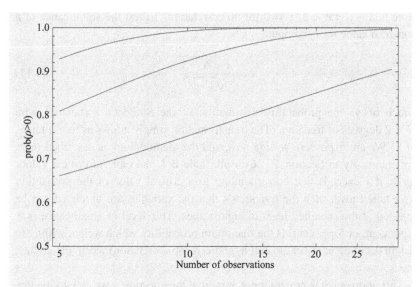

Figure 4.6 The probability of ρ being positive, as a function of sample size, for r values of 0.25 (lowest curve), 0.5 and 0.75 (uppermost curve).

4.2.2 The classical approach to correlation-testing

The alternative approach to the correlation problem starts by regarding ρ as a fixed quantity, not a variable about which probabilistic statements might be made. This approach therefore arrives at the probability of the *data*, given ρ (and, of course, the background hypothesis that a bivariate Gaussian is adequate). The result (Fisher, 1944) is

$$\text{prob}(r \mid \rho, H) \propto \frac{(1 - \rho^2)^{(N-1)/2}(1 - r^2)^{(N-4)/2}}{(1 - \rho r)^{N-3/2}} \left(1 + \frac{1}{N - 1/2} \frac{1 + r\rho}{8} + \cdots \right) \tag{4.9}$$

What can we do with this answer? The standard approach is to pick the easy 'null hypothesis' $\rho = 0$, compute r, and then compute the probability, under the null hypothesis, of r *being this big or bigger*. If this probability is very small, we may feel that the null hypothesis is rather unlikely.

The standard parametric test is to attempt to reject the hypothesis that $\rho = 0$ and we do this by computing r. The standard deviation in r is

$$\sigma_r = \frac{(1 - r^2)}{\sqrt{N - 1}} \tag{4.10}$$

Note that $-1 < r < 1$; $r = 0$ for no correlation. To test the significance of a non-zero value for r, compute

$$t = \frac{r\sqrt{N-2}}{\sqrt{1-r^2}} \tag{4.11}$$

which obeys the probability distribution of the Student's t statistic[1] with $N - 2$ degrees of freedom. (The transformation simply allows us to use tables of t.) We are *hypothesis testing* now, and the methodology is described more systematically in Section 5.1. Consult Table B.3, the table of critical values for t; if t exceeds that corresponding to a critical value of the probability (two-tailed test), then the hypothesis that the variables are unrelated can be rejected at the specified level of significance. This level of significance (say 1 per cent, or 5 per cent) is the maximum probability which we are willing to risk in deciding to reject the null hypothesis (no correlation) when it is, in fact, true.

This approach has probably not answered the question – we embark on this sort of investigation when it is apparent that the data contain correlations; we merely want some justification by knowing 'how much'. Also, the inclusion in the testing procedure of values of r that have not been observed poses the usual difficulties.

The test is widely used, and is formally powerful, but as one statistics book says, 'There are data to which this kind of correlation method cannot be applied.' This is a gross understatement. The data must be on continuous scales, obviously. The relation between them must be linear. (How would we know this? In many cases in astronomy we change the scales at will (log–log, log–linear, etc.) to give roughly linear appearance to our plots.) The data must be drawn from Normally distributed populations. (How would we know this? Certainly, if we have changed our data axes to log form, there must be doubt.) They must be free from restrictions in variability or groupings. There are parametric tests that help: the F test for non-linearity and the correlation ratio test which gets around non-linearity. However, to circumvent the problems, it is far better to go to a *non-parametric* test. These permit additional tests on data which are not numerically defined (binned data, or ranked data), so that, in some instances, they may be the only alternative.

---

[1] After its discoverer W. S. Gosset (1876–1937), who developed the test while working on quality control sampling for Guinness. For reasons of industrial secrecy, Gosset was required to publish under a pseudonym; he chose 'Student', which he used for years in correspondence with his (former) professor at Oxford, Karl Pearson.

4.2.3 Correlation-testing: classical, non-parametric

The best-known non-parametric test consists of computing the Spearman rank correlation coefficient (Conover, 1999; Siegel & Castellan, 1988):

$$r_s = 1 - 6\frac{\sum_{i}^{N}(X_i - Y_i)^2}{N^3 - N},$$
(4.12)

where there are N data pairs, and the N values of each of the two variables are ranked so that (X_i, Y_i) represents the ranks of the variables for the ith pair, $1 < X_i < N$, $1 < Y_i < N$.

The range is $0 < r_s < 1$; a high value indicates significant correlation. To find how significant, refer the computed r_s to Table B.5, a table of critical values of r_s applicable for $4 \leq N \leq 30$. If r_s exceeds an appropriate critical value, the hypothesis that the variables are unrelated is rejected at that level of significance. If N exceeds 30, compute

$$t_r = r_s\sqrt{\frac{N-2}{1 - r_s^2}},$$
(4.13)

a statistic whose distribution for large N asymptotically approaches that of the t statistic with $N - 2$ degrees of freedom. The significance of t_r may be found from Table B.3, and this represents the associated probability under the hypothesis that the variables are unrelated.

How does use of r_s compare with use of r, the most powerful parametric test for correlation? Very well: the efficiency is 91 per cent. This means that if we apply r_s to a population for which we have a data pair (X_i, Y_i) for each object and both variables are Normally distributed, we will need, on average, 100 (X_i, Y_i) for r_s to reveal that correlation at the same level of significance which r attains for 91 (X_i, Y_i) pairs. The moral is that if in doubt, little is lost by going for the non-parametric test.

The Kendall rank correlation coefficient does the same thing as r_s, and with the same efficiency (Siegel & Castellan, 1988).

Example A 'correlation' at the notorious 2σ level is shown in Figure 4.7. Here, $r_s = 0.28$, $N = 55$, and the hypothesis that the variables are unrelated is rejected at the 5 per cent level of significance. Here, we have no idea of the underlying distributions; nor are we clear about the nature of the axes. The assumption of a bivariate Gaussian distribution would be rash in the extreme, especially in view of a uniformly filled Universe producing a V/V_{max} statistic uniformly distributed between 0 and 1 (Schmidt, 1968). The V_{max} method is discussed in Section 8.3.

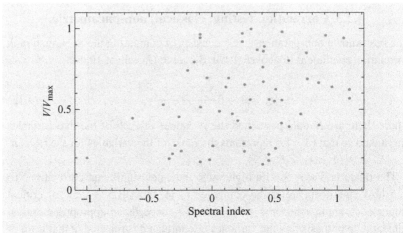

Figure 4.7 V/V_{max} as a function of high-frequency spectral index for a sample of radio quasars selected from the Parkes 2.7-GHz survey.

There is yet another way, the *permutation test*. In the case of correlation analysis, we have data $(X_1, Y_1), (X_2, Y_2), \ldots$ and we wish to test the null hypothesis that x and y are uncorrelated. In this regard if we have some home-made test statistic η, we can calculate its distribution, on the assumption of the null hypothesis, by simply calculating its value for many permutations of the X's amongst the Y's. For any reasonable data set, there will be far more possible permutations than we can reasonably explore, but choosing a random set will give an adequate estimate of the distribution of the test statistic.

If it turns out that the observed value of η is very improbable, under the null hypothesis, we may be interested in estimating the distribution for non-zero correlation. This is the route to useful Bayesian analysis, of the kind we described for the product–moment coefficient ρ. Here, the bootstrap method (Section 6.6) or the jackknife will be useful in estimating features of the test statistic.

These methods can be used to derive distributions of statistics such as Spearman's or Kendall's correlation coefficients in cases when a correlation is apparently present.

4.2.4 Correlation-testing: Bayesian versus non-Bayesian tests

Let us be clear: the non-parametric tests circumvent some of the issues involved in the non-Bayesian approach, but they have no bearing on the fundamental issue – what was the real question? However, the Bayesian approach, strong in answering the real question, forces reliance on a model.

There is rather little difference, in practice, between the Fisher test and results from Jeffreys' distribution. We can show this with some random Gaussian data with a correlation of zero. In the standard way, we can use the r distribution to find the probability of r being as large, or larger, than we observe, on the hypothesis that $\rho = 0$. If this probability is small, the test is hinting at the possibility that the correlation is actually positive. Therefore, we compare with the probability, from the Jeffreys' distribution, that ρ is positive. If the probability from Fisher's r distribution is small, we expect the probability from ρ to be large; and in fact we can see, either from simulations or from the algebraic form of the distributions, that the sum of these two probabilities is always very close to 1. In other words, interpreting the standard Fisher test (illegally!) to be telling us the chance that ρ is positive actually works very well.

4.3 Partial correlation

The 'lurking third variable' can be dealt with (provided that its influence is recognized in the first place) by *partial correlation*, in which the 'partial' correlation between two variables is considered by nullifying the effects of the third (or fourth or more) variable upon the variables being considered. Partial correlation is a science in itself; it is covered in both parametric and non-parametric forms by Macklin (1982), Siegel & Castellan (1988) and Stuart & Ord (1994). A bogus correlation, driven by correlation of the variables being studied with a third, unsuspected variable, is probably one of the commonest errors in statistical analysis (and political decision-making).

In the parametric form, consider a sample of N objects for which parameters x_1, x_2 and x_3 have been measured. The *first-order partial correlation coefficient* between variables x_1 and x_2 is

$$r_{12.3} = \frac{r_{12} - r_{13}r_{23}}{\sqrt{(1 - r_{13}^2)(1 - r_{23}^2)}}, \tag{4.14}$$

where the r are the product–moment coefficients defined in Section 4.2.2. If there are four variables, then the *second-order partial correlation coefficient* is

$$r_{12.34} = \frac{r_{12.3} - r_{14.3}r_{24.3}}{\sqrt{(1 - r_{14.3}^2)(1 - r_{24.3}^2)}}, \tag{4.15}$$

where the correlation is being examined between x_1 and x_2 with x_3 and x_4 held constant. Examination of the correlation between the other variables requires manipulation of the subscripts in the foregoing.

And so forth for higher-order partial correlations between more than four variables, with the the standard error of the partial correlation coefficients being given by

$$\sigma_{r_{12.34...m}} = \frac{1 - r^2_{12.34...m}}{\sqrt{N - m}},$$

where m is the number of variables involved. The significance then comes from the Student's t test as above.

Example Consider data from a sample of lads aged 12–19. The correlation between height and weight will be high because the older boys are taller on average. But with age held constant, the correlation would still be significantly positive because at all ages, taller boys tend to be heavier. In such a sample of 10, the correlation between height and weight (r_{12}) is calculated as 0.78; between height and age (r_{13}) 0.52, and between weight and age, $r_{23} = 0.54$. The first-order partial coefficient of correlation (Equation (4.14)) is thus $r_{12.3} = 0.69$; $\sigma_{r_{12.3}} = 0.198$; and the correlation is significant at the level of 0.2 per cent.

Consider further a measure of strength for each lad. The correlation between strength and height (r_{41}) is 0.58; between strength and weight (r_{42}) 0.72. Will lads of the same weight show a dependence of strength upon height? The answer is given by $r_{41.2} = 0.042$; the correlation between strength and height essentially vanishes and we would conclude that height, as such, has no bearing on strength; only by virtue of its correlation with weight does it show any correlation at all.

As for second-order partials, is there a correlation between strength and age if height and weight are held constant? The raw correlation between age and strength was 0.29; the second-order partial also yields 0.29. It seemingly makes little difference if height and weight are allowed to vary; the relation between age and strength is the same.

4.4 But what next?

If we have demonstrated a correlation, it is logical to ask what the correlation means, i.e. what is the law which relates the variables. It is common practice

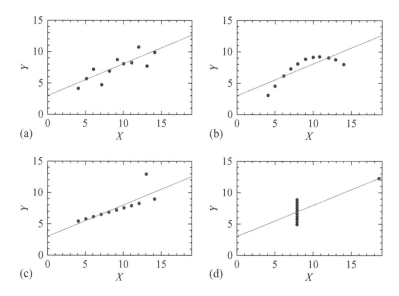

Figure 4.8   Anscombe's quartet: four fictitious sets of 11 (X_i, Y_i), each with the same $(\bar{X}, \bar{Y})$, identical coefficients of regression, regression lines, residuals in Y and estimated standard errors in slopes.

to dash off and fit a *regression*[2] *line*, usually applying the *method of least squares* (Section 6.2). It is essential to note that this is model-fitting now; the distinction between data modelling (Chapter 6) and hypothesis testing (here; and Chapter 5) is important.

Before doing so, there are several considerations, most of which are addressed in more detail in Section 6.2. Are there better quantities to minimize than the squares of deviations? What errors result on the regression-line parameters? Why should the relation be linear? And – most crucial of all – what are we trying to find out? If we have found a correlation between x and y, which variable is dependent; do we want to know x on y or y on x? The coefficients are generally completely different.

As an argument against blind application of correlation-testing and line-fitting, consider the famous Anscombe (1973) quartet, shown in Figure 4.8. Anscombe's point is the essential role of graphs in good statistical analysis. However, the examples illustrate other matters: the rule of thumb (Section 4.1),

---

[2] Galton (1889) introduced the term *regression*; it is from his examination of the inheritance of stature. He found that the sons of fathers who deviate x inches from the mean height of all fathers themselves deviate from the mean height of all sons by less than x inches. There is what Galton termed a 'regression to mediocrity'. The mathematicians who took up his challenge to analyse the correlation propagated his mediocre term, and we are stuck with it.

and the distinction between *independence* of data points and *correlation*. In more than one of Anscombe's data sets the points are clearly related. They are far from *independent*, while not showing a particularly strong (formal) *correlation*. Figure 4.8(b) is a case in which a linear fit is of indifferent quality, while choice of the 'right' relation between X and Y would result in a perfect fit. The quartet further emphasizes how dependent our analyses are on the assumption of normality: the covariance matrix, which intuitively we might expect to reflect some of the structure in the individual plots, is identical for each.

Note that X *independent* of Y means $\mathrm{prob}(X, Y) = \mathrm{prob}(X)\mathrm{prob}(Y)$, or $\mathrm{prob}(X \mid Y) = \mathrm{prob}(X)$; while X *correlated* with Y means $\mathrm{prob}(X, Y) \neq \mathrm{prob}(X)\mathrm{prob}(Y)$ in a particular way, giving $r \neq 0$. It is perfectly possible to have $\mathrm{prob}(X, Y) \neq \mathrm{prob}(X)\mathrm{prob}(Y)$ *and* $r = 0$ – the standard example being points distributed so as to form the Union Jack.

If we simply wish to map the dependence of variables on each other with minimal judgemental input, it strongly suggested, here and in Section 6.2, that principal component analysis is the appropriate technique.

4.5 Principal component analysis

Principal component analysis (PCA) is the ultimate correlation searcher when many variables are present. Given a sample of N objects with n parameters measured for each of them, how do we find what is correlated with what? What variables produce primary correlations, and what produce secondary, via the lurking third (or, indeed, $n - 2$) variables?

PCA is one of a family of algorithms (known as multivariate statistics; see, e.g., Kendall (1980), Manly (1994) Joliffe (2002)) designed for this situation. Its task is the following: given a sample of N objects with n measured variables x_n for each, find a new set of ξ_n variables that are orthogonal (independent), each one a linear combination of the original variables:

$$\xi_i = \sum_{j=1}^{n} a_{ij} x_j \qquad (4.16)$$

with values of a_{ij} such that the *smallest number* of new variables account for as much of the variance as possible. The ξ_i are the *principal components*. If most of the variance involves just a few of the n new variables, we have found a simplified description of the data. Finding which of the variables correlate (and how) may lead to that successful fishing expedition – we may have caught new physical insight.

Following (with permission) the excellent paper of Francis & Wills (1999), PCA may be described algebraically, through covariance matrices (Section 4.2), or geometrically. Taking the latter approach, consider the N objects represented by a large cloud in n-dimensional space. If two of the n parameters are correlated, the cloud is elongated along some direction in this space. PCA identifies these extension directions and uses them as a sequential set of axes, sequential in the sense that the most extended direction is identified first by minimizing the sums of squares of deviations. This direction forms the first principal component (or *eigenvector 1*), accounting for the largest single linear variation amongst the object properties. Then the $(n-1)$-dimensional hyperplane orthogonal to the first principal component is considered and searched for the direction representing the greatest variance in $(n-1)$ space; and so forth, defining a total of n orthogonal directions.

Example As an elementary PCA example via geometry, let us return to the early Hubble diagram of Figure 4.3: 24 galaxies with two measured variables, velocity of recession v and distance d. It is standard practice to normalize by subtracting the means from each variable and dividing by the standard deviation, i.e. to plot $v_i' = (v_i - <v>)/\sigma_v$ versus $d_i' = (d_i - <v>)/\sigma_v$, as shown in Figure 4.3(b). Then we find the first principal component by simply rotating the axis through the origin to align with maximum elongation, the direction of apparent correlation, and we do this with least squares (Section 6.2) – maximizing the variance along PC1 is equivalent to minimizing the sums of the squares of the distances of the points from this line through the origin. The distance of a point from the direction PC1 (shown dotted in Figure 4.3(b)) represents the value (score) of PC1 for that point. PC1 is clearly a linear combination of the two original variables; in fact it is $v' = d'$. Because the new coordinate system was found by simple rotation, distances from the origin are unchanged; the total variance of v' and d' is unchanged and is 2.0. The variance of PC1, the normalized distances squared from PC2, is 1.837. The remaining variance of the sample must be accounted for by the projection of data points onto the axis PC1, perpendicular to PC2; the length of these projections are the object's values or scores of the second principal component, and this is verified as 0.163, with the sum of these variances 2.0 as expected. Table 4.1 sets out the results in the standard way of PCA.

Table 4.1   *Principal components from Figure 4.3*

	PC1	PC2
Eigenvalue	1.837	0.163
Proportion	0.918	0.082
Cumulative	0.918	1.000
Variable	PC1	PC2
d (Mpc)	0.707	0.707
v (km s^{-1})	0.707	0.707

Now consider the matrix approach. In the process of PCA the usual methodology is to construct the error matrix (Section 4.2), e.g. for the two-variable case of the example, $a(1, 1) = \sum d'^2, a(1, 2) = a(2, 1) = \sum v'd', a(2, 2) = \sum v'^2$. We then seek a principal axis transformation that makes the cross-terms vanish; we seek an axis transformation to rotate the ellipses of Figure 4.4 so that the axes of the ellipses coincide with the principal axes of the coordinate system. This, of course, is simply done in matrix notation. We determine the eigenvalues of the error matrix and form its eigenvectors (readily shown for the example to be $v' = d'$ and $v' = -d'$ as seen in Figure 4.3(b)). These eigenvectors then form the transpose matrix T, for variable transformation and axis rotation. The axis rotation *diagonalizes the matrix*, i.e. in the new axis system, the cross-terms are zero; we have rotated the axes until there is no x, y covariance.

Note that for the purpose our set of data has been reduced from 48 numbers for the 24 galaxies to four numbers, a 2×2 matrix. How did this happen? PCA assumes that *the covariance (or error) matrix suffices to describe the data*; this is the case if the data are drawn from a multivariate Gaussian (Section 4.2, Figure 4.4), or in general when a simple quadratic form, using the covariance matrix, can describe the distribution of the data. It is far from generally true that the clouds of points in most n-variate hyperspaces will be so simply distributed – see the following example.

In multivariate data sets, the disparate units are taken care of by normalizing as in the above example, subtracting mean values and dividing by variances. This is not a prescription, however. For example, the variance for any particular variable might be dominated by a monstrous outlier which there are good grounds to reject. The choice of weights does, therefore, depend on familiarity with the data and preferences – there is plenty of room for subjectivity. It should also be noted that PCA is a linear analysis and tests need to be performed on the linearity of the principal components. For example, plotting the scores of PC1 versus PC2 should show a Gaussian distribution consistent with $\rho = 0$. It may be apparent how to reject outliers or to transform coordinates to reduce the problem to a linear analysis. In large data sets such processes can reveal unusual objects.

Example Some PCA problems have a larger number of variables than input observables, $p > n$, resulting in singular matrices requiring modifications to standard techniques to solve the eigenvector equations (Wilkinson, 1978; Mittaz *et al.*, 1990). This situation occurs in *spectral PCA* for which the p variables are fluxes in p wavelength or frequency bins (Francis *et al.*, 1992; Wills *et al.*, 1997). The technique is ideal for dealing with a huge sample and was therefore adopted in the 2dF survey which aimed to measure 250 000 galaxy spectra to provide a detailed picture of the galaxy distribution out to a redshift of 0.25 (see Section 11.1). The PCA approach to 2dF galaxy classification is discussed in detail by Folkes *et al.* (1999). Figure 4.9, drawn from this paper, shows examples of 2dF spectra prepared for PCA, the mean spectrum, and the first three principal components. These three components represent the eigenvectors of the covariance matrix of these prepared spectra. In this example, the first PC accounts for 49.6 per cent of the variance; the first three components account for 65.8 per cent of the variance. Much of the remainder is due to noise.

The key aspect Folkes *et al.* wished to address was how luminosity function depends on galaxy type. The objects in the PC1–PC2 plane form a single cluster (Figure 4.9, blue emission-line object to the left, red objects with absorption lines to the right, and strong emission-line objects straggling downward.) Five spectral classes were then adopted, shown by the slanted lines in this figure. Confirmation that these spectral classes correspond to morphological classification came from placing the 55 Kennicut (1992) standard galaxies into this plot; the five classes are roughly E/S0, Sa, Sb, Scd and Irr. The way ahead to use the PCA classes to work out luminosity functions for each is clear, and the punch line is that significantly different Schecter functions emerged for each class.

Note how asymmetrical the distribution looks. This need not invalidate the analysis – here primarily one of classification – but the effectiveness must, in general, be reduced. Asymmetrical shapes in the PC planes must result in unquantifiable errors in the classification.

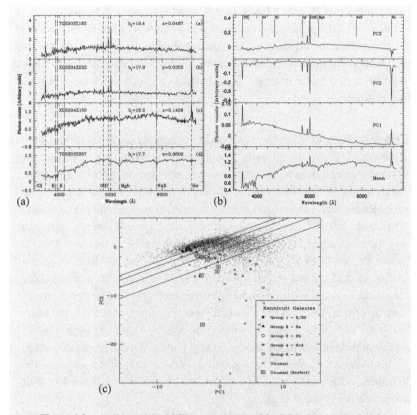

Figure 4.9   (a) Examples of 2dF spectra prepared for PCA. Instrumental and
atmospheric features have been removed, with the spectra transformed to the
rest frame, resampled to 4Å bins and normalized to unit mean flux. (b) The
mean spectrum and first three principal components; the sign of the PCs is
arbitrary. (c) Distribution of 2dF galaxy spectra in the PC1–PC2 plane. Slanted
lines divide the plane into the five spectral classes adopted by Folkes *et al.*
(1999); the positions of galaxies typed by Kennicut (1992) are shown.

In addition to spectral classification and analysis, spectral time variability is
amenable to PCA (Mittaz *et al.*, 1990; Turler & Courvoisier, 1998).

Of course we would like to know if the PCs are 'real' and so some indication
of the distribution of each one would be useful. This can be computed by a
bootstrap (Section 6.6) on the original data set. This will show how stable
the eigenvectors and eigenvalues actually are, in particular whether the largest
eigenvector is reliably detected.

Exercises

4.1 Correlation testing (D). Consider the Hubble plot of Figure 4.3. What is (a) the most likely value for ρ via the Jeffreys' test; (b) the significance of the correlation via the standard Fisher test and (c) the significance via the Spearman rank test? Estimate distributions for these statistics with a bootstrap (Section 6.6); and compare the results with the standard tests.

4.2 Multivariate random numbers. (a) Give the justification for why the prescription (Section 4.2) for generating (x, y) pairs following a Gaussian of given variance and correlation coefficient is correct. (b) Using a Gaussian Monte Carlo generator, find 1000 (x, y) pairs following a given prescription, i.e. σ_x^2, σ_y^2 and ρ. Plot these on contours of the bivariate probability distribution, as in Figure 6.2, to check roughly that the prescription works. (c) Find the error matrix for the (x, y) pairs to verify that the prescription works.

4.3 Permutation tests (D). (a) Take a small set of uncorrelated pairs (X, Y), preferably non-Gaussian. By permutation methods on the computer, derive distributions of Fisher r, Spearman's and Kendall's statistics. (b) Try the same numerical experiment with correlated data, using the bootstrap and the jackknife to estimate distributions (Section 6.6). Correlated non-Gaussian data are provided for the multivariate t distribution, which is Cauchy-like for one degree of freedom and becomes more Gaussian for larger degrees of freedom. How robust are the conclusions against outliers?

4.4 Principal component analysis (D). Carry through a PCA on the data of the quasar sample given in Francis & Wills (1999). Compute errors with a bootstrap analysis or jackknife (Section 6.6).

4.5 Lurking third variables. Consider the following correlations, and speculate on how a third variable might be involved. (a) During the Second World War, J. W. Tukey discovered a strong positive correlation between accuracy of bombing and the presence of enemy fighter planes. (b) There is a well-known correlation between stock market indices and the sunspot cycle. (c) The apparent angular size of radio sources shows a strong inverse correlation with radio luminosity.

In the exercises denoted by (D), data sets are provided on the book's website; or create your own.

5

Hypothesis testing

How do our data look?
I've carried out a Kolmogorov–Smirnov test . . .
Ah. **That** *bad.*
> *(interchange between Peter Scheuer and his then student, CRJ)*

(The) premise that statistical significance is the only reliable indication of causation is flawed.
> *(US Supreme Court, Matrixx Initiatives, Inc. vs. Siracusano,*
> *22 March 2011)*

It is often the case that we need to do sample comparison: we have someone else's data to compare with ours; or someone else's model to compare with our data; or even our data to compare with our model. We need to make the comparison and to *decide something*. We are doing *hypothesis testing* – are our data consistent with a model, with somebody else's data? In searching for correlations as we were in Chapter 4, we were *hypothesis testing*; in the model-fitting of Chapter 6 we are involved in *data modelling* and *parameter estimation*.

A frequentist point of view might be to consider the entire science of *statistical inference* as *hypothesis testing* followed by *parameter estimation*. However, if experiments were properly designed, the Bayesian approach would be right: it answers the sample-comparison questions we wished to pose in the first place, namely what is the probability, given the data, that a particular model is right? Or: what is the probability, given two sets of data, that they agree? The two-stage process should be unecessary at best. Indeed, the two-step process can be integrated into a Bayesian method of model choice, which we discuss in Chapter 7.

A frequentist works from the probability distribution of a statistic; a Bayesian works from the probability distribution of a hypothesis or model. Of course, it is the latter that we really want to know, but there are times when it has to be done the classical way, as, for example, when a model is not available. This chapter is basically concerned with those times.

Classical methods of hypothesis testing may be either *parametric* or *non-parametric, distribution-free* as it is sometimes called.

There are four reasons why statistical inference based on known probability distributions does not work, or limits our possibilities severely.

(i) We are measuring in experiments being run out there in the Universe, not by us. The underlying distributions may be far from known or understood; no averaging may be going on to lead us towards the central limit theorem and Gaussian distributions (see Chapter 2); yet we still wish to draw inferences about the underlying population. We only do so safely with *non-parametric statistics*, methods that do not require knowledge of the underlying distributions.

(ii) We may have to deal with small-number samples, such as $N = 3$. Non-parametric techniques have the ability to do this.

(iii) The range of observation scales available to us is given in Table 1.3. Each such scale has formal definition and formal properties. Each has admissible operations. Suffice it to say here that use of scales other than numerical ('interval' and 'ratio') requires in most (but not all) cases that we use non-parametric methods. *We may wish to make statistical inference without recourse to numerical scales.*

(iv) Others use such methods to draw inference. We need to understand what they are doing.

Non-parametric methods thus enormously increase the possibilities in decision-making via classical testing and form an essential part of our process.

Bayesian methods necessarily involve a known distribution, and hence, non-parametric methods do not apply. We briefly described the concepts of Bayesian versus frequentist and parametric versus non-parametric in the introductory Chapters 1 and 2. Table 5.1 summarizes these apparent dichotomies and indicates appropriate usage.

That non-parametric Bayesian tests do not exist appears self-evident, as the key Bayesian feature is the probability of a particular model in the face of the data. However, it is not quite this clear-cut, and there has been consideration of non-parametric methods in a Bayesian context (Gull & Fielden, 1986). If we understand the data so that we can model its collection process, then the Bayesian route beckons (see Chapters 2, 7 and their examples).

Table 5.1 *Bayes/frequentist/parametric/non-parametric usage*

	Parametric	Non-parametric
Bayesian testing	Model known. Data gathering and uncertainty understood.	Such tests do not exist.
Classical testing	Model known. Underlying distribution of data known. Large enough numbers. Data on ordinal or interval scales.	Small numbers. Unknown model. Unknown underlying distributions or errors. Data on nominal or categorical scales.

The classical tests involve us in 'rejecting the null hypothesis', i.e. in rejecting rather than accepting a hypothesis at some level of significance. The hypothesis we reject may not be one in which we have the slightest interest. This is a *process of elimination*.

5.1 Methodology of classical hypothesis testing

Classical hypothesis testing, as developed by Neyman and Pearson, follows these steps.

(i) Set up two possible and exclusive hypotheses, each with an associated *terminal action*:

H_0, the *null hypothesis* or hypothesis of no effect, usually formulated to be rejected, and

H_1, an alternative, or *research hypothesis*.

(ii) Specify a priori the *significance level* α; choose a test which (a) approximates the conditions and (b) finds what is needed; obtain the *sampling distribution* and the *region of rejection*, whose area is a fraction α of the total area in the sampling distribution.

(iii) Run the test; reject H_0 if the test yields a value of the statistic whose probability of occurrence under H_0 (usually called p) is $\leq \alpha$.

(iv) Carry out the terminal action.

It is vital to emphasize (ii). The significance level has to be chosen before the value of the test statistic is glimpsed; otherwise, some arbitrary convolution of the data plus the psychology of the investigator is being tested. This is not a game. You must be prepared to carry out the terminal action on the stated terms. There is no such thing as an inconclusive hypothesis test!

There are two types of error involved in the process, traditionally referred to (surprisingly enough) as Types I and II. *Type I error* occurs when H_0 is, in fact, true, and the probability of a Type I error is the probability of rejecting H_0 when it is, in fact, true, i.e. α. The *Type II error* occurs when H_0 is false, and the probability of a Type II error is the probability β of the failure to reject a false H_0; β is not related to α in any direct or obvious way. The *power* of a test is the probability of rejecting a false H_0, or $1 - \beta$.

This approach is conceptually distinct from Fisher's original method, which focuses on the p value, not considering alternative hypotheses. In the Neyman–Pearson approach, α is treated as a design parameter of the experimental procedure. Fisher disagreed violently with Neyman and Pearson, partly on the grounds that in science there is often no well-defined alternative hypothesis. The interpretation of p continues to provoke discussion and controversy (Berger & Sellke, 1987). We quoted Jeffreys' trenchant opinion on p in Section 3.5. Note also the final paragraph of this chapter (Section 5.6).

The *sampling distribution* is the probability distribution of the test statistic, i.e. the frequency distribution of area unity including all values of the test statistic under H_0. The probability of the occurrence of any value of the test statistic in the *region of rejection* is less than p, by definition; but where the region of rejection lies within the sampling distribution depends on H_1. If H_1 indicates *direction*, then there is a single region of rejection and the test is *one-tailed*; if no direction is indicated, the region of rejection is comprised of the two ends of the distribution and we are dealing with a *two-tailed* test. This is the only use we make of H_1; the testing procedure can only convince us to accept H_1 if it is the sole alternative to H_0. Beware – it is human nature to think that your H_1 is the only possible alternative to H_0.

If we do not have a well-defined H_1, then setting a threshold α for action makes less sense; this was Fisher's objection. A more nuanced view than Neyman–Pearson might be, calculate p, notice that it is uncomfortably small, be provoked into thinking about other possibilities. There is nothing wrong with this as long as we do not ascribe a probability to our rejection of H_0.

A simple table helps to clarify the notions of Type I and Type II errors. Suppose that the critical value of our test statistic is (say) t_c, and further assume that the chance of *exceeding* t_c under H_0 is α. (It might equally be a critical value of chi-square, or the F statistic, or whatever, depending on our exact method.) Our procedure is then: compute the test statistic T, compare with

t_c, reject H_0 (and hence accept H_1) if $T > t_c$. The table then illustrates the possibilities.

	H_0 true	H_1 true
$T \geq t_c$	A: Type I error	B: correct
$T < t_c$	C: correct	D: Type II error

Notice that the probabilities of cells A and C (or of cells B and D) add up to unity; but there is no corresponding relationship along the rows, so that there is no relationship between the Type I and the Type II error rates. The probability of B being occupied is the power. Generally, the power and the Type I error rate have to be traded off against each other for the specific alternative H_1.

The notion of 'power' is used much more in disciplines other than astronomy, where alternatives to the (usually simple) null hypothesis are more often available. Both *parametric* and *non-parametric* (classical) tests follow this procedure; both use a test statistic with a known sampling distribution. The non-parametric aspect arises because the test statistic does not itself depend upon properties of the population(s) from which the data were drawn. There are persuasive arguments for following the non-parametric testing in using classical methods, as outlined at the head of Chapter 5. But first we consider the parametric route in some detail in order to establish methodology.

5.2 Parametric tests: means and variances, t and F tests

A very common question arises when we have two sets of data (or one set of data and a model) and we ask if they differ in location or spread. The best-known parametric tests for such comparisons concern samples drawn from Normally distributed parent populations; these tests are, of course, the Student's t test (comparison of *means*) and the F test (comparison of *variances*), and are discussed in most books on statistics, e.g. Martin (1971), Stuart & Ord (1994). The t and F statistics have been introduced already (Section 3.4)

To contrast the classical and Bayesian methods for hypothesis testing, we look at the simple case of comparison of means. We deal with a Gaussian distribution, because its analytical tractability has resulted in many tests being developed for Gaussian data; and then, of course, there is the central limit theorem.

Let us suppose that we have n data X_i drawn from a Gaussian of mean μ_x, and m other data Y_i, drawn from a Gaussian of identical variance but a different mean μ_y. Call the common variance σ^2.

The Bayesian method is to calculate the joint posterior distribution

$$\text{prob}(\mu_x, \mu_y, \sigma) \propto \frac{1}{\sigma^{n+m+1}} \exp\left[-\frac{\sum_i (x_i - \mu_x)^2}{2\sigma^2}\right] \exp\left[-\frac{\sum_i (y_i - \mu_y)^2}{2\sigma^2}\right] \tag{5.1}$$

in which we have used the Jeffreys' prior (Exercise 2.6) for the variance. Integrating over the 'nuisance' parameter σ, we would get the joint probability $\text{prob}(\mu_x, \mu_y)$ and could use it to derive, for example, the probability that μ_x is bigger than μ_y.

From this we can calculate the probability distribution of $(\mu_x - \mu_y)$ (see, e.g., Lee, 2004, Chapter 5). The result depends on the data via a quantity

$$t' = \frac{(\mu_x - \mu_y) - (\overline{X} - \overline{Y})}{s\sqrt{m^{-1} + n^{-1}}}, \tag{5.2}$$

where

$$s^2 = \frac{nS_x + mS_y}{\nu},$$

with the usual mean squares $S_x = \sum(X_i - \overline{X})^2/n$, similarly for S_y, and $\nu = n + m - 2$.

The distribution for t' is

$$\text{prob}(t') = \frac{\Gamma[\frac{\nu+1}{2}]}{\sqrt{\pi\nu}\,\Gamma[\frac{\nu}{2}]}\left(1 + \frac{t'^2}{\nu}\right)^{-(\nu+1)/2}. \tag{5.3}$$

By this route we do not really hypothesis test. We regard the data as fixed and $(\mu_x - \mu_y)$ as the variable, simply computing the probability of any particular difference in the means. We might alternatively work out the range of differences which are, say, 90 per cent probable, or we might carry the distribution of $(\mu_x - \mu_y)$ on into a later probabilistic calculation.

If we instead follow the classical line of reasoning, we do not treat the μ's as random variables. Instead, we guess that the difference in the averages $\overline{X} - \overline{Y}$ will be the statistic we need; and we calculate its distribution on the null hypothesis that $\mu_x = \mu_y$. We find that

$$t = \frac{\overline{X} - \overline{Y}}{s\sqrt{m^{-1} + n^{-1}}} \tag{5.4}$$

follows a t distribution with ν degrees of freedom. This is the classical Student's t. Critical values are given in Table B.3.

This gives the basis of a classical hypothesis test, the t *test* for means. Assuming that $(\mu_x - \mu_y) = 0$, (the null hypothesis) we calculate t. If it (or some greater value) is very unlikely, we think that the null hypothesis is ruled out.

The t statistic is heavy with history and reflects an era when analytical calculations were essential. The penalty is the total reliance on the Gaussian. However, with cheap computing power we may expect to be able to follow the basic Bayesian approach outlined above for any distribution.

By analogous calculations, we can arrive at the F *test* for variances. Again, Gaussian distributions are assumed. The null hypothesis is $\sigma_x = \sigma_y$, the data are X_i ($i = 1, \ldots, N$) and Y_i ($i = 1, \ldots, M$) and the test statistic is

$$\mathcal{F} = \frac{\sum_i (X_i - \overline{X})^2 / (N-1)}{\sum_i (Y_i - \overline{Y})^2 / (M-1)}. \tag{5.5}$$

This follows an F distribution with $N - 1$ and $M - 1$ degrees of freedom (Table B.4) and the testing procedure is the same as for Student's t. Clearly this statistic will be particularly sensitive to the Gaussian assumption.

Example Suppose that we have two small sets of data, from Gaussian distributions of equal variance: -1.22, -1.17, 0.93, -0.58, -1.14 (mean -0.64), and 1.03, -1.59, -0.41, 0.71, 2.10 (mean 0.37), with a pooled standard deviation of 1.2. The standard t statistic is 1.33. If we do a two-tailed test (so being agnostic about whether one mean is larger than another), we find a 22 per cent chance that these data would arise if the means were the same. The one-tailed test (testing whether one mean is larger) gives 11 per cent. From a Bayesian point of view, we can calculate the distribution of $(\mu_x - \mu_y)$ for the same data. In Figure 5.1 we can see clearly that one mean is smaller; the odds on this being so are about 10 to 1, as can be calculated by integrating the posterior distribution of the difference of means. We can also check that the variances are indeed the same for the two samples by using the F test. Here we get a one-tailed significance of 21 per cent, so this justifies (to some extent!) treating the variances as the same in the t test.

We see that the Bayesian approach, and the classical t test, give the same *numbers* in this case. Indeed, this is necessarily the case: Equation (5.3) *is the t* distribution. Despite this, the *interpretation* of the classical and Bayesian results is very different.

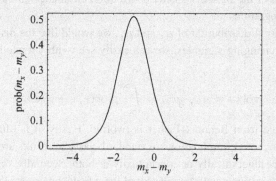

Figure 5.1   The distribution of the difference of means for the example data.

5.2.1 The Behrens–Fisher Test

Relaxing the assumption of equal variances may be important. It is indeed possible to derive the distribution of the difference in means without the assumption of equal variances in the two samples; the resulting distribution is called the Behrens–Fisher distribution. It is of great interest in statistics because it is a rare example of a Bayesian analysis having no classical analogue; there is no classical test for the case of possibly unequal variances. Lee (2004) discusses this in some detail.

The analytical form of the Behrens–Fisher distribution is complicated (see, e.g., Lee, 2004, Chapter 5) and involves a numerical integration anyway, so we may as well resort to a computer right away to calculate it from Bayes' theorem. We suppose that our data are drawn from Gaussians with means μ and standard deviations σ. The joint posterior distribution (using Jeffreys' prior on the σ) is

$$\text{prob}(\mu_x, \mu_y, \sigma_x, \sigma_y) \propto \frac{1}{\sigma_x^{n+1}} \exp\left[-\frac{\sum_i (x_i - \mu_x)^2}{2\sigma_x^2}\right]$$
$$\times \frac{1}{\sigma_y^{n+1}} \exp\left[-\frac{\sum_i (y_i - \mu_y)^2}{2\sigma_y^2}\right] \quad (5.6)$$

We have a multidimensional integration to do in order to get rid of the two *nuisance parameters* (σ_x and σ_y) and to ensure that the resulting joint distribution $\text{prob}(\mu_x, \mu_y)$ is properly normalized. This is now not much of a problem, although until recently these integrations (for anything other than Gaussians) were a formidable obstacle to Bayesian methods – see Section 7.7. The analytical derivation of the Behrens–Fisher distribution eliminates all but one of the numerical integrations.

Given the joint distribution of μ_x and μ_y, we would like the distribution of $\mu_y - \mu_x$. By changing variables, we can easily see (with yet another integration!) that

$$\text{prob}(u = \mu_y - \mu_x) = \int_\infty^\infty \text{prob}(v, v + u)\, dv.$$

The message from Behrens–Fisher is twofold. Firstly, it is often the case that quite innocuous complications to standard classical results are very complicated, either algebraically or conceptually or both. Secondly, the rescue by brute-force Bayes may require quite a lot of force to do the necessary numerical integrations. This will be a theme of our Bayesian examples. With this caveat, this general type of Bayesian test can be followed for any distribution – as long as we know what it is and can do the integrations.

Example Consider the same example data as before, relaxing the assumption that the variances are equal. (The sample standard deviations are 0.9 and 1.4, not significantly different, according to the F test.) We see from Figure 5.2 that the distributions of $\mu_y - \mu_x$ are similar to the t distribution, although, as we might expect, the distribution is a little wider without the assumption that the variances are equal. Although we cannot tell (classically) that the variances differ, we will obtain different results by *not* assuming that they are the same.

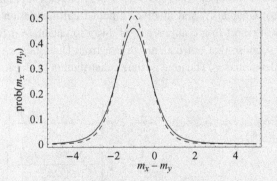

Figure 5.2   Distribution of the difference of means assuming equal variances (dashed) and without this assumption (solid).

We have now been introduced to the classical *t* test, the Neyman–Pearson concepts of error rates and power, and the complications of simple extensions to classical parametric methods. In the following example we will see all of these at work.

Example We generate two sets of 10 random data X_i and Y_i, drawn from Gaussian distributions with unit standard deviation and means zero and μ. These values correspond to hypotheses H_0 and H_1, respectively.

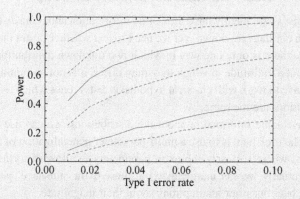

Figure 5.3   Power versus Type I error rate for a very simple *t* test, where the true differences in the means range from 1 unit (bottom) to 3 units (top). The dashed lines are for an increase of 25 per cent in the standard deviation of the Gaussian from which the samples are drawn.

By computing the *t* statistic, and comparing it to the critical value for a given significance level α, a large number of repetitions allows us to find the Type I error rate and the power, following the method outlined in Section 5.1. The results are shown in Figure 5.3, and we see the differing trade-offs between power and Type I error rates, depending on the value of μ that defines the alternative hypothesis H_1. It is salutary to attempt to do this calculation analytically; calculation of the sampling distribution of *t* when H_1 applies requires a non-central *t* distribution, which is quite complicated.

In all the examples in this chapter, we are dealing with cases where the data are free of measurement noise. So when we compare empirical distributions like the ones in this example, we are assuming that the data are accurately known so that the observed distributions reflect the true ones. If the X_i and Y_i were stellar magnitudes, for example, they might be subject to measurement error which would blur out the distributions and reduce the difference in the true distribution of magnitudes. We can see the effect of an additive Gaussian error of this kind simply by increasing the standard deviation of the Gaussian distributions from which X_i and Y_i are drawn. Increasing it from 1.0 to 1.25 gives the dashed lines in Figure 5.3, where we see how noise decreases the power of the test – an intuitive and general result, illustrating the utility of the concept of power.

5.2.2 Non-Gaussian parametric testing

In astronomy we frequently have little or no information about the distributions from which our data are drawn, yet we need to test whether or not they are the same. Since there is only one way in which two unknown distributions can be the same, but a multitude in which they may differ, it is not surprising that we currently have to work with classical hypothesis tests – ones which assume that the distributions are the same.

If we have some information about the distributions, we can use Bayesian methods. The trick here is to use a multi-parameter generalization of a familiar distribution, where we carry the extra parameters to allow distortions in the shape. Eventually we can marginalize out these extra 'nuisance' parameters, integrating over our prior assumptions about their magnitude.

The most common example of this sort of generalization is the Gram–Charlier series:

$$\exp\left(-\frac{x^2}{2\sigma^2}\right)\left(1 + \sum_i a_i H_i(x)\right) \tag{5.7}$$

in which the H's are the Hermite polynomials. The coefficients a_i are the free parameters we need. (Because the Hermite polynomials are orthogonal with respect to Gaussian weights, these coefficients are also related to the moments of the distribution we are trying to create.) The effect of these extra terms is to broaden and skew a Gaussian, and so for some data a few-term Gram–Charlier series may give quite a useful basis for a parametric analysis. Priors on the coefficients have to be set by judgement. The even Hermite polynomials have the effect of changing scale, and so should follow the same Jeffreys' prior as the standard deviation. The odd polynomials will change both scale and location and here setting the prior is less obvious.

There are two other variants on the Gram–Charlier series. For a distribution allied to the exponential $\exp(-x/a)$, a Laguerre series will function in the same way as a Gram–Charlier series, except the distorting functions are the Laguerre polynomials. The Gamma series is based on the distribution $x^\alpha(1 - x)^\beta$, defined on the interval from 0 to 1; the distorting functions are the even less

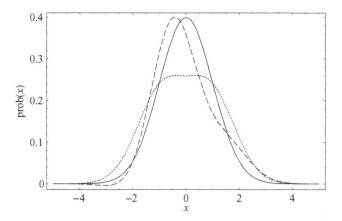

Figure 5.4   Various distributions resulting from using just two terms in a Gram–Charlier distribution; the solid curve is a pure Gaussian.

familiar Jacobi polynomials. However, computer algebra packages such as MATHEMATICA give comprehensive support for special functions and make the application of these series rather straightforward (Reinking, 2002). A useful summary and critique of approximations to weakly non-Gaussian distributions is given by Blinnikov & Moessner (1998).

This approach clarifies the problem that a non-parametric test has to solve. Suppose we fix on a two-term Gram–Charlier expansion as a realistic representation of our data; the versatility is demonstrated in Figure 5.4. For data set 1, we then get the posterior $\text{prob}(\mu^{(1)}, \sigma^{(1)}, a_1^{(1)}, a_2^{(1)})$, and similarly for data set 2. If we ask the apparently innocuous question 'are these data drawn from different distributions?' we see that there are many possibilities (in fact, 2^4) of the form, for instance, $\mu^{(1)} > \mu^{(2)}$ and $\sigma^{(1)} < \sigma^{(2)}$ and $a_1^{(1)} > a_1^{(2)}$ and $a_2^{(1)} < a_2^{(2)}$. Working through these possibilities could be quite tedious. A different question might be 'are these distributions at different locations, regardless of their widths?', in which case we could marginalize out the σ's and a_2's (Section 2.2); the location, in a Gram–Charlier expansion, is a simple combination of μ and a_1.

5.2.3   Which model is better? The Bayes factor

This does suggest that comparison of models in the sense 'are these data drawn from the same distribution?' might be a more tractable question. Notice that we are not asking if $\mu^{(1)} = \mu^{(2)}$, etc. as the probability of exact equality is zero. With the Bayesian approach, often knowing the posterior distribution of the parameter of interest is enough; we might be making a comparison with an exactly known quantity, perhaps derived from some theory. However, we may wish to compare with an experimental determination of some other parameter $\vec{\beta}$. A typical case, for scalar parameters α and β, would be to ask for the probability that, say, α is bigger than β.

Suppose, therefore, that we have derived two distributions $\text{prob}(\alpha) = p_A(\alpha)$ and $\text{prob}(\beta) = p_B(\beta)$ from independent samples. The probability that α is larger than β is

$$p(\alpha > \beta) = \int_{-\infty}^{\infty} p_B(y) \, dy \int_{y}^{\infty} p_A(x) \, dx$$

and the double integral simplifies to

$$p(\alpha > \beta) = \int_{-\infty}^{\infty} [1 - C_A(x)] p_B(x) \, dx$$

in which C_A is the cumulative distribution corresponding to p_A. If p_A and p_B are the same distribution, this becomes $p(\alpha > \beta) = 1/2$, as expected. Usually these integrals have to be done numerically case by case, but are worth the effort.

We may express posterior probabilities by using the notion of *odds*, a handy way of expressing probabilities when we have only two possibilities (see Section 2.5). The odds on event A are just

$$\frac{\text{prob}(A)}{\text{prob}(\text{not } A)}.$$

For instance, the odds on throwing a six with a fair die are 5:1 (probability of $1/6$ for throwing a six, $5/6$ for anything else). From a betting point of view, the odds on a bet give the profit that might be made on a stake; in the case of our example with dice, being offered 5:1 odds for a 6 means that we would get \$5 profit (\$6 payout) on a stake of \$1, if a six comes up. Of course a bookie will offer slightly different odds, to be sure of a profit in the long run. If we have two exclusive possibilities for a prior, say A and not A, then the posterior odds are given by the ratio of the posterior probabilities with each prior, and give an indication of which prior to bet on, given the available data.

An important method of model choice involves the *Bayes factor*. We will deal with this in much more detail in Section 7.1 (and see Lee, 2004), but here a flavour of the method is instructive.

Suppose we try to describe all of the example data X_i, Y_i we have been discussing with just one distribution G. This distribution may have parameters, so let us denote this hypothesis by (G, θ). Alternatively (and by hypothesis exhaustively) we may use (G_x, θ_x) for the data X_i and (G_y, θ_y) for the data

Y_i. This hypothesis is $(G_x, \theta_x, G_y, \theta_y)$. We will need prior probabilities for our two options, G or $G_x G_y$. Bayes' theorem then tells us

$$\text{prob}(G, \theta \mid X, Y) \propto \text{prob}(X, Y \mid G, \theta)\text{prob}(G, \theta) \tag{5.8}$$

with a similar (but longer!) equation for the paired hypothesis involving two distributions. In this equation, the prior $\text{prob}(G, \theta)$ tells us both the prior distribution of the parameters and the overall prior probability of the model. The term $\text{prob}(X, Y \mid G, \theta)$ is shorthand for the likelihood. The proportionality symbol is used because we will not need the normalizing constant, which is common to this equation and the one for the paired distributions.

To make a model choice we want the probability of the *model*, averaged over the possible values for the parameters; in other words, we want to convert the left-hand side of Equation (5.8) to a number. We do this by integrating it over the parameters. So we have

$$\text{prob}(G) \propto \int \text{prob}(G, \theta \mid X, Y)\,\mathrm{d}\theta$$

with again a similar equation for the pair of distributions, which is explicitly

$$\text{prob}(G_x, G_y) \propto \int \text{prob}(G_x, \theta_x \mid X)\,\mathrm{d}\theta_x \int \text{prob}(G_y, \theta_y \mid Y)\,\mathrm{d}\theta_y.$$

The ratio of probabilities can now be taken, cancelling out the normalizing factor and giving the odds on G:

$$\mathcal{B} = \frac{\text{prob}(G)}{\text{prob}(G_x, G_y)}.$$

This ratio $\mathcal{B}$ is the Bayes factor (Section 7.1).

The odds on the distinct distributions are, in detail,

$$\frac{\int \text{prob}(G_x, \theta_x \mid X)\,\mathrm{d}\theta_x \int \text{prob}(G_y, \theta_y \mid Y)\,\mathrm{d}\theta_y}{\int \text{prob}(G, \theta \mid X, Y)\,\mathrm{d}\theta}. \tag{5.9}$$

To work out these odds, we integrate the likelihood functions, weighted by the priors, over the range of parameters of the distributions.

The Bayes factor can be used for much more complex model choices, as we shall see in Chapter 7. In this chapter we will be most often concerned with the significance level p as the output of a hypothesis-testing procedure where *no* alternative is specified. It is a remarkable fact that knowing only p actually gives a bound on the Bayes factor for a very wide class of alternative hypotheses (Sellke *et al.*, 2001). If p is the significance level obtained on testing

H_0, a lower bound on the Bayes factor or odds on the alternatives is given by

$$\mathcal{B} \geq \frac{-1}{ep \log p},\tag{5.10}$$

a strange connection between the Bayesian and non-Bayesian worlds.

Example Suppose, we have the following two data sets: $X_i = -0.16, 0.12,$
0.44, 0.60, 0.70, 0.87, 0.88, 1.44, 1.74, 2.79 and $Y_i = 0.89, 0.99, 1.29,$
1.73, 1.96, 2.35, 2.51, 2.79, 3.17, 3.76. The means differ by about one
standard deviation. We consider two a-priori equally likely hypotheses. One
is that all 20 data are drawn from the same Gaussian. The other is that
they are drawn from different Gaussians. In the first case, the likelihood
function is

$$\frac{1}{(\sqrt{2\pi}\sigma)^{20}} \exp\left[-\frac{\sum_i(X_i - \mu)^2 + \sum(Y_i - \mu)^2}{2\sigma^2}\right]$$

and we take the prior on σ to be $\frac{1}{\sigma}$. We also assume a uniform prior for the
μ's. In the second case, the likelihood is

$$\frac{1}{(\sqrt{2\pi}\sigma_x)^{10}} \exp\left[-\frac{\sum_i(X_i - \mu_x)^2}{2\sigma_x^2}\right] \frac{1}{(\sqrt{2\pi}\sigma_y)^{10}} \exp\left[-\frac{\sum_i(Y_i - \mu_y)^2}{2\sigma_y^2}\right]$$

and the prior on the standard deviations is $\frac{1}{\sigma_x\sigma_y}$. Integrating over the range of
the μ's and σ's, the odds on the data being drawn from different Gaussians
are about 40:1 – a good bet. Taking a significance level from a t test on
these data, we also find from Equation (5.10) that the Bayes factor should
be bigger than about 16.

The priors need some care in this type of calculation because they have to
be normalizable. Thus, in this example we are assuming that the priors on
the μ's and σ's extend over some wide but not infinite range – wide enough
not to affect the answer, but finite. In the example, we have assumed for
simplicity that the ratio of the normalizing factors on these priors is simply
unity. Better and more interesting assumptions are possible, as we will see
in Chapter 7.

In the exercises, we suggest following a classical t and F test on these
data, and contrasting to the Bayes factor in further detail.

5.3 Non-parametric tests: single samples

We now leave Bayesian methods and return to classical territory for the remainder of this chapter.

'Non-parametric tests' implies that 'no distribution is assumed', but something must be assumed, to make any progress. It may not be the Gaussian, but what is it? Various tests exploit different things, but a common method is to use counting probabilities. Take as an example the chi-square test (Section 5.3.1). The number of items in bin i is N_i, and we expect E_i. For smallish numbers, Poisson statistics tells us that the variance is also E_i. So $(N_i - E_i)^2/E_i$ should be roughly a squared Gaussian variable, of unit variance. As another example, the runs test (Section 5.3.3) is just using the assumption that each successive observation is equally likely to be 'up' or 'down', so a binomial distribution applies. The assumptions underlying non-parametric tests are weaker, and so more general, than for the parametric tests.

It is worth emphasizing again why we are going to advocate the non-parametric tests.

- These make fewer assumptions about the data. If, indeed, the underlying distribution is unknown, *there is no alternative*.
- If the sample size is small, probably we must use a non-parametric test.
- The non-parametric tests can cope with *data in non-numerical form*, e.g. ranks, classifications. There may be no parametric equivalent.
- Non-parametric tests can treat samples of observations from several different populations.

What are the counter-arguments? The main one concerns *binning* – binning is bad; it loses information and therefore loses *efficiency*. The *power* of non-parametric tests may be somewhat less, but typically no more than 10 per cent less than their parametric equivalents.

5.3.1 Chi-square test

Pearson's paper (Pearson, 1900) in which the chi square was introduced, is a foundation stone of modern statistical analysis[1]; a comprehensive and readable review (plus bibliography) is given by Cochran (1952).

Consider observational data which can be binned, and a model/hypothesis which predicts the population of each bin. The chi-square statistic describes

---

[1] Pearson's paper is entitled *On the criterion that a given system of deviations from the probable in the case of a correlated system of variables is such that it can be reasonably supposed to have arisen from random sampling*. It is wonderful polemic and gives several examples of the previous abuse of statistics, covering the frequency of buttercup petals to the incompetence of Astronomers Royal. ('Perhaps the greatest defaulter in this respect is the late Sir George Biddell Airy...') He demonstrates, for extra measure, that a run of bad luck at his roulette wheel, Monte Carlo, in July 1892 had one chance in 10^{29} of arising by chance; he avoids libel by phrasing his conclusion '... it will be more than ever evident how little chance had to do with the results...'

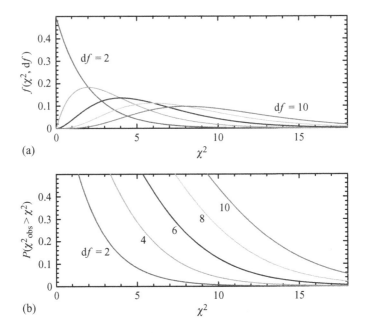

Figure 5.5   The chi-square distribution. (a) $f(\chi^2, df)$, the probability density function of χ^2 for df degrees of freedom. (b) The distribution function $\int_{\chi^2}^{\infty} f(\chi^2, df)d\chi^2$ of Table B.6, consulted to determine if χ^2 is 'large enough' to reject H_0.

the goodness-of-fit of the data to the model. If the *observed* numbers in each of k bins are O_i, and the *expected* values from the model are E_i, then this statistic is

$$\chi^2 = \sum_{i=1}^{k} \frac{(O_i - E_i)^2}{E_i}. \tag{5.11}$$

The null hypothesis H_0 is that the number of objects falling in each category is E_i; the chi-square procedure tests whether the O_i are sufficiently close to E_i to be likely to have occurred under H_0. The sampling distribution under H_0 of the statistic χ^2 follows the chi-square distribution (Figure 5.5) with $\nu = (k - 1)$ degrees of freedom. One degree of freedom is lost because of the constraint that $\sum_i O_i = \sum_i E_i$. The chi-square distribution is given by

$$f(x) = \frac{2^{-\nu/2}}{\Gamma[\nu/2]} x^{\nu/2 - 1} e^{-x/2} \tag{5.12}$$

(for $x \geq 0$), the distribution function of the random variable $Y^2 = Z_1^2 + Z_2^2 + \cdots + Z_\nu^2$ where the Z_i are independent random variables of standard Normal distribution. Table B.6 presents critical values; if χ^2 exceeds these values, H_0 is rejected at that level of significance.

The premise of the chi-square test, then, is that the deviations from E_i are due to statistical fluctuations from limited numbers of observations per bin, i.e. 'noise' or Poisson statistical variation, and the chi-square distribution simply gives the probability of the chance deviations of O_i from E_i. As we shall see, we need enough data per bin to ensure that each term in the chi-square summation is approximately Gaussian. To use the chi-square statistic, we ask the question 'What is the chance that this value of chi square, or a bigger one, arises by chance?' and so we use the cumulative distribution of Figure 5.5. We have to do this because the probability of any particular value of the chi-square statistic is, of course, infinitesimally small.

There is good news and bad news about the chi-square test. Firstly the good: it is a test of which most scientists have heard, with which many are comfortable, and from which some are even prepared to accept the results. Moreover, because χ^2 is additive, the results of different data sets which may fall in different bins, bin sizes, or which may apply to different aspects of the same model, may be tested all at once. The contribution to χ^2 of each bin may be examined and regions of exceptionally good or bad fit delineated. In addition, χ^2 is easily computed, and its significance readily estimated as follows. The *mean* of the chi-square distribution equals the *number of degrees of freedom*, while the *variance* equals *twice the number of degrees of freedom*; see plots of the function in Figure 5.5. So, as another rule of thumb, if χ^2 should come out (for more than four bins) as $\sim$(number of bins $-$ 1) then accept H_0, but if χ^2 exceeds twice (number of bins $-$ 1), probably H_0 will be rejected. Finally, minimizing χ^2 is an exceptionally common method of *model fitting* (see Section 6.3); and an example of the chi-square test (and model-fitting) is shown as Figure 6.3.

Because (for more than four bins) chi square is expected to be $\sim$ the number of degrees of freedom, it is frequent practice to work in terms of *reduced chi square*, namely χ^2/ν. If all is well, this should come out near unity; values of 4 or more are highly improbable on the premise that deviations are due to random fluctuations around the model in question. The use of reduced chi square is so common that at times researchers forget to mention the little fact that they have divided the observed χ^2 by ν – beware of this.

Now the bad news: the data must be binned to apply the test, and the bin populations must reach a certain size because it is obvious that instability results as $E_i \rightarrow 0$. As another rule of thumb then: >80 per cent of the bins must have $E_i > 5$. Bins may have to be combined to ensure this, an operation which is perfectly permissible for the test. However, the binning of data in general, and certainly the binning of bins, results in loss of efficiency and information, resolution in particular.

Thus, the advantages of the chi-square test are its general acceptance, the ease of computation, the ease of guessing significance, and the fact that model testing is for free. The disadvantages are the loss of power and information via binning, and the lack of applicability to small samples, in particular the serious instability at <5 counts per bin. Moreover, the chi-square test cannot tell *direction*, i.e. it is a ' two-tailed' test; it can only tell whether the differences between sample and prediction exceed those which can be reasonably expected on the basis of statistical fluctuations due to the finite sample size. There must be something better, and indeed there is:

5.3.2 Kolmogorov–Smirnov one-sample test

The test is extremely simple to carry out.

(i) Calculate $S_e(x)$, the *predicted* cumulative (integral) frequency distribution of the model under H_0.

(ii) Consider the sample of N observations, and compute $S_o(x)$, the *observed* cumulative distribution, the sum of all observations to each x divided by the sum of all N observations.

(iii) Find

$$D = \max \mid S_e(x) - S_o(x) \mid \qquad (5.13)$$

(iv) Consult the known sampling distribution for D under H_0, as given in Table B.7, to determine the fate of H_0. If D exceeds a critical value at the appropriate N, then H_0 is rejected at that level of significance.

Thus, as for the chi-square test, the sampling distribution indicates whether a divergence of the observed magnitude is 'reasonable' if the difference between observations and prediction is due solely to statistical fluctuations.

The Kolmogorov–Smirnov test has some enormous advantages over the chi-square test. Firstly, it treats the individual observations separately, and no information is lost because of grouping. Secondly, it works for small samples; for very small samples it is the only alternative. For intermediate sample sizes it is more powerful. Finally, note that as described here, the Kolmogorov–Smirnov test is non-directional or two-tailed, as is the chi-square test. However, a method of finding probabilities for the one-tailed test does exist (Birnbaum & Tingey, 1951; Goodman, 1954), giving the Kolmogorov–Smirnov test yet another advantage over the chi-square test.

Then why not always use it? There are perhaps two valid reasons, in addition to the invalid one (that it is not so well known). Firstly, the distributions must be continuous functions of the variable to apply the Kolmogorov–Smirnov test. The chi-square test is applicable to data which can be simply binned, grouped, categorized – there is no need for measurement on a numerical scale. Secondly, in model-fitting/parameter estimation, the chi-square test is readily adapted (Section 6.3) by simply reducing the number of degrees of freedom according to the number of parameters adopted in the model. The Kolmogorov–Smirnov test cannot be adapted in this way, since the distribution of D is not known when parameters of the population are estimated from the sample.

5.3.3 One-sample runs test of randomness

This delightfully simple test is contingent upon forming a binary (1–0) statistic from the sample data, e.g. heads–tails, or the sign of the residuals about the mean, or a best-fit line. It is to test H_0 that the sample is random; that successive observations are independent. Are there too many or too few *runs*?

Determine m, the number of *heads* or 1's; n, the number of *tails* or 0's, $N = n + m$; and r, the number of runs.

Look up the level of significance from the tabled probabilities (Table B.8) for a one or two-tailed test – depending on H_1, which can specify (as the *research hypothesis*) how the non-randomness might occur. In general we are concerned simply with the one-tail test, asking whether or not the number of runs is too few, the issue being independence or otherwise of data in a sequence. Situations giving rise to too many runs are infrequent; but if, indeed, there are significantly too many runs, it does say something serious about the data structure – probably in the sense that we do not understand it.

In fact for m 'heads' and n 'tails' with N data, the expectation value of number of runs is

$$\mu_r = \frac{2\,mn}{m+n} + 1 \tag{5.14}$$

and in the large N approximation this is asymptotically Gaussian with

$$\sigma_r = \sqrt{\frac{2\,nm(2\,nm - N)}{N^2(N-1)}}. \tag{5.15}$$

For large samples, then, it is possible to use the Normal distribution in the standard way by forming

$$z = \frac{r - \mu_r}{\sigma_r}$$

and consulting Table B.1, the integral Gaussian or erf function. This is the procedure when the numbers exceed 20 and run off the end of Table B.8.

Example Figure 5.6 shows the optical spectrum of quasar 3C207. The baseline has been estimated by the method of minimum Fourier components (Section 9.3.2). Does it fit properly? Is there low-level signal present in broad emission lines? Carefully selected regions of the spectrum are examined with the runs test.

The runs test is applied by using one-bit digitization – is the datum above or below the fitted baseline? The lower-wavelength region has enhanced continuum, a quasar 'blue bump', where the likelihood of line emission is significantly reduced. The runs test yields concordance, 36 positive deflections, 29 negative, 31 runs against an expectation of 32.1 runs, $z = -0.28$. The second region lies in the range of the hydrogen Balmer-line series, and several members are clearly present in emission. The result, a foregone conclusion here, is rejection of randomness by the runs test at about 4σ: 31 positives, 32 negatives, 16 runs against an expectation of 31.5, $z = -3.94$. The broad emission lines yield the contiguous regions that decrease the number of runs to a highly significant degree.

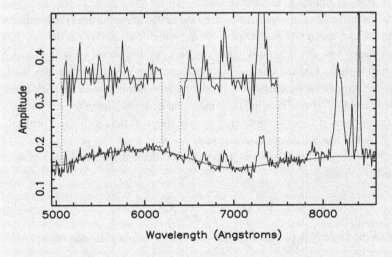

Figure 5.6   A spectrum of the quasar 3C207, taken with the 4.2-m William Herschel telescope. The solid curve is a baseline fitted by a Fourier minimum-component technique. The regions considered for runs testing are shown in the separated sections, each with the baseline subtracted and magnified by a factor of 3.

The test is at its most potent in looking for independence between adjacent sample members, e.g. in checking sequential data of *scan* or *spectrum* type as in the above example. It is frequently used for checking sequences of residuals, scatter of data about a model line, and in this guise it can give a straightforward answer as to whether a model is a good representation of the data.

5.4 Non-parametric tests: two independent samples

Now suppose that we have two samples; we want to know whether they could have been drawn from the same population, or from different populations, and if the latter, whether they differ in some predicted direction. Again assume that we know nothing about probability distributions, so that we need non-parametric tests. There are several.

5.4.1 Fisher exact test

The test is for two independent *small* samples for which discrete *binary* data are available, e.g. scores from the two samples fall in two mutually exclusive bins yielding a 2×2 *contingency table* as shown in Table 5.2. H_0: the assignment of 'scores' is random.

Table 5.2 2×2 *contingency table*

Sample =	1	2
Category = 1	A	C
= 2	B	D

Compute the following statistic:

$$p = \frac{(A + B)!(C + D)!(A + C)!(B + D)!}{N!A!B!C!D!} \tag{5.16}$$

This is the probability that the total of N scores could be as they are *when the two samples are in fact identical*. But in fact the test asks: What is the probability of occurrence of the observed outcome *or one more extreme* under H_0? Hence, by the laws of probability (see, e.g., Stuart & Ord 1994), $p_{tot} = p_1 + p_2 + \cdots$; computation can be tedious. Nevertheless this is the best test for small samples; and if $N < 20$, it is probably the only test to use.

5.4.2 Chi-square two-sample (or k-sample) test

Again the much-loved chi-square test is applicable. All the previous short-comings apply, but for data which are not on a numerical scale, there may be

Table 5.3 *Multi-sample contingency table*

Sample: $j =$	1	2	3
Category: $i = 1$	O_{11}	O_{12}	O_{13}
2	O_{21}	O_{22}	O_{23}
3	O_{31}	O_{32}	O_{33}
4	O_{41}	O_{42}	O_{43}
5	O_{51}	O_{52}	O_{53}
.	...	...	...

no alternative. To begin, each sample is binned in the same r bins (a $k \times r$ *contingency table*) – see Table 5.3. H_0 is that the k samples are from the same population.

Then compute

$$\chi^2 = \sum_{i=1}^{r} \sum_{j=1}^{k} \frac{(O_{ij} - E_{ij})^2}{E_{ij}}. \tag{5.17}$$

The E_{ij} are the *expectation* values, computed from

$$E_{ij} = \frac{\sum_{j=1}^{k} O_{ij} \sum_{i=1}^{r} O_{ij}}{\sum_{i=1}^{r} \sum_{j=1}^{k} O_{ij}}. \tag{5.18}$$

Under H_0 this is distributed as χ^2, with $(r-1)(k-1)$ degrees of freedom.

Note that there is a modification of this test for the case of the 2×2 contingency table (Table 5.2) with a total of N objects. In this case,

$$\chi^2 = \frac{N(|AD - BC| - N/2)^2}{(A+B)(C+D)(A+C)(B+D)} \tag{5.19}$$

has just one degree of freedom.

The usual chi-square caveat applies – beware of the lethal count of 5, below which the cell populations should not fall in any number. If they do, combine adjacent cells, simulate the distribution of the test statistic under the null hypothesis, or abandon the test. And if there are only 2×2 cells, the total (N) must exceed 30; if not, use the Fisher exact probability test.

There is one further distinctive feature about the chi-square 2×2 contingency-table test. It may be used to test a directional alternative to H_0, i.e. H_1 can be that the two groups differ in some predicted sense. If the alternative to H_0 is directional, then use Table B.6 in the normal way and halve the probabilities at the heads of the columns, since the test is now one-tailed. For degrees of freedom > 1, the chi-square test is insensitive to order, and another test thus may be preferable.

5.4.3 Wilcoxon–Mann–Whitney U test

This test is usually preferable to chi square, mostly because it avoids binning. There are two samples, A (m members) and B (n members); H_0 is that A and B are from the same distribution or have the same parent population, while H_1 may be one of three possibilities:

(i) that A is stochastically larger than B;
(ii) that B is stochastically larger than A;
(iii) that A and B differ in some other way, perhaps in spread or skewness.

The first two hypotheses are directional, resulting in one-tailed tests; the third is not and correspondingly results in a two-tailed test. To proceed, first decide on H_1 and, of course, the significance level α. Then

(i) *Rank* in ascending order the combined sample A + B, preserving the A or B identity of each member.
(ii) (Depending on the choice of H_1) sum the number of A-rankings to get U_A, or vice versa, the B-rankings to get U_B. Tied observations are assigned the average of the tied ranks. Note that if $N = m + n$,

$$U_A + U_B = \frac{N(N + 1)}{2},$$

so that only one summation is necessary to determine both – but a decision on H_1 should have been made a priori.
(iii) The sampling distribution is known (of course, or there would not be a test); Table B.9, columns labelled c_u (upper-tail probabilities), presents the exact probability associated with the occurrence (under H_0) of values of U greater than that observed. The table also presents exact probabilities associated with values of U less than those observed; entries correspond to the columns labelled c_l (lower-tail probabilities). The table is arranged for $m \leq n$, which presents no restriction in that group labels may be interchanged. What does present a restriction is that the table presents values only for $m \leq 4$ and $n \leq 10$. For samples up to $m = 10$ and $n = 12$, see Siegel & Castellan (1988). For still larger samples, the sampling distribution for U_A tends to Normal with mean $\mu_A = m(N + 1)/2$ and variance $\sigma_A^2 = mn(N + 1)/12$. Significance can be assessed from the Normal Distribution, Table B.1, by calculating

$$z = \frac{U_A \pm 0.5 - \mu_A}{\sigma_A},$$

where $+0.5$ corresponds to considering probabilities of $U \leq$ that observed (lower-tail), and -0.5 for $U \geq$ that observed (upper-tail). If the *two-tailed* ('the samples are distinguishable') test is required, simply double the probabilities as determined from either Table B.9 (small samples) or the Normal-distribution approximation (large samples).

Example An application of the test is shown in Figure 5.7, which presents magnitude distributions for flat and steep spectrum quasars from a complete sample of quasars in the Parkes 2.7-GHz survey (Masson & Wall, 1977). H_1 is that the flat-spectrum quasars extend to significantly lower (brighter) magnitudes than do the steep-spectrum quasars, a claim made earlier by several observers. The eye agrees with H_1, and so does the result from the U test, in which we found $U = 719, z = 2.69$, rejecting H_0 in favour of H_1 at the 0.004 level of significance.

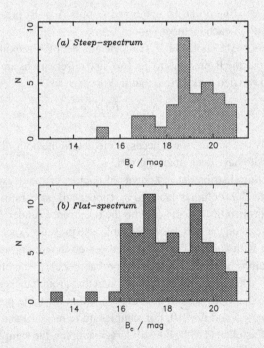

Figure 5.7 Magnitude histograms for a complete sample of quasars from the Parkes 2.7-GHz survey, distinguished by radio spectrum. H_0, that the magnitude distributions are identical, is rejected using the Wilcoxon–Mann-Whitney U test at the 0.004 level of significance.

In addition to this versatility, the test has a further advantage of being applicable to small samples. In fact it is one of the most powerful non-parametric tests; the efficiency in comparison with the Student's t test is ≥ 95 per cent for even moderate-sized samples. It is therefore an obvious alternative to the chi-square test, particularly for small samples where the chi-square test is illegal, and when directional testing is desired.

An alternative is the following test.

5.4.4 Kolmogorov–Smirnov two-sample test

The formulation parallels the Kolmogorov–Smirnov one-sample test; it considers the maximum deviation between the normalized cumulative distributions of two samples with m and n members. H_0 is (again) that the two samples are from the same population, and H_1 can be that they differ (two-tailed test), or that they differ in a specific direction (one-tailed test).

To implement the test, refer to the procedure for the one-sample test (Section 5.3.2); merely exchange the cumulative distributions S_e and S_o for S_m and S_n corresponding to the two samples. Critical values of D are given in Tables B.10 and B.11. Table B.10 gives the values for small samples, one-tailed test, while Table B.11 is for the two-tailed test. For large samples, two-tailed test, use Table B.12. For large samples, one-tailed test, compute

$$\chi^2 = 4D^2 \frac{mn}{m+n}, \tag{5.20}$$

which has a sampling distribution approximated by chi square with two degrees of freedom. Then consult Table B.6 to see if the observed D results in a value of χ^2 large enough to reject H_0 in favour of H_1 at the desired level of significance.

The test is extremely powerful with an efficiency (compared to the t test) of >95 per cent for small samples, decreasing somewhat for larger samples. The efficiency always exceeds that of the chi-square test, and slightly exceeds that of the U test for very small samples. For larger samples, the converse is true, and the U test is to be preferred.

Note that the Kolmogorov–Smirnov test can also be used to compare *two-dimensional* distributions (Peacock, 1983).

Example Two examples, drawn from an investigation of flattening and radio emission among elliptical galaxies (Disney *et al.*, 1984), are shown in Figure 5.8.

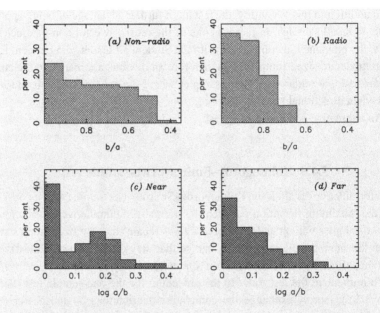

Figure 5.8   Kolmogorov–Smirnov tests on subsamples of ellipticals from the Disney–Wall (1977) sample of bright ellipticals. Upper panels – distribution functions in b/a, minor to major axis, for (a) the 102 undetected and (b) the 30 radio-detected ellipticals in the sample. The Kolmogorov–Smirnov two-sample test rejects H_0, that the subsamples are drawn from the same population, at a significance level of <1 per cent. Lower panels – distribution functions in *log a/b* for (c) the 51 ellipticals closer than 30 Mpc, (d) 76 bright ellipticals in the sample more distant than this. The Kolmogorov–Smirnov test indicates no significant difference between these latter subsamples.

The upper diagrams compare the axial ratio *b/a* (minor to major axis) for (a) 102 bright ellipticals for which no radio emission was detected and (b) 30 ellipticals for which emission was detected. The Kolmogorov–Smirnov test rejects H_0, that the two distributions are from the same parent population, at the 1 per cent level of significance. The lower pair, to do with ascertaining whether seeing is affecting measurement of axial ratio (the radio ellipticals are on average more distant), shows some difference by eye, but no significant difference when the Kolmogorov–Smirnov test is carried out.

These and tests on additional subsamples were used to show that there is a strong correlation between radio activity and flattening, in the sense that radio ellipticals are both inherently and apparently rounder than the average elliptical.

5.5 Summary, one- and two-sample non-parametric tests

Tables 5.4, 5.5 and 5.6, adapted from Siegel & Castellan (1988), attempt a summary, demonstrating an apparent wide world of non-parametric tests available for sample comparison. But is this really so? In deciding which test(s), the following points should be noted; the decision may be made for you.

(i) The two-sample and *k*-sample cases each contain columns of tests for *related samples*, i.e. matched-pair samples, or samples of paired replicates. This is common experimental practice in biological and behavioural sciences, where the concept of the control sample is highly developed. It is not so common in astronomy for obvious reasons, but has been exploited on occasion. The powerful tests available to treat such experiments are listed in Table 5.4, and are described by Siegel & Castellan.

(ii) Table 5.4 runs downward in order of increasing sophistication of measurement level, from *Nominal* (in which the test objects are simply dumped into classes or bins) through *Ordinal* (by which objects are ranked or ordered) to *Interval* (for which objects are placed on a scale, not necessarily numerical, in which distance along the scale matters). None of the tests requires measurement on a *Ratio* scale, the strongest scale of measurement in which to the properties of the Interval scale a true zero point is added. An important feature of test-selection lies in the level of measurement required by the test; the table is cumulative downward in the sense that at any level of measurement, all tests above this level are applicable.

(iii) The efficiency of a particular test depends very much on the individual application. Is the search for goodness-of-fit and general *difference*, i.e. is this sample from a given population? Are these samples from the same population? Or is it a particular property of the distribution which is of interest, such as the *location*, e.g. central tendency, mean or median; or the *dispersion*, e.g. extremes, variance, rms? For instance in the two-sample case, the chi-square and the Kolmogorov–Smirnov (two-tailed) tests are both sensitive to any type of difference in the two distributions, location, dispersion, skewness, while the *U* test is reasonably sensitive to most properties, but is particularly powerful for location discrimination. To aid the process of choice, Tables 5.5 (single samples) and 5.6 (two samples) summarize the attributes of the one- and two-sample tests.

The choice of test may thus come down to Hobson's. However, if it does not, and two (or more) alternatives remain – beware of this plot of the Devil. It might be possible to 'test the tests' in searching for support of a point of

Table 5.4 *Non-parametric tests for comparison of samples*

Level of measurement	One-sample case	Two-sample case		k-sample case	
		Related	Independent	Related	Independent
Nominal or categorical	Binomial test *chi-square test	McNemar change test	*Fisher exact test for 2 × 2 tables *chi-square test for r × 2 tables	Cochran Q test	*chi-square test for r × k tables
Ordinal or ordered	*Kolmogorov–Smirnov one-sample test *One-sample runs test Change-point test	Sign test Wilcoxon signed-ranks test	Median test *U (Wilcoxon–Mann–Whitney) test Robust rank-order test *Kolmogorov–Smirnov two-sample test Siegel–Tukey test for scale-differences	Friedman two-way analysis of variance by ranks Page test for ordered alternatives	Extension of median test Kruskal–Wallis one-way analysis of variance Jonckheere test for ordered alternatives
Interval		Permutation test for paired replicates	Permutation test for two independent samples Moses rank-like test for scale differences		

*Described in this chapter; Siegel & Castellan (1988) discuss the other tests.

Table 5.5   *Single-sample non-parametric tests*

Test	Applicability[2]	$N < 10$?	Comment
Binomial test	Goodness-of-fit (N)	Yes	Appropriate for two-category (dichotomous) data; do *not* dichotomize continuous data.
[1]Chi-square test	Goodness-of-fit (N)	No	For testing categorized, pre-binned, or classified data; choose categories with expected frequencies 6–10.
[1]Kolmogorov–Smirnov one-sample test	Goodness-of-fit (O)	Yes	The most powerful test for data from a continuous distribution; may always be more efficient than chi-square test.
[1]One-sample runs test	Randomness of event sequences (O)	Yes	Does not estimate differences between groups.
Change-point test	Change in the distribution of an event sequence (O)	Yes	Robust with regard to changes in distributional form; efficient.

[1] Described in this chapter; Siegel & Castellan (1988) discuss the other tests.
[2] *Goodness-of-fit* indicates general testing for any type of difference, i.e. H_o is that the distribution is drawn from the specified population. The level of measurement required is indicated by N – Nominal, O – Ordinal, or I – Interval.

view. If such a procedure is followed, quantification of the amount by which significance is reduced must be considered: for a chosen significance level p in a total of N tests, the chance that one test will (randomly) come up significant is $Np(1 - p)^{N-1} \simeq Np$ for small p. The application of efficient statistical procedure has power; but the application of common sense has more.

5.6 Statistical ritual

We pursue the common-sense theme. We draw the reader's attention to a salutary paper entitled *Mindless Statistics* (Gigerenzer, 2004), whose abstract starts 'Statistical rituals largely eliminate thinking in the social sciences.' Before we nod knowingly, let us look in the mirror. Why are we doing these tests; and what do we take from them? Here is a (paraphrased) sanity check under the section title in Gigerenzer (2004) 'Collective Illusions', which has a long history (Oakes, 1986; Haller & Krauss, 2002). The issue is this. You have done an experiment: you have a 'result' sample and a 'control' sample, each of 20.

Table 5.6   *Two-sample non-parametric tests*

Test	Applicability[2]	$N < 10$?	Comment
[1]Fisher exact test for 2×2 tables	Difference (N)	Yes	The most powerful test for dichotomous data.
[1]Chi-square test for $r \times 2$ tables	Difference (N)	No	Best for pre-binned, classified, or categorized data.
Median test	Location (O)	Yes	Best for small numbers; efficiency *decreases* with N.
[1]U (Wilcoxon–Mann–Whitney) test	Location (O)	Yes	One of the most efficient non-parametric tests.
Robust rank-order test	Location (O)	Yes	Efficiency similar to U test.
[1]Kolmogorov–Smirnov two-sample test	Two-tailed: Difference One-tailed: Location (O)	Yes	The most powerful test for data from a continuous distribution.
Siegel–Tukey test for scale-differences	Dispersion (O)	Yes	The medians must be the same (or known) for both distributions. Low efficiency.
Permutation test	Location (I)	Yes	Very high efficiency.
Moses rank-like test for scale-differences	Dispersion (I)	(No)	Does not require identical medians; valid for small samples, but efficiency increases with sample size.

[1] Described in this chapter; Siegel & Castellan (1988) discuss the other tests.

[2] *Difference* signifies sensitivity to any form of difference between the two distributions, i.e. H_0 is that the two distributions are drawn from the same population; *Location* indicates sensitivity to the position of the distributions, e.g. means or medians; and *Dispersion* indicates sensitivity to the spread of the distributions, i.e. variance, rms extremes. The level of measurement required is indicated by N – Nominal, O – Ordinal or I – Interval.

You test the means via a Student's t test and the difference is significant ($t = 2.7$, $\nu = 18$, $p = 0.01$). Recall (Section 5.1) that a p value is the probability of the observed data (or of more extreme data points), given that the null hypothesis H_0 is true, defined in symbols as $p(D \mid H_0)$.

Do you judge the following six statements as *TRUE* or *FALSE*?

(i) You have absolutely disproved the null hypothesis (that is, there is no difference between the population means).

(ii) You have found the probability of the null hypothesis being true.

(iii) You have absolutely proved your experimental hypothesis (that there is a difference between the population means).

(iv) You can deduce the probability of the experimental hypothesis being true.

(v) You know, if you decide to reject the null hypothesis, the probability that you are making the wrong decision.

(vi) You have a reliable experimental finding in the sense that if, hypothetically, the experiment were repeated a great number of times, you would obtain a significant result 99 per cent of the time.

Statements (i) and (iii) are clearly false, because a significance test can never disprove the null hypothesis or the (undefined) experimental hypothesis.

Statements (ii) and (iv) are also false. The probability $p(D \mid H_0)$ is not the same as $p(H_0 \mid D)$ and, more generally, a significance test does not provide a probability for a hypothesis. (Bayesian statistics (Section 5.2.3) does provide just such a tool, of course.) Statement (v) also refers to a probability of a hypothesis. This is because if one rejects the null hypothesis, the only possibility of making a wrong decision is if the null hypothesis is true. Thus it makes the same claim as Statement (ii), and both are incorrect. Statement (vi) amounts to the 'replication fallacy'. Here, $p = 1$ per cent is taken to imply that such significant data would reappear in 99 per cent of the repetitions. Statement (vi) could be made only if one knew that the null hypothesis was true. In formal terms, $p(D \mid H_0)$ is confused with $1 - p(D)$.

All six statements are incorrect.

Exercises

5.1 Kolmogorov–Smirnov (D). Use the data provided, two data sets, one with a total of $m = 290$ observations, the other with 385 measurements. The former is of flux densities measured at random positions in the sky; the latter of flux densities at the positions of a specified set of galaxies. Using the Kolmogorov–Smirnov two-sample test, examine the hypothesis that there is excess flux density at the non-random positions.

5.2 Wilcoxon–Mann–Whitney (D). Repeat the test with the Wilcoxon–Mann–Whitney statistic. Is the significance level different? How would you combine the results from these two tests, plus the chi-square test in the text?

5.3 *t* **test and outliers (D).** Create two data sets, one drawn from a Gaussian of unit variance, the other drawn from a variable combination of two Gaussians, the dominant one of unit variance and the other three times wider. All Gaussians are of zero mean. Perform a *t* test on sets of 10 observations and investigate what happens as contamination from the wide Gaussian is increased. Compare the effect on the posterior distribution of the difference of the means. Now shift the narrow Gaussian by half a unit, and repeat the experiment. What effect do the outliers have on our ability to refute the null hypothesis? How does the Bayesian approach compare?

5.4 *F* **test (D).** Create some random data, as in the first part of Exercise 5.3. Investigate the sensitivity of the standard *F* test to a small level of contamination by outliers.

5.5 Non-parametric alternatives (D). Repeat the analysis of the last two exercises, using a non-parametric test; the Wilcoxon–Mann–Whitney test for the location test, and the Kolmogorov–Smirnov test for the variance test. How do the results compare with the parametric tests? Can you detect genuine differences in variance, apart from the outliers?

5.6 Several data sets, one test. Suppose you have N independent data sets, and with a certain test you obtain a significance level of p_i for each one. A useful overall significance is given by the W statistic (Peacock, 1985) which is

$$W = \prod_{i=1}^{N} p_i.$$

Find the distribution of log W and describe how it could be used. Note that this contrasts to the case discussed in the text, where we might perform several different tests on the *same* data set. (Each p_i will be uniformly distributed between zero and one, under the null hypothesis. The distribution of log W is the sum of these uniformly distributed numbers, and tends to a Gaussian of mean N and variance N.)

5.7 Gram–Charlier (D). Take some data drawn from a Gaussian and investigate the posterior likelihood if just one term (the quadratic) is used in a Gram–Charlier expansion as an assumption for the 'true' distribution. Take the location as known. Find the distribution of the variance, marginalizing out the Gram–Charlier parameter. Also, find the odds on including the parameter in the model. What does this tell you about assuming a Gaussian distribution when the amounts of data are limited?

5.8 Odds versus classical tests. Use the small data set from the example in Section 5.2.3. Perform a classical analysis, using t and F tests. Compare and contrast to the odds calculated in the text. Does the Behrens–Fisher distribution give a better answer than either or both? See Jaynes' comments on confidence intervals (Jaynes, 1983).

In the exercises denoted by (D), data sets are provided on the book's website; or create your own.

6

Data modelling and parameter estimation: basics

But what are the errors on your errors?
(Graham Hine at a Mark Birkinshaw colloquium, Cambridge 1979)

Many pages of statistics textbooks are devoted to methods of estimating parameters, and calculating confidence intervals for them. For example, if our N data Z_i follow a Gaussian distribution

$$\text{prob}(z) = \frac{1}{\sigma \sqrt{2\pi}} \exp \left[-\frac{(z - \mu)^2}{2\sigma^2} \right],$$

then the statistic

$$m = \frac{1}{N} \sum_i Z_i$$

is a good estimator for μ and has a known distribution (a Gaussian again) which can be used for calculating confidence limits. Or, from the Bayesian point of view, we can calculate a probability distribution for μ, given the data.

Any data-modelling procedure is just a more elaborate version of this, assuming that we know the relevant probability distributions. Suppose that our data Z_i were measured at various values of some independent variable X_i, and we believed that they were scattered, with Gaussian errors, around the underlying functional relationship

$$\mu = \mu(x, \alpha_1, \alpha_2, \ldots),$$

in which $\alpha_1, \alpha_2, \ldots$ are unknown parameters (slopes, intercepts, ...) of the relationship. We then have

$$\text{prob}(z \mid \alpha_1, \alpha_2, \ldots) = \frac{1}{\sigma \sqrt{2\pi}} \exp \left[-\frac{(z - \mu(x, \alpha_1, \alpha_2, \ldots))^2}{2\sigma^2} \right],$$

and, by Bayes' theorem, we have the posterior probability distribution for the parameters

$$\text{prob}(\alpha_1, \alpha_2, \ldots \mid Z_i, \mu)$$
$$\propto \prod_i \frac{1}{\sigma \sqrt{2\pi}} \exp\left[-\frac{(Z_i - \mu(X_i, \alpha_1, \alpha_2, \ldots))^2}{2\sigma^2} \right] \text{prob}(\alpha_1, \alpha_2, \ldots) \quad (6.1)$$

including, as usual, our prior information. We have included μ as one of the 'givens' to emphasize that everything depends on it being the correct model.

This, at least formally, completes our task; we have a probability distribution for the parameters of our model, given the data.

This is a very general approach. In the limiting case of uninformative or diffuse priors, it is very closely related to the method of maximum likelihood; if the distribution of the residuals from the model is indeed Gaussian, it is closely related to the method of least squares. Moreover, it can be used in a clear way to update models as new data arrive; the posterior from one stage of the experiments becomes the prior for the next.

We can also deal nicely with unwanted parameters ('nuisance' parameters). Typically we will end up with a probability distribution for various parameters, some of interest (say, cosmological parameters) and some not (say, instrumental calibrations). We can 'marginalize out' the unwanted parameters by an integration, leaving us with the distribution of the variable of interest that takes account of the range of plausible values of the unwanted variables. Later examples will develop these ideas.

Modelling can be a very expensive part of any investigation. Analytic approximations were developed in past years for very good reasons. Modelling processes always involve finding an extreme value, a maximum or minimum, of some merit function. Without help from an analytic solution, this means evaluating the function, and perhaps its derivatives, many times. The model itself may be the result of a complex and time-consuming computation, so evaluating it over a range of parameters is even worse.

Another difficulty that arises in the Bayesian approach is numerical integration. Interesting problems have many parameters; operations such as marginalizing these out or discriminating between models involve multidimensional integrals. These are often very time-consuming, and laborious to check. Any analytical help we can get is especially welcome in doing integrations. We will see the relevance of this in the next section, where powerful theorems may allow great simplifications.

Perhaps the most important thing to remember about models is blindingly obvious; they may be wrong. The most insidious case of this is a mistake in the assumed distribution of residuals about the model. Inevitably, the parameters deduced from the model will be wrong. Worse, the inferred errors on these parameters will be wrong too, often giving a quite false sense of security. It is important to have a range of models available, and always to check optimized models against the data, inspecting the residuals for strange outliers or clusters of positive or negative residuals. The *chi-square test* (Section 5.3.1), or the *runs test* (Section 5.3.3) are helpful in this respect.

6.1 The maximum-likelihood method

Maximum likelihood (ML) has a long history; it was derived by Bernoulli in 1776 and Gauss around 1821, and worked out in detail by Fisher in 1922.

We have met the likelihood function several times already; together with the prior probabilities, it makes up the posterior probability from Bayes' theorem. Suppose that our data are described by the probability density function $f(x, \alpha)$, where x is a variable, α is a parameter (maybe many parameters) characterizing the known form of f. We want to estimate α. If $X_1, X_2, \ldots, X_N$ are data, presumed independent and all drawn from f, then the likelihood function is

$$\mathcal{L}(X_1, X_2, \ldots, X_N) = \prod_{i}^{N} f(X_i \mid \alpha) \qquad (6.2)$$

From the classical point of view this is the probability, given α, of obtaining the data. From the Bayesian point of view it is *proportional to* the probability of α, given the data and assuming that the priors are 'diffuse'. Practically speaking, this means that they change little over the peaked region of the likelihood function. Finding the constant of proportionality involves the troublesome integrals we referred to before.

If the priors are not diffuse, this means that they are having as strong an effect on our conclusions as the data. This is not an unlikely situation, but it does rule out the handy analytical approximations we will describe later.

From either point of view, more intelligibly from the Bayesian, the peak value of $\mathcal{L}$ seems likely to be a useful choice of the 'best' estimate of α. This does rather depend on what we want to do next with our estimate, however.

Formally, the maximum-likelihood estimator (MLE) of α is $\hat{\alpha} =$ (that value of α which maximizes $\mathcal{L}(\alpha)$ for all variations of α). Often we can find this from

$$\frac{\partial}{\partial \alpha} \ln \mathcal{L}(\alpha) \mid_{\alpha = \hat{\alpha}} = 0 \qquad (6.3)$$

but sometimes we cannot – an example of this will be given later.

Maximizing the *logarithm* is often convenient, both algebraically and numerically. The MLE is a *statistic* – it depends only on the data, and not on any parameters.

Example Consider our old friend the regression line, for which we have values of Y_i measured at given values of the independent variable X_i. Our model is

$$y(a, b) = ax + b$$

and assuming that the Y_i have a Gaussian scatter, each term in the likelihood product is

$$\mathcal{L}_i(y \mid (a, b)) = \exp\left[-\frac{(Y_i - (aX_i + b))^2}{2\sigma^2}\right]$$

i.e. the residuals are $(y_i - model)$, and our model has the free parameters (a, b). Maximizing the log of the likelihood products then yields

$$\frac{\partial \mathcal{L}}{\partial a} = -2\Sigma X_i(Y_i - aX_i - b) = 0$$

$$\frac{\partial \mathcal{L}}{\partial b} = -2\Sigma(Y_i - aX_i - b) = 0$$

from which two equations in two unknowns we get the well-known

$$a = \frac{\overline{XY} - \overline{X} \cdot \overline{Y}}{\overline{X^2} - (\overline{X})^2}$$

$$b = \overline{Y} - a\overline{X}.$$

With this simple maximum-likelihood example, we have derived the standard OLS, the ordinary least squares estimate of y on the independent variable x. But note how this happened. We were given the fact that the Y_i were Normally distributed with their scatter described by a single deviation σ; and, of course, we were given the fact that a straight-line model was correct. It need not be this way; we could have started knowing that each Y_i had an associated σ_i, or even that the distribution in y about the line was not Gaussian, perhaps say uniform, or dependent on $|y_i - model|$ rather than $(y_i - model)^2$. The formulation is identical, although the algebra may not work out as neatly as it does for an OLS regression line. This of course is another advantage of maximum likelihood – the likelihood function can be computed and the maximum found without calculus.

Example Jauncey (1967) showed that maximum likelihood was an excellent way of estimating the slope of the number–flux-density relation, the dependence of source surface density on intensity, for extragalactic radio sources. The source count is assumed to be of the power-law form

$$N(> S) = kS^{-\gamma},$$

where N is the number of sources on a particular patch of sky with flux densities greater than S, k is a constant and $-\gamma$ is the exponent, or slope in the $\log N - \log S$ plane, which we wish to estimate; see Figure 6.1.

The probability distribution for S (the chance of getting a source with a flux density near S) is then

$$\text{prob}(S) = \gamma k S^{-(\gamma+1)}$$

and k is determined by the normalization to unity (survey limit S_0).

$$\int_{S_0}^{\infty} \text{prob}(S) \, dS = 1.$$

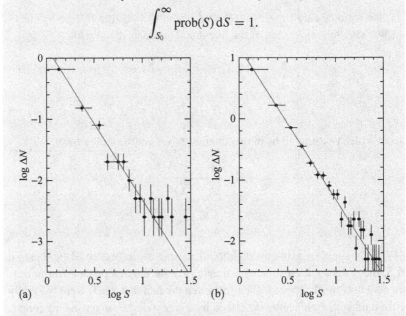

(a)   (b)

Figure 6.1   A maximum-likelihood application. The figures show differential source counts generated via Monte Carlo sampling with an initial uniform deviate (see Section 2.6) obeying the source-count law $N(> S) = kS^{-1.5}$. The straight line in each shows the anticipated count with slope -2.5. (a) $k = 1.0$, 400 trials, (b) $k = 10.0$, 4000 trials. The ML results for the slopes are -2.52 ± 0.09 and -2.49 ± 0.03, the range being given by the points at which the log-likelihood function has dropped from its maximum by a factor of 2. The anticipated errors in the two exponents, given by $|\text{slope}|/\sqrt{\text{trials}}$ (see the next-but-one example), are 0.075 and 0.024.

(We have taken the maximum possible flux density to be infinity, with small error for steep counts.) k is then S_0^γ and the log likelihood is, dropping constants,

$$\ln \mathcal{L}(\gamma) = M \ln \gamma - \gamma \sum_i \ln \frac{S_i}{S_0},$$

where we have observed M sources with flux densities S brighter than S_0. Differentiating this with respect to γ to find the maximum then gives the equation for $\hat{\gamma}$, the MLE of γ:

$$\hat{\gamma} = \frac{M}{\sum_i \ln \frac{S_i}{S_0}}$$

a nicely intuitive result. This application of ML makes optimum use of the data in that the sources are not grouped and the loss of power which always results from binning is avoided.

The MLE cannot always be obtained by differentiation, as the following example shows.

Example A supernova produces an intense burst of neutrinos. The intensity of this burst decays exponentially after the core collapse of the precursor star. A few neutrinos (say N in number) were detected from supernova 1987a with arrival times (in order) $T_1, T_2, \ldots$ The probability of a neutrino arriving at time t is

$$\text{prob}(t) = \exp[-(t - t_0)]$$

for $t > t_0$ and zero otherwise. Times are measured in units of the half-life and t_0 is the parameter we want, the start of the burst.
    The log likelihood is just

$$\ln \mathcal{L}(t_0) = N t_0 - \sum_i T_i$$

and this does not appear to have a maximum. However, clearly $t_0 < T_1$ and so the likelihood is maximized, within the allowable range of t_0, at $\hat{t}_0 = T_1$.

After the MLE estimate has been obtained, it is essential to perform a final check: does the MLE model fit the data reasonably? If it does not, then the data are erroneous when the model is known to be right; or, the adopted or assumed model is wrong; or (most commonly) there has been blunder of some kind. There are many ways of carrying out such a check; two of these, the chi-square test and the Kolmogorov–Smirnov test, are described in Sections 5.3.1 and 5.3.2, respectively.

If the deviations between the best-fit model and the data (the residuals) are Gaussian, the log-likelihood function becomes a sum of squares of residuals and we have the famous method of least squares (Section 6.2).

Now for those theorems. The strongest reason for picking the MLE of a parameter is that it has desirable properties – it has minimum variance compared to any other estimate, and it is asymptotically distributed around the true value. An MLE is not always unbiased, however.

If we estimate a vector $\hat{\alpha}$ by the maximum-likelihood method, then the components of the estimated vector are asymptotically distributed around the true value like a multivariate Gaussian (Section 4.2). 'Asymptotically' implies that we have lots of data, strictly speaking infinite amounts. The covariance matrix that describes this Gaussian can be derived from the second derivatives of the likelihood with respect to the parameters. This involves a famous matrix called the Hessian, which is

$$
\mathcal{H} = \begin{bmatrix} \frac{\partial^2 \ln \mathcal{L}}{\partial \alpha_1^2} & \frac{\partial^2 \ln \mathcal{L}}{\partial \alpha_1 \partial \alpha_2} & \frac{\partial^2 \ln \mathcal{L}}{\partial \alpha_1 \partial \alpha_3} & \cdots \\ \frac{\partial \ln \mathcal{L}}{\partial \alpha_2 \partial \alpha_1} & \frac{\partial^2 \ln \mathcal{L}}{\partial \alpha_2^2} & \frac{\partial^2 \ln \mathcal{L}}{\partial \alpha_2 \partial \alpha_3} & \cdots \\ \frac{\partial^2 \ln \mathcal{L}}{\partial \alpha_3 \partial \alpha_1} & \frac{\partial^2 \ln \mathcal{L}}{\partial \alpha_3 \partial \alpha_2} & \frac{\partial^2 \ln \mathcal{L}}{\partial \alpha_3^2} & \cdots \\ \vdots & \vdots & \vdots & \end{bmatrix}. \tag{6.4}
$$

This matrix of course depends on the data. Taking its expectation value (the 'average' value of each component of the matrix, $E[\mathcal{H}]$ for short, Section 3.1), we have a simple expression for the covariance matrix of the multivariate Gaussian distribution of the MLEs of the parameters:

$$
C = (E[\mathcal{H}])^{-1}, \tag{6.5}
$$

the $(\ldots)^{-1}$ signifying the inverse matrix. The (negative) average of the Hessian is important enough to have a name; it is called the Fisher information matrix (Fisher, 1935). It is important because it describes the width of the likelihood function and hence the scatter in the maximum-likelihood estimators, as we now see. The Fisher matrix can be calculated for various experimental designs as a metric of how well the experiment will perform (Tegmark *et al.*, 1997).

The probability distribution of our N MLEs $\hat{\alpha}$ is then

$$
\text{prob}(\hat{\alpha}_1, \hat{\alpha}_2, \ldots) = \frac{1}{\sqrt{(2\pi)^N \mid \det C \mid}} \exp\left[-\frac{1}{2}(\hat{\alpha} - \alpha) \cdot C^{-1} \cdot (\hat{\alpha} - \alpha)^T \right] \tag{6.6}
$$

so that, as stated, the MLE ($\hat{\vec{\alpha}}$) is distributed around the true value $\vec{\alpha}$ with a spread described by the covariance C, or equivalently by the Fisher matrix. ($|\det C|$ is the absolute value of the determinant of C.)

Taking the expectation value is obviously important, as otherwise the matrix would be different for each set of data. Sometimes we can carry out the expectation, or averaging, operation analytically in terms of $\vec{\alpha}$, the parameters of the original model. Sometimes the matrix does not involve the data at all. Most commonly, we just have to take the single matrix, given by our one set of data, as the best estimate we can make of the average value.

Why should the MLEs obey this theorem? Take a simple case, a Gaussian of true mean μ and variance σ^2. If we have N data X_i, the log likelihood is (dropping constants)

$$\log \mathcal{L} = \frac{-1}{2\sigma^2} \sum_i (X_i - \mu)^2 - N \log \sigma$$

and

$$\frac{-\partial^2 \log \mathcal{L}}{\partial \mu^2} = \frac{N}{\sigma^2}.$$

This is the Hessian 'matrix' for our simple problem. Taking its expectation and then inverse, not too hard in this case, gives us the variance on the estimate of the mean as σ^2/N, the anticipated result.

This example provides some justification for the theorem. In the exercises we set the somewhat more complicated case of estimating μ and σ together. This gives a matrix problem rather than a scalar one, and some real expectations have to be evaluated.

Example In the source-count example, we have just one parameter. The variance on $\hat{\gamma}$ is then

$$\frac{-1}{E\left[\frac{\partial^2 \mathcal{L}(\gamma)}{\partial \gamma^2}\right]}$$

which is $\frac{\gamma^2}{M}$. The expectation is easy in this case. However, we see that the error is given in terms of the thing we want to know, namely γ. As long as the errors are small we can approximate them by $\frac{\hat{\gamma}^2}{M}$.

6.2 The method of least squares: regression analysis

Least squares is a famous old method of dealing with noisy data; it was invented, for astronomical use, by Gauss and Laplace at the beginning of the nineteenth century. There is a huge literature, e.g. Williams (1959); Linnik (1961); Montgomery & Peck (1992). The justification for the method follows immediately from the method of maximum likelihood. For data Y_i measured at independent variable values X_i, if the distribution of the residuals is Gaussian, then the log likelihood is a sum of squares of the form

$$\log \mathcal{L} = \text{constant} - \sum_{i=1}^{N} \xi_i (Y_i - \mu(X_i, \alpha_1, \alpha_2, \ldots))^2, \tag{6.7}$$

where the ξ are the weights, usually inversely proportional to the variance on the measurements. Usually the weights are assumed equal for all the data, and least squares is just that; we seek the model parameters which minimize

$$\log \mathcal{L} = \text{constant} - \frac{1}{2\sigma^2} \sum_{i=1}^{N} (Y_i - \mu(X_i, \alpha_1, \alpha_2, \ldots))^2.$$

These will just be the MLEs, and everything we have said before about them carries over. In particular, they are asymptotically distributed like a multivariate Gaussian. If we do not know the error level (the σ) we do not need to use it, but we will not be able to infer errors on the MLE; we will get a model fit, but we cannot know how good or bad the model is.

The matrix of second derivatives defining the covariance matrix of the estimates, the Hessian matrix (Section 6.1), takes on a particular significance in the method of least squares because it is often used by the numerical algorithms which find the minimum. There are many powerful variations on these algorithms – see *Numerical Recipes* (Press *et al.*, 2007) for details. Typically, the value of the Hessian matrix, at the minimum, pops out as a by-product of the minimization. We can use this directly to work out the covariance matrix, as long as our model is linear in the parameters; in this case, the expectation operation is straightforward and the matrix does not depend on any of the parameters. We saw before why this is a problem (in the source-count example) – we want to *find* the parameters, and using the estimates in the covariance matrix is not an ideal procedure.

The notion of a linear model is worth clarifying. Suppose that our data X_i are measured as a function of some independent variable Z_i. Then a linear model – linear in the *parameters* – might be $\alpha z^2 + \beta \exp(-z)$, whereas $\alpha \exp(-\beta z)$ is not a linear model. Of course a model may be approximately linear near the MLE. However, how close must it be? This illustrates again the general feature of the asymptotic Normality of the MLE – we can use the approximation, but we cannot tell how good it is. Usually things will start to go wrong first in the wings of the inferred distributions (we have seen this in a previous example) and so high degrees of significance usually cannot be trusted unless they have been calculated exactly, or simulated by Monte Carlo methods.

Example In the notation we used before, suppose that our model is

$$\mu(x, \alpha, \beta) = \alpha x + \beta x^2,$$

a simple polynomial. The covariance matrix can be calculated from M, the matrix of derivatives of the log likelihood; it is just

$$C = \frac{1}{\sigma^2} \begin{bmatrix} \sum_i X_i^2 & 0 \\ 0 & \sum_i X_i^4 \end{bmatrix}$$

so the variance on β, for example, is $\sigma^2 / \sum_i X_i^4$. Evidently, where we make the measurements (the X_i) will affect the variance. The effects are obvious enough in this simple case, but in more complicated cases it may be worth examining the experimental design, via the covariance matrix, to minimize the expected errors.

Quite often we will not be confident that we are dealing with Gaussian residuals, and usually this is because of outliers – residuals which are extremely unlikely on the Gaussian hypothesis. One convenient distribution which has 'fat' tails, and is a useful contrast to a Gaussian, is the simple exponential

$$\text{prob}(x) = \frac{1}{2a} \exp \left[-\frac{|x - \mu|}{a} \right].$$

A t distribution may also be a helpful model. If the residuals are distributed exponentially, then it is easy to see that ML leads to the minimization of the sum of the absolute values of the residuals. Working out an MLE in this way will give some indication of whether outliers are driving the answer. The only problem may be that relatively slow numerical routines have to be used; least-squares minimization routines are highly developed, by comparison.

Let us return for the last time to our simple regression line, the least-squares fit of model $y = ax + b$ through N pairs of (X_i, Y_i) by minimizing the squares of the residuals. This yields the well-known expressions for slope and intercept (see Section 6.1):

$$a = \frac{N \sum_{i}^{N} X_i Y_i - \sum_{i}^{N} X_i \sum_{i}^{N} Y_i}{N \sum_{i}^{N} X_i^2 - \left(\sum_{i}^{N} X_i\right)^2} \tag{6.8}$$

$$\text{and } b = \left(\sum_{i}^{N} Y_i - a \sum_{i}^{N} Y_i\right) / N \tag{6.9}$$

In the absence of knowledge of the how and why of a relation between the x_i and the y_i (Section 4.4), any two-parameter curve may be fitted to the data pairs just with simple coordinate transformations; for example

(i) an exponential, $y = b \exp a$ requires y_i to be changed to $\ln y_i$ in the above expressions;
(ii) a power law, $y = bx^a$; change y_i to $\ln y_i$ and x_i to $\ln x_i$;
(iii) a parabola, $y = b + ax^2$; change x_i to $\sqrt{x_i}$.

Note that the residuals cannot be Gaussian for *all* of these transformations (and may not be Gaussian for any). Of course it is always possible to minimize the squares of the residuals, but it may well not be possible to retain the formal justification for doing so. The tests of Chapter 5 can be revealing as to which (if any) model fits, particularly the runs test.

This simple formulation of the least-squares fit for y on x represents the tip of an iceberg – there is an enormous variety of least-squares linear regression procedures. Amongst the issues involved in choosing a procedure:

- Are the data to be treated weighted or unweighted?
- (And the related question) do all the data have the same properties, e.g. in the simple case of y on x, is one σ_y^2 applicable to all y? Or does σ_y^2 depend on y? In the uniform σ case, the data are described as *homoskedastic*, and in the opposite case, *heteroskedastic*.
- Is the right fit the standard ordinary least-squares solution y on x (OLS(Y/X)) or x on y (OLS(X/Y)? Or something different, as discussed below?
- If we *know* we have heteroskedasticity, with the uncertainty different but known in each Y_i and perhaps also in each X_i, how do we use this information to estimate the uncertainty in the fit?
- Are the data truncated or censored; do we wish to include upper limits in our fit? This is perfectly possible; see Section 8.5.

The thorough papers of Feigelson and collaborators (Isobe *et al.*, 1990; Babu & Feigelson, 1992; Feigelson & Babu, 1992) consider these issues, describe the complexities, indicate how to find errors with bootstrap and jackknife resampling (Section 6.6), and identify appropriate software routines. In the astronomical context, Feigelson & Babu (1992) emphasize that much of the proliferation of linear regression methods in the cosmic-distance-scale literature is due to *lack of precision in defining the scientific question*. The question defines the statistical model. The serious fitter must consult the Feigelson references. In the interim and as an indication of why you must, consider the following example.

Example Return to our bivariate Gaussian of Section 4.2 and Figure 4.4; and now consider random variates (x_i, y_i) selected (a) in accord with $\rho = 0.05$ (little correlation) and (b) with $\rho = 0.95$ (strongly correlated), following the prescription of Section 4.2. The ellipses of the contours are shown in Figure 6.2. For the case of little correlation, the two OLS lines are stunningly different, almost orthogonal; for the relatively strong correlation, the lines are very similar.

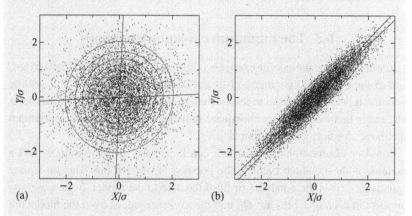

Figure 6.2   Linear contours of the bivariate Gaussian probability distribution. (a) $\rho = 0.05$, a Gaussian distribution with weak connection between x and y; (b) $\rho = 0.95$, indicative of a strong connection between x and y. In each case 5000 (x, y) pairs have been plotted, selected at random from the appropriate distribution as described in Section 2.6. Two lines are shown as fits for each distribution, the OLS(X/Y) and the OLS(Y/X).

The point is that we know the answer here for the relation: it is a line of slope unity, $45°$. With little (yet formally significant) correlation, the OLS lines mislead us dramatically. Of course the so-called bisector line (the average of the two OLS) would get it right, as would the orthogonal regression line which minimizes the perpendicular distances, but for the former – if the points were not Gaussian in distribution, would you trust it? A few outliers (mistakes?) would soon wreck it. The latter is principal component analysis (Section 4.5) precisely. It has already been emphasized that when the dependencies of variables on each other are not understood, PCA is the way to go. It gives the right answer in this example; it tells us what the relation between x and y is, without us assuming which variable is in control.

Frequently, we have data where there are errors in *both* x and y and we wish to fit a straight line. This is surprisingly difficult to do, although there are many ad-hoc solutions. For further help, see Gull (1989, Bayesian data analysis: straight-line fitting) and (for the full treatment) Zellner (1987); see also Exercise 6.5.

We continue to follow classical lines in our discussion of likelihood. The method is very useful; the main limitation is the difficulty in calculating the parameters of the asymptotic distribution of the MLE. And, of course, without an exact solution, it is difficult to be sure how useful this asymptotic distribution is anyway.

6.3 The minimum chi-square method

A dominant classical modelling process is minimum chi-square, a simple extension of the chi-square goodness-of-fit test described in Section 4.2. It will be seen that it is closely related to least-squares and weighted least-squares methods, and in fact the minimum chi-square statistic is closely related to maximum likelihood, for normally distributed errors.

Consider observational data which can be (or are already) binned, and a model/hypothesis which predicts the population of each bin. The chi-square statistic describes the goodness of fit of the data to the model. If the *observed* numbers in each of k bins are O_i, and the *expected* values from the model are E_i, then this statistic is

$$\chi^2 = \sum_{i=1}^{k} \frac{(O_i - E_i)^2}{E_i} \qquad (6.10)$$

(The parallel with weighted least squares is evident: the statistic is the squares of the residuals weighted by what is effectively the variance if the procedure is governed by Poisson statistics.) The minimum chi-square method of model-fitting consists of minimizing the χ^2 statistic by varying the parameters of the model. The premise on which this technique is based is simply that the model is assumed to be qualitatively correct, and is adjusted to minimize (via χ^2) the differences between the E_i and O_i which are deemed to be due solely to statistical fluctuations. In practice, the parameter search is easy enough as long as the number of parameters is less than four; if four or more, then sophisticated search procedures may be necessary (Press *et al.*, 2007). The appropriate number of degrees of freedom to associate with χ^2 for k bins and N parameters is $v = k - 1 - N$.

Table 6.1 *Chi-square differences*
($\Delta(v, \chi^2)$) above minimum

Confidence	Number of parameters		
c	1	2	3
0.68	1.00	2.30	3.50
0.90	2.71	4.61	6.25
0.99	6.63	9.21	11.30

The essential issue, having found appropriate parameters, is to estimate confidence limits (Section 3.1) for them. The answer (Avni, 1976; Press *et al.*, 2007) is that the region of confidence is defined by a zone in parameter space, generally elongated and approximately elliptical, with

$$\chi^2 = \chi^2_{min} + \Delta(\chi^2),$$

where $\Delta(\chi^2)$ is from Table 6.1. The table signifies that there is probability c that this region will contain the true values of the parameters; it is calculated from

$$c(v, \Delta\chi^2) = P(v/2, \Delta\chi^2/2), \tag{6.11}$$

with P the incomplete Gamma function (Press *et al.*, 2007). Equation (6.11) enables extension of the table to meet your requirements (as well as enabling conversion of any $\Delta\chi^2$ to a probability). It is interesting to note that the $\Delta(v, \chi^2)$ probability depends only on the number of parameters involved, and not on the goodness of fit (via χ^2_{min}) actually achieved.

Example The model to describe an observed distribution (Figure 6.3(a)) requires two parameters, γ and K. Contours of χ^2 resulting from the parameter search are shown in Figure 6.3(b). Consulting Table 6.1 gives $\chi^2_{0.68} = \chi^2_{min} + 2.30$, for the value corresponding to 1σ (significance level $= 0.68$); the contour $\chi^2_{0.68} = 6.2$ defines a region of confidence in the (γ, K) plane corresponding to the 1σ level of significance. (Because the range of interest for γ was limited from other considerations to $1.9 < \gamma < 2.4$, the parameter search was not extended to define this contour fully.)

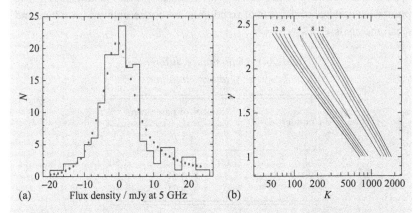

Figure 6.3   An example of model fitting via minimum chi-square. The object of the experiment was to estimate the surface-density count (the $N(S)$ relation, see Section 6.1, Figure 6.1) of faint extragalactic sources at 5 GHz, assuming a power law $N(> S) = KS^{-(\gamma-1)}$, γ and K to be determined from the distribution of background deflections, the so-called $p(D)$ *method*, Section 8.6. The histogram of measured deflections is shown (a), together with the curve representing the optimum model from minimizing χ^2. Contours of χ^2 in the (γ, K) plane are shown (b), with χ^2 indicated for every second contour.

There are three good features of the minimum chi-square method, and two bad ones. The good:

(i) Because χ^2 is additive, the results of different data sets that may fall in different bins, bin sizes or that may apply to different aspects of the same model, may be tested all at once.

(ii) The contribution to χ^2 of each bin may be examined and regions of exceptionally good or bad fit delineated.

(iii) One of the best features of the method is that you get goodness of fit for free. Table B.6 indicates probabilities of χ^2 for given degrees of freedom. It is to be hoped that the model comes out with a probability of order 0.50. The peak of the χ^2 distribution is at $\sim$ (number of degrees of freedom) when $\nu \geq 4$ (Figure 5.5). So, in the example above, there are 7 bins, 2 parameters, and the appropriate number of degrees of freedom is therefore 4. The value of $\chi^2_{\min}$ is about 4, just as one would have hoped. The probability of this value being exceeded, if the model is correct, is about 0.4 and the optimum model is thus a satisfactory fit.

The bad and downright ugly:

(i) Low bin-populations in the chi-square sums will cause severe instability. As a rule of thumb, 80 per cent of the bins must have $E_i > 5$. As for the chi-square test, it does not work for small numbers.

(ii) Finally it is important to repeat the mantra: data-binning is bad. In general, it loses information and efficiency. What is worse is the bias it can cause. Just consider a skewed distribution with rather few data defining it – the consequent need for wide bins may 'erase' the skewness entirely.

6.4 Weighting combinations of data

The form of the chi-square terms shows that the effect of data on the fitted parameters is weighted by the variance on the data. If an observation has large variance, its effect on the solution is correspondingly reduced.

A simple case of this is the common one of estimating a mean, when the data in the sum have different variances. Working this through does make the form of chi square more plausible.

We may look for a 'weighted mean' X_w

$$X_w = \alpha_1 X_1 + \alpha_2 X_2 + \cdots + \alpha_N X_N$$

in which we determine the coefficients α_i by the requirement that the variance in X_w is minimized. This is a particular example of a general technique for constructing linear estimators. Subject to the requirement that the X_i's are uncorrelated (so, excluding the possibility of common systematic errors) the result is

$$X_w = \frac{\frac{X_1}{\sigma_1^2} + \frac{X_2}{\sigma_2^2} + \cdots + \frac{X_N}{\sigma_N^2}}{\frac{1}{\sigma_1^2} + \frac{1}{\sigma_2^2} + \cdots + \frac{1}{\sigma_N^2}}.$$

The optimum weight for an observation of standard deviation σ is just $1/\sigma^2$. If all the variances are the same, we recover the simple average; if one is much smaller than the others, it dominates the estimate. These are nice intuitive properties. However, the fact that we need the squares of the quoted standard deviations places a premium on their accuracy. An estimate with too low a standard deviation can easily dominate the results.

The standard deviation of this weighted mean is

$$\sqrt{\left(\frac{1}{\left(\frac{1}{\sigma_1^2} + \frac{1}{\sigma_2^2} + \cdots + \frac{1}{\sigma_N^2} \right)} \right)}$$

These results also follow from a maximum-likelihood analysis, assuming Gaussian distributions, but the results are more general than that. Hints for the derivation are in Exercise 6.2.

6.5 Bayesian likelihood analysis

We now move on to Bayesian aspects of model-fitting. Bayes' theorem says, for model parameters (a vector, in general) $\vec{\alpha}$ and data X_i,

$$\text{prob}(\vec{\alpha} \mid X_i) \propto \mathcal{L}(\vec{\alpha} \mid X_i)\text{prob}(\vec{\alpha}) \qquad (6.12)$$

so the likelihood function is important here too. However, given the posterior probability of $\vec{\alpha}$, we may choose to emphasize properties other than the most probable $\vec{\alpha}$ – we may only be interested in the probability that it exceeds a certain value, for example.

Two great strengths of the Bayesian approach are the ability to deal with nuisance parameters via marginalization, and the use of the evidence or Bayes factor to choose between models. Another useful product of the Bayesian approach is the asymptotic distribution of the likelihood function itself. $\mathcal{L}(\vec{\alpha})$ is asymptotically a multivariate Gaussian distributed around the MLE $\hat{\vec{\alpha}}$, with covariance matrix given by the inverse of the matrix

$$F' = - \begin{bmatrix} \frac{\partial^2 \ln \mathcal{L}}{\partial \alpha_1^2} & \frac{\partial^2 \ln \mathcal{L}}{\partial \alpha_1 \partial \alpha_2} & \frac{\partial^2 \ln \mathcal{L}}{\partial \alpha_1 \partial \alpha_3} & \cdots \\ \frac{\partial \ln \mathcal{L}}{\partial \alpha_2 \partial \alpha_1} & \frac{\partial^2 \ln \mathcal{L}}{\partial \alpha_2^2} & \frac{\partial^2 \ln \mathcal{L}}{\partial \alpha_2 \partial \alpha_3} & \cdots \\ \frac{\partial^2 \ln \mathcal{L}}{\partial \alpha_3 \alpha_1} & \frac{\partial^2 \ln \mathcal{L}}{\partial \alpha_3 \partial \alpha_2} & \frac{\partial^2 \ln \mathcal{L}}{\partial \alpha_3^2} & \cdots \\ \vdots & \vdots & \vdots & \end{bmatrix} \qquad (6.13)$$

evaluated at the peak, namely the MLE of $\vec{\alpha}$. (The expectation or average over many realizations of F' is the Fisher information matrix.) The likelihood then takes the form

$$\mathcal{L}(\vec{\alpha} \mid X_i) = \mathcal{L}(\hat{\vec{\alpha}} \mid X_i) \exp\left(-\frac{1}{2}(\hat{\vec{\alpha}} - \vec{\alpha})^T F(\hat{\vec{\alpha}} - \vec{\alpha})\right), \qquad (6.14)$$

where T denotes the matrix transpose. This is called the Laplace approximation (MacKay, 2003, Chapter 26). The integral of this form, over the whole parameter space, is useful:

$$\int d\vec{\alpha}\, \mathcal{L}(\vec{\alpha} \mid X_i) = \mathcal{L}(\hat{\vec{\alpha}} \mid X_i)\sqrt{\frac{(2\pi)^k}{\mid \det F \mid}}, \qquad (6.15)$$

where det F is the matrix determinant of F and the likelihood is maximized at $\hat{\vec{\alpha}}$. This will come into its own in the next chapter; in this one, integrals over *some* of the components of $\vec{\alpha}$ will be potentially more useful – see Jaynes (2003) for the full marginalization integral procedure.

We will illustrate this approach by developing a simple two-parameter example, fitting a power law to some radio flux density data. This example will appear in various guises in this chapter, but each time we will assume Gaussian statistics and uniform priors. These assumptions do not simplify the calculations, which were all done numerically in any case; they do simplify the presentation. Use the error distribution and prior that fits *your* problem.

Example Let us suppose that we have flux density measurements at 0.4, 1.4, 2.7, 5 and 10 GHz. The corresponding data are 1.855, 0.640, 0.444, 0.22 and 0.102 flux units – see Figure 6.4.

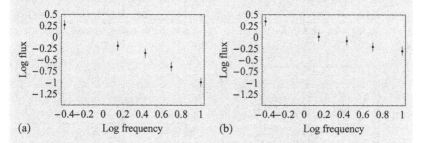

Figure 6.4   The two experimental spectra we will examine; (b) contains an offset error as well as random noise.

Let us label the frequencies as f_i and the data as S_i. These follow a power law of slope -1, but have a 10 per cent Gaussian noise added. The noise level is denoted ϵ, and the model for the flux density as a function of frequency is $Kf^{-\gamma}$. Assuming that we know the noise level and distribution, each term in the likelihood product is of the form

$$\frac{1}{\sqrt{2\pi}\epsilon Kf_i^{-\gamma}} \exp\left[-\frac{(S_i - Kf_i^{-\gamma})^2}{2(\epsilon Kf_i^{-\gamma})^2}\right].$$

The likelihood is therefore a function of K and γ. A contour map of the log likelihood is in Figure 6.5. We can calculate the Gaussian approximation to the likelihood with Equation (6.14), also shown in Figure 6.5.

At this point, there are at least two possibilities for further analysis. We may wish to know which pairs of (K, γ) are, say, 90 per cent probable. This, in general, involves a very awkward integration of the posterior probabilities. The multivariate Gaussian approximation to the likelihood is much easier to use; it is automatically normalized and there are analytic forms for its integral over any number of its arguments, as mentioned before.

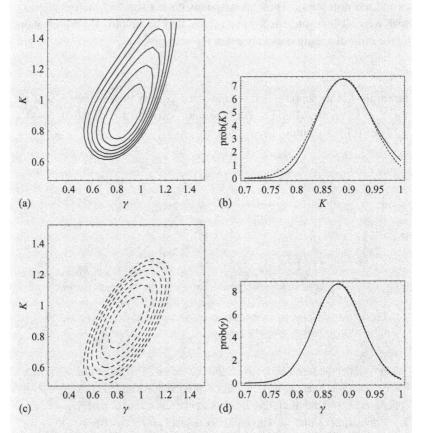

Figure 6.5   (a) A contour plot of the log likelihood function for our spectral parameters (K, γ); (c), the Gaussian approximation; (b) and (d), the marginal distributions of K and γ, comparing the Gaussian approximation to the full likelihood.

As can be seen in Figure 6.5, the areas defined by a particular probability requirement are simple ellipses.

Another possibility is to ask for the probability of, say, K *regardless* of γ. Getting rid of γ in this context is called *marginalization*. So we have a posterior probability prob($K, \gamma \mid S_i$) and we form

$$\text{prob}(K \mid S_i) = \int \text{prob}(K, \gamma \mid S_i) \, d\gamma.$$

The probability distributions for K and γ are also shown in Figure 6.5, along with the distributions deduced from the Gaussian approximation. As we can see, the agreement is quite good.

Marginalization (Section 2.2) can be a very useful technique. Often we are not interested in all the parameters we need to estimate to make a model. If we were investigating radio spectra, for instance, we would want to marginalize out K in our example. We may also have to estimate instrumental parameters as part of our modelling process, but at the end we marginalize them out in order to get answers which do not depend on these parameters. Of course, the marginalization process will always broaden the distribution of the parameters we do want, because it is absorbing the uncertainty in the parameters we do not want – the *nuisance parameters*.

Example In our radio spectrum example we will add (somewhat artificially) an offset of 0.4 flux units to each measurement. This has the effect of flattening the spectrum quite markedly. We calculate two possibilities. Model A is the simple one we assumed before, with no offsets built in. Model B uses a model for the flux densities of the form $\beta + Kf^{-\gamma}$. Each likelihood term is then

$$\frac{1}{\sqrt{2\pi}\epsilon K f_i^{-\gamma}} \exp\left[-\frac{(S_i - (\beta + Kf_i^{-\gamma}))^2}{2(\epsilon K f_i^{-\gamma})^2}\right].$$

We also suppose that we have some suspicion of the existence of this offset, so we place a prior on β of mean 0.4, standard deviation ϵ. Model B therefore returns a posterior distribution for K, γ and β. We are not actually interested in β (although an instrument scientist might be) so we marginalize it out. The likelihoods from the two models are shown in Figure 6.6, and it is clear that the more complex model does a better job of recovering the true parameters. The procedure works because there is information in the data about both the instrumental and the source parameters, *given the model of the spectrum*. If our model for the spectrum had a 'break' in it, we would not be able to recover much information about β, if any. If our fluxes had a pure scale error, we would not have been able to recover this either.

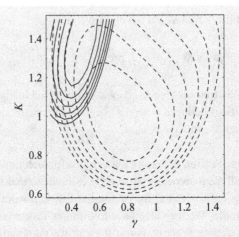

Figure 6.6   The log likelihoods for the two models of radio spectra; the black contours are for Model A and the dashed contours are for Model B.

In the real world, of course, we do not have the truth available to guide us as to our choice of Model A or Model B. As remarked before, we ought to check the 'fit' of the two models. In one dimension there are various ways to do this, as discussed in Chapter 5. In many dimensions things are harder. If at least one of the models appears to fit reasonably well, we may use the Bayes factor, introduced in Section 5.2.3. As discussed there, suppose we are choosing between Model A and Model B and we believe they are the only possibilities. The prior probability of A is, say, p_A and of B is p_B.

The posterior odds on Model A, compared to Model B, are then

$$\mathcal{P} = \frac{\int_\alpha p_A \mathcal{L}(X_i \mid \alpha, A) \mathrm{prob}(\alpha \mid A)}{\int_\alpha p_B \mathcal{L}(X_i \mid \alpha, B) \mathrm{prob}(\alpha \mid B)} \qquad (6.16)$$

in which we have to integrate over the range of parameters appropriate to each model. This is worth the effort because we get a straightforward answer to the question: which of A or B would it be better to bet on?

Example In the previous two examples, we have worked out the likelihood functions, which we abbreviate as $\mathcal{L}(X_i \mid K, \gamma, A)$ for Model A and similarly for Model B. In Model B we also have a prior on the offset β, which is

$$\mathrm{prob}(\beta \mid B) = \frac{1}{\sqrt{2\pi}\,\epsilon} \exp \frac{-(\beta - 0.4)^2}{2(\epsilon)^2}.$$

We then form the ratio of the integrals

$$p_A \int dK \int d\gamma \, \mathcal{L}(X_i \mid K, \gamma, A)$$

and

$$p_B \int dK \int d\gamma \int d\beta \, \mathcal{L}(X_i \mid K, \gamma, B) \mathrm{prob}(\beta \mid B).$$

Take $p_A = p_B$, an agnostic prior state; also, we have implicity assumed uniform priors on K and γ. Cranking through the integrations numerically, we get *odds on* B *compared to* A: about 8 to 1.

6.6 Bootstrap and jackknife

In some data-modelling procedures, confidence intervals for the parameters fall out of the procedure. But are these realistic? And what about the procedures where they do not? Computer power can provide the answer, with the *Bootstrap method* method invented by Efron (1979); see also Diaconis & Efron (1983) and Davison & Hinkley (1997). It apparently gives something for nothing, and Efron so named it from the image of lifting oneself up by one's own bootstraps.

The method is so blatant (described, e.g., in *Numerical Recipes* as 'quick-and-dirty Monte Carlo') that it took some time to gain respectability, but the foundations are now secure (see, e.g., LePage & Billiard, 1993; Efron & Tib-shirani, 1993). Suppose the sample consists of N data points, each consisting of one or more numbers (e.g. single measurements, or x, y pairs), and we wish to ascertain the error on a parameter estimated from these data points (e.g. mean, or slope of a best fit). We calculate the parameter using a modelling process such as one of those described above. We then 'bootstrap' to find its uncertainty, as follows.

(i) *Label* each data point;
(ii) *Draw at random* a sample of N with replacement (simply done by computer with a random-number generator);
(iii) *Recalculate* the parameter.
(iv) *Repeat* this process as many times as possible.

That's it. Provided that the data points are independent (in distribution and in order), the distribution of these recalculated parameters maps the uncertainty in the estimate from the original sample.

Example Bhavsar (1990) described how well suited the bootstrap method is to estimating uncertainty in measuring the slope of the angular two-point correlation function for galaxies. This function $w(\theta)$ (Section 10.4) measures the excess surface density over that expected from a uniform independent and random distribution at angular scales θ. The data points are the (x, y) pairs of galaxy coordinates on the sky, and the difficulty in estimating the accuracy of this slope is even more notorious than that of estimating the slope of the counts of radio sources. The reason is similar: $\sqrt{N}$ error bars are readily assigned, but they are not independent, and unlike the case of source counts for which a differential version is possible, there is no ready way of assessing the significance of the correlated errors in a correlation function. Figure 6.7 shows an example of such a two-point correlation function estimate, part of a search for clustering in the distribution of radio sources on the sky (Wall *et al.*, 1993).

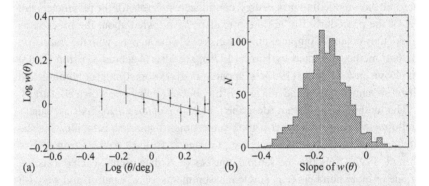

Figure 6.7   A bootstrap method application. (a) The two-point correlation function for 2812 radio sources with extended radio structure, from the White–Becker catalogue of the NRAO 1.4-GHz survey of the northern sky. A least-squares fit gives a slope of -0.19. (b) The distribution of slopes obtained in bootstrap method-testing the sample with 1000 trials. The mean slope is -0.157, while the rms scatter is ± 0.082; the slope is less than zero (i.e. signal is present) for 96.8 per cent of the trials.

The bootstrap method is ideal for computing errors in a PCA analysis. It is a good way of telling if any of the principal components have been detected above the sampling error. The bootstrap method takes us back to the quotation starting this chapter. If errors are not well known, it is still possible to ascertain errors on a model. Moreover, the errors may be known well; but, as in the above example, their significance in terms of defining a model may not be understood. In either case it is possible to bootstrap one's way to safety.

The *jackknife* is a rather similar technique to the bootstrap, but much older, first described by Tukey (one of the inventors of the FFT) in 1958.

The algorithm is again quite simple. Suppose we are interested in some function $f(X_1, X_2, \ldots)$ which depends on the N observations X_i. Usually this will be because f is a useful estimator of a parameter α. Thus, we have

$$\hat{\alpha} = f(X_1, X_2, \ldots).$$

The jth partial estimate is obtained by deleting the jth element of the data set:

$$\hat{\alpha}_j = f(X_1, X_2, \ldots, X_{j-1}, X_{j+1}, \ldots, X_N),$$

giving N partial estimates. The next step (and the crucial one) is to define the pseudo-values

$$\hat{\alpha}_j^* = N\hat{\alpha} - (N-1)\hat{\alpha}_j,$$

and finally the *jackknifed estimate* of α is the simple average of the pseudo-values

$$\hat{\alpha}^* = \frac{1}{N} \sum_{i=1}^{N} \hat{\alpha}_j^*. \tag{6.17}$$

The great merit of the jackknife is that it removes bias. Often the bias will depend inversely on the sample size (a simple example of this being the MLE for the variance of a Normal distribution, Section 3.2) and the jackknifed estimate will not contain this bias. In general, we can construct an mth order jackknifed estimate by removing m observations at a time, and this will eliminate bias that depends on $1/N^m$.

For estimators which are asymptotically Normal (e.g. MLEs) it is useful to calculate the sample variance on the pseudo-values, which is

$$(\sigma^*)^2 = \frac{1}{N(N-1)} \sum_j (\hat{\alpha}_j^* - \hat{\alpha}^*)^2. \tag{6.18}$$

This can be used to give a confidence interval on $\alpha - \alpha^*$, which is distributed according to $\sigma^* t$ with t having $N - 1$ degrees of freedom. This works to the extent that Normality has been obtained. In practice it is easier to use a bootstrap method for confidence intervals, because the assumption of Normality is not needed. If the jackknife intervals can be checked with a bootstrap method, they are, of course, much less computationally intensive to calculate.

The jackknife is related to the technique called *cross-validation*, in which the data set is divided into parts and the results of analysing the parts separately are compared for statistical consistency. There are a number of sophisticated variants of this idea.

Exercises

6.1 Covariance matrix. Consider N data X_i, drawn from a Gaussian of mean μ and standard deviation σ. Use ML to find estimators of both μ and σ, and find the covariance matrix of these estimates.

6.2 Weighting data. Show that the optimum weight for an observation of standard deviation σ is just $1/\sigma^2$. This weight turns up naturally in modelling using minimum χ^2.

6.3 MLE and power laws. In the example Section 6.1 we fit a power law truncated at the faint end, and assume that we know where to cut it off. What happens if you try to infer the faint-end cutoff by ML as well? Formulate this problem at least.

6.4 MLEs. Find an estimator of μ when the distribution is (a)

$$\text{prob}(x) = \exp(-|x - \mu|)$$

and (b) the Poisson

$$\text{prob}(n) = \mu^n \frac{e^{-\mu}}{n!}.$$

6.5 Least-squares linear fits. Derive the 'minimum distance' OLS for errors in both x and y, assuming Gaussian errors.

6.6 Marginalization. Using the data supplied, use ML to find the distribution of the parameters of a fitted Gaussian plus a baseline. Test to see how the estimates are affected by marginalizing out the baseline parameters.

6.7 The jackknife. Using the MLE for a power-law index (Section 6.1), work out and compare the confidence intervals with the analytic result from that section using the jackknife and bootstrap tests. Check how the results depend on sample size.

7

Data modelling and parameter estimation:
advanced topics

Frustra fit per plura quod potest fieri per pauciora – it is futile to do with
more things that which can be done with fewer.

(William of Ockham, c.1285–1349)

Nature laughs at the difficulties of integration.

(Pierre-Simon de Laplace, 1749–1827; Gordon & Sorkin, 1959)

One of the attractive features of the Bayesian method is that it offers a principled
way of making choices between models. In classical statistics, we may fit to a
model, say by least squares, and then use the resulting χ^2 statistic to decide if
we should reject the model. We would do this if the deviations from the model
are unlikely to have occurred by chance. However, it is not clear what to do if
the deviations are likely to have occurred, and it is even less clear what to do
if several models are available. For example, if a model is in fact correct, the
significance level derived from a χ^2 test (or, indeed, any significance test) will
be uniformly distributed between zero and one (Exercise 7.1).

The problem with model choice by χ^2 (or any similar classical method) is that
these methods do not answer the question we wish to ask. For a model H and
data D, a significance level derived from a minimum χ^2 tells us about the conditional probability, prob($D \mid H$). (In fact, it usually tells us not even this, but a
probability involving data we did not observe – see Section 3.5.) But what we
want is prob($H \mid D$), and we cannot get this from prob($D \mid H$) without an application of Bayes' theorem and consequent involvement with prior probabilities.

7.1 Model choice and Bayesian evidence

To develop the Bayesian ideas, let us suppose that we have just two models H_1
and H_2, associated with parameters $\vec{\alpha}$ and $\vec{\beta}$, respectively. For a set of data D,

Bayes' theorem gives the posterior probabilities:

$$\text{prob}(H_1, \vec{\alpha} \mid D) = \frac{\text{prob}(D \mid H_1, \vec{\alpha})\,\text{prob}(\vec{\alpha} \mid H_1)\text{prob}(H_1)}{E} \qquad (7.1)$$

and

$$\text{prob}(H_2, \vec{\beta} \mid D) = \frac{\text{prob}(D \mid H_2, \vec{\beta})\,\text{prob}(\vec{\beta} \mid H_2)\text{prob}(H_2)}{E}. \qquad (7.2)$$

In these equations the terms are the following. Firstly, there is 'the prior probability of the model', for example $\text{prob}(H_1)$. Here, 'the model' means a statement like 'this spectral line is Lorentzian' and so the prior probability of the model is a *number* which reflects your prior belief that the line in question is, indeed, Lorentzian. Next comes the prior probabilities of the *parameters* in the model, given that we have accepted it for the time being. This is the term $\text{prob}(\vec{\alpha} \mid H_1)$. If the Lorentzian model has a scale parameter σ to which Jeffreys' prior applies, then this term would be proportional to $1/\sigma$, for instance. Lastly, we have $\text{prob}(D \mid H_1, \vec{\alpha})$, which is simply the familiar likelihood of the data. E is the necessary normalizing factor to make the left-hand side $\text{prob}(H_1, \vec{\alpha} \mid D)$ a probability; the components of E will soon assume great significance.

Assuming (a key assumption) that our set of possible models is exhaustive, we can find E by integrating out the parameters to obtain the posterior probabilities of H_1 and H_2 and requiring that these sum to unity:

$$\int \text{prob}(H_1, \vec{\alpha} \mid D)\,d\vec{\alpha} + \int \text{prob}(H_2, \vec{\beta} \mid D)\,d\vec{\beta} = 1. \qquad (7.3)$$

This integration may be formidable in multidimensional spaces, but it gives us E, which is

$$E = \int \text{prob}(D \mid H_1, \vec{\alpha})\,\text{prob}(\vec{\alpha} \mid H_1)\,d\vec{\alpha}\,\text{prob}(H_1)$$
$$+ \int \text{prob}(D \mid H_2, \vec{\beta})\,\text{prob}(\vec{\beta} \mid H_2)\,d\vec{\beta}\,\text{prob}(H_2). \qquad (7.4)$$

Putting together Equations (7.1) and (7.4) gives the posterior probability of model H_1

$$\text{prob}(H_1) = \frac{1}{1 + \mathcal{BP}} \qquad (7.5)$$

in which $\mathcal{B}$ is the *Bayes factor* (Section 5.2.3)

$$\mathcal{B} = \frac{\int \text{prob}(D \mid H_2, \vec{\beta})\text{prob}(\vec{\beta} \mid H_2)\,d\vec{\beta}}{\int \text{prob}(D \mid H_1, \vec{\alpha})\text{prob}(\vec{\alpha} \mid H_1)\,d\vec{\alpha}} \qquad (7.6)$$

multiplied by the *prior odds*. Given the posterior probabilities of the competing models, we then also have the *posterior odds* as the ratio of those posterior probabilities.

$$\mathcal{P} = \frac{\text{prob}(H_2)}{\text{prob}(H_1)}. \tag{7.7}$$

Equations (7.7), (7.6), and (7.5) encapsulate the Bayesian model choice method. The key ingredient is the Bayes factor, which is a ratio of terms sometimes called *evidence*. These evidence terms are *the average of the likelihood function over the prior on the parameters* and it is the relative magnitude of the evidence for each model that determines its posterior probability. Our normalizing term E we now see to be the sum of the evidence terms, each weighted by the prior probability of the relevant model.

The method of Bayes' factors was developed by the astronomer and geophysicist Sir Harold Jeffreys (Jeffreys, 1961). A good early review is by Kass & Raftery (1995).

Example Suppose that the prior odds are large (you think that H_2 is the right model). If the ratio of the evidence integrals is small – this means that the data are much more likely under H_1 – then we will get a small value for the Bayes factor $\mathcal{B}$. In fact, if $\mathcal{B}$ is small enough it will outweigh the large value for the prior odds $\mathcal{P}$ and thus $\text{prob}(H_1) \simeq 1$. Your view of what is likely to be the case will have been modified by the data. The posterior probability of H_1 is large, whereas before using the data, your assigned probability for H_2 was large.

This is the Bayesian model choice method; the prior odds on the models are modified by the data to give the posterior probabilities.

7.2 Model simplicity and the Ockham factor

The priors on the parameters are within the evidence integrals. What does this mean? Take a simple case where the priors are rather flat and the likelihood functions narrow and peaked, and we have just one parameter for each model instead of a vector. Suppose we know in advance that α is somewhere inside a range $\sim\Delta\alpha$, and β is within a range $\sim\Delta\beta$. Then $\text{prob}(\alpha) = 1/\Delta\alpha$ and $\text{prob}(\beta) = 1/\Delta\beta$. In Equation (7.5) we then have

$$\mathcal{B}\mathcal{P} = \text{ratio of integrals of likelihoods} \times \frac{\Delta\alpha}{\Delta\beta} \times \text{prior odds.} \tag{7.8}$$

In our previous example where the prior odds were large (H_2 favoured), suppose that our prior ideas about the parameters of H_2, namely β, are much vaguer than they are for the competing model's parameters α. This means that $\Delta\alpha/\Delta\beta \ll 1$, which will reduce the effect of the prior odds. The data then have to work strongly in favour of H_2 for that model to come out with a larger posterior probability, more strongly than they would if $\Delta\alpha \simeq \Delta\beta$.

What we have seen here is the so-called Ockham factor at work. The model H_1 is 'simpler' than H_2 because it is more specific (uses less prior range on its parameter) and there is an automatic boost to the posterior probability of the simpler model, unless the data work in favour of the more complex model. This built-in mechanism that favours simplicity, in the very specific sense of prior volumes or ranges, is named after the medieval philosopher William of Ockham whose opinions on complexity head this chapter.

This is quite a narrow definition of simplicity. For example, in any particular case a model may appear more complex than its competitors, but with a wider view may be seen to have more explanatory power (and therefore be 'simpler'). So a judgement about simplicity has to take into account the scope over which a model is to be judged. Nonetheless, the ability to penalize models automatically for gratuitous complexity is a very attractive feature of the Bayesian model choice method. In general there is not a multiplicative Ockham factor (we were only able to get a factor by considerable simplifications) but the effect is present in any calculation built on the Bayes factor.

For more details on Ockham factors, see Gregory (2004), and MacKay (2003, Chapter 28).

7.3 The integration problem

This attractive Bayesian formalism has one major problem: the integrals. For most problems of interest, there are many parameters – the vector $\vec{\alpha}$ may have many components. Performing the integrals in the resulting many-dimensional spaces is a formidable numerical problem, one that has only recently been solved (or at least, made much easier) by Markov chain Monte Carlo (MCMC) techniques. We outline these techniques later in this chapter.

Simpler methods than MCMC can be useful. The heart of the matter is the integration of a likelihood function over the prior on the parameters:

$$\int \text{prob}(D \mid H_1, \vec{\alpha})\text{prob}(\vec{\alpha} \mid H_1)\,\mathrm{d}\vec{\alpha}.$$

If the residuals from the model are Normally distributed, the likelihood term involves the sum of squares of the residuals, as usual. We can write it in terms of chi square:

$$\text{prob}(D \mid \vec{\alpha}, H_1) = \exp\left(-\frac{1}{2}\chi^2(\alpha)\right), \tag{7.9}$$

where the residual term depends, of course, on the parameters, and the chi-square term is the sum of the squared residuals, each divided by the variance of that residual.

This is just one form of the likelihood, although a common one. In any case, if the likelihood is much more peaked (as a function of $\vec{\alpha}$) than the prior, the integration simplifies considerably by use of the Laplace approximation or a steepest-descent method (see Section 6.5 and also p. 341 of MacKay, 2003). In this approximation we replace the likelihood term by a Gaussian near its peak, and integrate over it using standard methods. The result is

$$\int \text{prob}(D \mid H_1, \vec{\alpha})\text{prob}(\vec{\alpha} \mid H_1)\,d\vec{\alpha} \simeq \frac{(2\pi)^{m/2}}{\sqrt{|\det(\mathcal{H})|}} L(\vec{\alpha}^*)\,P(\vec{\alpha}^*), \tag{7.10}$$

where $\vec{\alpha}^*$ is the value at the peak of the likelihood, $\mathcal{H}$ is the Hessian matrix of second derivatives of the log of the likelihood at the peak, and m is the number of parameters (Section 6.1). The value of the likelihood function at its maximum has been abbreviated as $L(\vec{\alpha}^*)$ and the value of the prior on the parameters at this point is abbreviated as $P(\vec{\alpha}^*)$.

The integration of the likelihood term in the Bayesian method then reduces to the less laborious task of finding the maximum posterior probability, and evaluating the matrix $\mathcal{H}$ and its determinant. Averaging $\mathcal{H}$ over many realizations of the data yields the Fisher matrix (Section 6.1), which may be inverted to predict the covariance matrix of the components of $\vec{\alpha}$ (e.g. Tegmark *et al.*, 1997). This is an extremely useful way of approaching a Bayesian analysis, as its approximations may well be satisfied. The computational load may then be small enough to justify the use of the method on relatively small but common and interesting model choice problems.

It is vital in this method to keep track of the normalizing factors in the priors; as written above, $P(\vec{\alpha})$ must be normalized to unity over the allowed range of the parameters. It is this normalization that brings the Ockham factor into play. (It is also effected through the determinant term to some extent).

Example Suppose that we are trying to decide if a single spectral line is Gaussian or Lorentzian. The parameters in either case are the baseline, the height, width and centre. Call the Gaussian model H_1 and the Lorentzian model H_2. The input Gaussian profile is

$$\alpha_1 + \alpha_2 \exp\left(-\frac{1}{2\alpha_3^2}(x - \alpha_4)^2\right) \tag{7.11}$$

with initial values $\alpha_1 = 0$, $\alpha_2 = 1$, $\alpha_3 = 1$, $\alpha_4 = 0$. The Lorentzian profile is

$$\beta_1 + \beta_2 \frac{1}{1 + \frac{(x-\beta_4)^2}{\beta_3^2}}. \tag{7.12}$$

We will be modelling the case where we are able to look at several realizations of the data while the spectral line remains the same. This might be the case, for instance, in which we are able to observe the object in question several times.

We generate data from the Gaussian profile model by adding Gaussian random noise, and then fit Gaussian and Lorentzian profiles to these data by least squares. We thus get sums of squared residuals, normalized to the assumed known standard deviations that were fed into the random-number generators. (In real life we would probably handle the noise level as another free parameter.) This means we are using a likelihood function of the form of Equation (7.9). This can be integrated using the Laplace approximation (Equation (7.10)).

However we need to consider the vital matter of the priors and their normalizations.

For simplicity we will assume that flat priors, of the same ranges, apply to the four parameters of each of the models. We could do better than this; a Jeffreys' prior, for instance, should be used for the line width since it is a scale parameter. However, since the prior ranges are all of order unity, for the accuracy we are needing we can, in effect, ignore the priors in what follows since there is a similar number, type and range of variable in both our competing models. We are also assuming that each model is a priori equally likely. The priors thus cancel out in the Bayes factor and prior odds.

Armed with all this, we may run a Monte Carlo simulation which creates some data that is Normally distributed around our Gaussian-shaped spectral line, fit our two candidate models by maximum likelihood, and finally compute the Bayes factor by using the Laplace approximation.

Examples of data and fits at two input noise levels are in Figure 7.1. The quoted odds, by eye, seem to be in the right sense, but the magnitudes are surprising. This leads us on to ask about the reliability of the model-choice method.

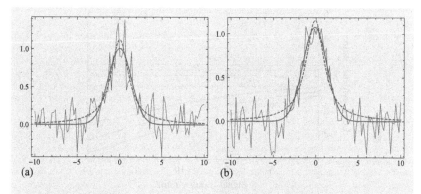

Figure 7.1 Spectral-line simulations. These are at a signal-to-noise ratio, in the peak of the line, of 5. (a) The Lorentzian fit is favoured at odds of 30:1, whereas (b) the Gaussian is favoured at odds of 100:1.

7.4 Pitfalls in model choice

The previous example set out to illustrate the method of model choice via Bayes factors, and used a Monte Carlo approach to generate data for the simulations. In the Bayesian paradigm we tend to think of data as being fixed and parameters as being the variables, but of course in many cases we can access many realizations of data for the same problem. Simply reobserving a galaxy on our list achieves this. A lesson from the Monte Carlo simulations in the example is that the Bayes factor is a *statistic* and shows random variation; that is, the odds on the Lorentzian, determined from *this* spectrum, will not be same as the odds determined from *that, repeated* spectrum.

In fact, the odds show considerable variation. By running the procedure described in the example many times, we can build up a picture of the variability in the derived odds, as a function of the noise level in the simulated data. This is shown in Figure 7.2. The most important feature of this graph is the *very* wide spread in the odds at a given signal-to-noise ratio: generally about a factor of 100.

We should be clear that while this aspect is analysed here for a Bayesian method, it is by no means specific to that approach. For example, in a classical test against a null hypothesis, the significance level will be a statistic and will be uniformly distributed between 0 and 1 (see Exercise 7.1). In the Bayesian case we are forced to consider alternative as well as null hypotheses and so it becomes clearer, as in Figure 7.2, that we need to be aware of the power of the method we are using. In our case the ability to pick the Gaussian profile when

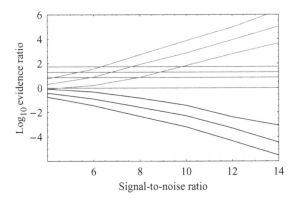

Figure 7.2   The odds for the Gaussian profile, plotted against signal-to-noise ratio. The light curve is calculated for starting data that has a Gaussian profile; the black curve for initial data that has a Lorentzian profile. The central curve is the median and the flanking solid lines are the 25th and 75th percentiles. 500 iterations were used at each noise level. The horizontal lines mark odds on the Gaussian that are even, e^2:1, e^3:1 and e^4:1.

it is actually the case is evidently quite poor at low signal-to-noise level. This, of course, is common sense as no model-choice method can conjure reliable results out of poor data.

Another feature of Figure 7.2 is the *size* of the odds. They seem improbably good in many cases, and indeed they are. This can be traced to three factors that will not apply in the real world outside simulations. Firstly, one of the models is exactly right, the one that was used as the basis of the fake data in the simulation. Secondly, the noise is known exactly; no outliers, no drifts, none of the usual problems of actual data. Finally, there is only one alternative; having several more or less similar alternatives obviously makes it harder to get good odds on one of them. Again, none of this is specific to Bayesian methods but it is important to be aware of the issues.

One potential pitfall which, however, is rather Bayesian relates to goodness of fit and the model suite. If we give our model-choice algorithm only two choices, it *will* choose one of them and it may give it quite high odds. This does not mean that the 'best' model would pass a goodness-of-fit test and it is vital to apply the basic sanity check of asking, does this model, with the odds in its favour, actually fit the data? Or is my set of models insufficiently comprehensive?

Issues like these, and especially the power of the Bayes factor in model choice, are explored in more detail in Exercises 7.6 and 7.7.

7.5 The Akaike and Bayesian information criteria

Two other commonly used criteria for the choice of models are the Akaike information criterion (AIC) and the Bayesian information criterion (BIC; Liddle *et al.*, 2006). They are defined by

$$\text{AIC} = -2 \log \mathcal{L}_{\text{max}} + 2k \tag{7.13}$$

and

$$\text{BIC} = -2 \log \mathcal{L}_{\text{max}} + k \log N \tag{7.14}$$

in which the first term is the ML of the model, and the second is a penalty term involving the number of adjustable parameters k and the number of data points N. Only the relative values of these criteria are used; that is, we pick the model with the smallest value of the AIC or BIC.

In general these are easier to use than a Bayesian analysis that requires a full integration to obtain the evidence. The BIC is actually an approximation to the full evidence. With the same number of parameters in both models, and a diffuse prior (broader than the likelihood function) the effect of full Bayesian methods is not very apparent. For Gaussian errors, choosing on the basis of the AIC or BIC is just minimum χ^2.

Once we do not have the same number of parameters in the model, differences emerge with even very simple priors. The AIC and BIC have somewhat different 'Ockham' penalties for adding parameters, but these are not large. If we have many data points, and Gaussian errors again, the AIC and BIC are pretty much minimum χ^2. The Bayes factor, however, includes the priors on the extra parameters, and, depending on this prior, may actually reverse the decision compared to minimum χ^2.

Example We can highlight the effect of the Ockham factor by adding some parameters to one of the models we considered before in the line-fitting example. Adding some curvature to the baseline of the Lorentzian model might offset its wider wings and make it more competitive to the Gaussian model. There may well be instrumental reasons to admit such curvature to the model, and knowledge of how big it can reasonably be. On the same assumption as before about the line-shape parameters (that they have similar flat priors that cancel out in the Bayes factor) we do need to account explicitly for the priors on the baseline parameters. If the additional baseline term in the Lorentzian model is $\beta_5 x^2 + \beta_6 x^4$ and the priors on these parameters are flat, with widths $\Delta\beta_5$ and $\Delta\beta_6$, we will need to account for these widths in the Bayes factor. A rough assignment of the prior width might result from requiring that the maximum, of both the quadratic and quartic terms separately, should not exceed the line height. Since the maximum value of the independent variable is 10 in the line-fitting example, this means that $\Delta\beta_5 = 10^{-2}$ and $\Delta\beta_6 = 10^{-4}$. This assignment of the prior could immediately shorten the odds on the Gaussian profile by six orders of magnitude! Following through the detailed calculation does not introduce quite such a large change – more like a mere three orders of magnitude – but the effect is considerable and places the odds much more in the real world.

Hence, now that we have an unequal number of parameters, we see very clearly how the Bayes factor approach differs fundamentally from minimum χ^2 – the prior really matters, as it should.

7.6 Monte Carlo integration: doing the Bayesian integrals

Many-dimensional numerical integration is often needed to complete the Bayesian solution of a problem. Two huge advantages of Bayesian methods are marginalization – integrating out 'nuisance parameters' to get the posterior distribution of the parameter of interest – and computing the evidence to make model choices, another integration. This is a technical subject, covered, for example, in Evans & Swartz (1995, 1996); Chib & Greenberg (1995); O'Ruanaidh & Fitzgerald (1996). Useful recent references are Gregory (2004) and MacKay (2003, Chapters 29 and 30).

To introduce the ideas, suppose that we have a function $f(x)$ defined for $a \leq x \leq b$. If we draw N random numbers X, uniformly distributed between a and b, then we have

$$\int_a^b f(x)\,\mathrm{d}x \simeq \frac{b-a}{N} \sum_i f(X_i). \tag{7.15}$$

This is Monte Carlo integration in its simplest form. A few numerical experiments will show that it is not very efficient, but it comes into its own for many-dimensional numerical integration because the alternatives are enormously time-consuming.

If f is a probability distribution, and the X_i are drawn from the distribution f itself, then obviously they will sample the regions where f is large more often and the integration will be more accurate. This technique is called *importance sampling*. For example, suppose that f is the posterior distribution of some interesting parameter and we want to know the expectation value of some function g of this parameter; it might be something as simple as the mean. If we can get random numbers X_i that are drawn from f, then the Monte Carlo integral is neatly

$$\int g(x) f(x) \, dx \simeq \frac{1}{N} \sum_i g(X_i) \qquad (7.16)$$

because if the X_i are indeed drawn from f, then $f(X_i)$ is uniformly distributed between zero and one, and we can use the basic Monte Carlo formula. This will work in the multivariate case as well.

Often, it is hard to get the needed random numbers. If we can get a random number Y_i from a distribution h that is rather like the f in which we are interested, then we can write

$$\int g(x) f(x) \, dx \simeq \frac{1}{N} \sum_i \frac{f(Y_i)}{h(Y_i)} g(Y_i). \qquad (7.17)$$

We want the surrogate function h to 'cover' f so that the denominator on the right-hand side does not explode.

Subject to this reservation, we can see that we have a route to evaluating expectations, or even the normalization of f (by setting $g = 1$). Thus, if f is a posterior from a Bayesian solution, we can estimate the Bayesian evidence integral:

$$\int f(x) \, dx \simeq \frac{1}{N} \sum_i \frac{f(Y_i)}{h(Y_i)}. \qquad (7.18)$$

This is sometimes useful if we can find a suitable distribution h, and in many dimensions really the only one available off the shelf is a multivariate Gaussian and its close cousins like the multivariate t. The real problem, however, is dimensionality. In many cases, we are working in quite high dimensions. Choosing a function h that adequately 'covers' f in many dimensions means that, because of the way volume depends on dimensions, a very large fraction of the random numbers we generate are virtually wasted ($f(Y_i)$ is very small in the above numerator). We need a better way because this inefficiency is exponential in the dimensionality of our problem.

7.7 The Metropolis–Hastings algorithm

In a Bayesian context, we would like to be able to generate random numbers from a probability distribution f/C, where C is an unknown normalizing factor.

Further, f will in general be a multivariate distribution (if it was not, we could use deterministic numerical integration).

The workhorse method for obtaining random numbers in this situation is the *Metropolis algorithm* or its cousin, the *Metropolis–Hastings algorithm*. This is a very simple method, which copies the way in which physical systems, in thermal equilibrium, will populate their distribution function. The enormous advantage of the method is that it works when we do not know the normalization, which is usually the case in Bayesian problems. A good overview of Metropolis–Hastings methods is given by MacKay (2003).

The Metropolis algorithm (Metropolis *et al.*, 1953) was invented in physics to compute the equation of state of interacting particles in a box. The idea is neat; at a current position of the particles, shuffle them about randomly in space and work out the resulting energy change ΔE of the system. If the energy has fallen, accept the new configuration. If it has not, accept the configuration with probability $\exp(-\Delta E/kT)$, the Boltzmann factor, or stay where we are and shuffle again. Metropolis *et al.* prove that the distribution of particles must approach thermal equilibrium, essentially because detailed balance applies in the way the particles are shuffled and because the system is ergodic (every state can be reached from every other state).

Translating to our statistical problem, the simple Metropolis algorithm looks like this (where f, as before, is an un-normalized distribution of interest, the 'target' and h is a suitable transition probability distribution, the 'proposal').

At step i:

- draw a random number X_i from h;
- draw a random number U_i, uniformly distributed between zero and one;
- compute α, the minimum of 1 and $f(X_i)/f(X_{i-1})$;
- if $U_i < \alpha$, then accept X_i;
- otherwise set $X_i = X_{i-1}$.

The set of random numbers delivered by this algorithm will be random numbers drawn from f/C. Successive numbers will not, however, be independent of each other and can even be the same.

The random numbers are generated sequentially, with each depending on its predecessors. The Metropolis–Hastings (Metropolis *et al.*, 1953; Hastings, 1970) algorithm thus produces a string of numbers called a *Markov chain*. Using these random numbers in Monte Carlo applications leads to the acronym MCMC, for Markov chain Monte Carlo. Because of the correlations in the chain, it is often useful to thin out the series of numbers by choosing only every nth number, where n may be quite large (tens or hundreds!).

In this original form of the algorithm the 'proposal' distribution h must be symmetric in one of the following two senses. Clearly, h engineers the jump from position x_{i-1} to position x_i. If the probability of a reverse jump is the same, h is symmetrical in the required sense. The other symmetrical possibility is that f depends only on the absolute value of the difference $(x_i - x_{i-1})$ – a Gaussian proposal distribution would be of this type.

The generalization to other types of proposal distribution leads to the full Metropolis–Hastings algorithm as discussed, for example, by Chib & Greenberg (1995). Choosing the proposal well is the key to efficient generation of random numbers by this method. Acceptance rates of around 0.25–0.5 seem to give a good balance between correlation and excessive rejection rates. If the proposal is too narrow, there is a lot of correlation and the chain has to be thinned a good deal for use. The whole structure of the target may not be explored. If the proposal is too wide, many jumps are rejected, and the creation of the chain takes a long time.

As remarked, the numbers generated by a Metropolis–Hastings algorithm will show serial correlation. This means that the numbers will depend on how the algorithm is started, although after a while this memory will be lost. The period where the initial conditions still matter is called the burn-in period.

Two basic checks on an MCMC chain are to be sure that burn-in has been accomplished, and also that the chain explores the whole space where the target has appreciable probability. The intuition here is that if a set of chains is started from initial values well spread over the target, then after burn-in the variation between chains will more or less match the variation within a chain. The spread of initial values gives some assurance that the chains have not become stuck from the outset on a local maximum in the target.

Formally, suppose that we have l chains each n long. (Probably this is the last n numbers from a significantly longer chain, and the numbers may come from a thinned chain). We first of all look at the within-chain standard deviations, derived from each of the l chains. These should not be evolving as the chain lengthens, and should not show systematic trends (for example, one chain always noisier than the others). We then look at the standard deviation of the l means, compared to the pooled in-chain standard deviations. This ratio should be close to $1/\sqrt{n}$, and should likewise not be evolving with chain length.

This simple check arises as follows. If burn-in has happened, each of the l chains must, by definition, be statistically identical. Each chain should have a standard deviation close to the pooled standard deviation of the l chains, defined by

$$\sigma^2 = \frac{1}{l} \sum_j \sigma_j^2.$$ (7.19)

Each of the chains also has a mean value m_j, $j = 1, \ldots, l$; the standard deviation of this set of means should be close to the standard deviation of any one mean, which is $\sigma_j/\sqrt{n}$, which, in turn, should be close to $\sigma/\sqrt{n}$. Thus, the ratio of the standard deviation of the means to the pooled standard deviation is $1/\sqrt{n}$.

Of course this is never exactly the case for finite-length chains and tests exist which give more formal acceptance criteria (see Gelman & Rubin (1992) for the classic, but somewhat complex, tests). However, if this simple ratio shows no trends with chain length, then you have done as well as you are going to do.

If the chains do not pass these basic tests, a few things can be tried. The effect of the width of the proposal should be examined, checking the rejection rate and the extent of correlation; the thinning can also be varied. The convergence of MCMC chains continues to be a research topic; see Dunkley *et al.* (2005) for a Fourier-based technique. It is often useful to look at the power spectrum of the numbers in a chain, as this shows immediately how severe correlations might be, and gives guidance as to the amount of thinning that is useful.

Example We will use the Metropolis algorithm to draw samples from the distribution that has the form

$$\text{prob}(x) \propto \frac{1}{1 + x^4}.$$ (7.20)

This could be integrated to get the cumulative distribution and so we could get random numbers directly from the uniform distribution, as described earlier, but we can use it to illustrate the Metropolis method. As a 'proposal' we use a simple Gaussian, centred at the current value in the chain, and take the width of this Gaussian as a free parameter in our experiment. We tried two cases. One was a narrow proposal (standard deviation = 1). Here, as expected, we were usually successful in making a transition to a new number in the chain (about 60 per cent of the time). The second case used a wide proposal (standard deviation = 10). Here, the algorithm often failed to make a transition to a new number, about 90 per cent of the time. However, we suspect that these numbers are less correlated with their predecessors than in the first, apparently more efficient, case.

To check this, we form the (average) power spectra (Section 9.2) of the two chains, initially 10 000 long, discarding the first quarter of the chain and thinning by extracting only every 15th number for use. We then also repeat the experiment several times to look at the scatter in the standard deviations that can be calculated from the chain. The exact standard deviation for this distribution is just 1. The chain generated by the wide proposal shows a scatter of only 4 per cent, despite the inefficiency of making a transition. The chain from the narrow proposal shows a scatter of 18 per cent.

The power spectra (Figure 7.3) show why this happens; as expected, the chain from the narrow proposal shows a good deal of correlation, even after thinning, and this outweighs having more distinct numbers in the chain. Clearly the Metropolis algorithm is inefficient, as with 100 000 independent samples we would expect a scatter of only 1 per cent in the standard deviation. However, in many dimensions it is exponentially more efficient than simple Monte Carlo integrations.

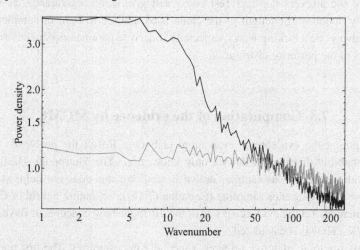

Figure 7.3 The power spectra of chains generated by a narrow proposal (black line) and a wide proposal (grey). The wavenumbers extend to the Nyquist frequency and the normalization of the power density is arbitrary.

The simplest implementation of the Metropolis algorithm is one-dimensional. What if we want random numbers from a multivariate $f(\alpha, \beta, \gamma, \ldots)$? This is a much more likely application in a Bayesian context. The arguments for the Metropolis algorithm generalize to many dimensions but a suitable proposal distribution may be harder to find.

Here we may use the *Gibbs sampler*. We guess a starting vector $(\alpha_0, \beta_0, \gamma_0, \ldots)$ and then draw α_1 from $f(\alpha_0, \beta_0, \gamma_0, \ldots)$. Next we draw β_1 from $(\alpha_1, \beta_0, \gamma_0, \ldots)$ and then γ_1 from $(\alpha_1, \beta_1, \gamma_0, \ldots)$; and so on. After we have cycled through all the variables once, we have our first multivariate sample. We may use one iteration of Metropolis–Hastings to make each of the draws, or it may happen that we can sample from the conditional distributions some other, easier way.

All the same, caution is necessary as in the 1D case, to ensure that burn-in has happened. An additional complication is that burn-in can be slowed considerably if there is correlation between the variables. It can be useful to change variables to combinations which are less correlated (e.g. approximations to principal components).

The combination of the Metropolis algorithm and the Gibbs sampler equips us to perform the multidimensional integrations we often need in Bayesian problems; for instance, we can perform marginalizations of uninteresting parameters and derive statistics such as means or percentiles, using Equation (7.16).

Are the answers correct? Test cases will give some reassurance, as will simply doubling the length of the chain used. Failure to burn in completely will always be a lurking worry, especially if derived parameters depend on the wings of the posterior distribution.

7.8 Computation of the evidence by MCMC

Computing the evidence is even more elaborate. Recall the problem: f is a probability distribution but we only know f/C. The Metropolis–Hastings algorithm will give us samples drawn from f but this does not help, as we cannot use importance sampling (Equation (7.16)): we cannot get rid of C by this method. Yet the evidence is one of the distinguishing concepts of Bayesian analysis. How is it calculated?

This is a topic of active research, especially in cosmology. The first port of call is probably 'thermodynamic integration'. This is a method which derives from an analogy to the partition function in statistical mechanics.

The evidence E is, by definition, a number which is the integral over parameters of the product of the likelihood and the prior on the parameters; we can write this in shorthand as

$$E = \int \mathcal{L}(\vec{\theta}) \, p(\vec{\theta}) \, \mathrm{d}\vec{\theta}. \tag{7.21}$$

This is exactly what we need to integrate in Bayesian applications.

An analogy from thermodynamics reminds us that we can calculate the partition function by an integration if we know the dependence of the mean energy of a system on its temperature; and the partition function looks formally like the integral we want to get the evidence. We introduce a parameter γ which is going to play a similar role to the inverse temperature in thermodynamics:

$$E(\gamma) = \int \mathcal{L}(\vec{\theta})^\gamma \, p(\vec{\theta}) \, d\vec{\theta} \qquad (7.22)$$

so that $E(0) = 1$, because our prior is normalized to unity; if $\gamma = 0$, then the above integral is simply $E(\gamma = 0) = \int p(\vec{\theta}) d\vec{\theta}$. Pursuing the analogy with physics, we expect the rate of change with γ to be interesting. It is

$$\frac{\partial \ln E(\gamma)}{\partial \gamma} = \frac{1}{E(\gamma)} \int \ln \mathcal{L}(\vec{\theta}) \, \mathcal{L}(\vec{\theta})^\gamma \, p(\vec{\theta}) \, d\vec{\theta}. \qquad (7.23)$$

The right-hand side is just the expectation of the log likelihood, with respect to the probability distribution that is proportional to $\mathcal{L}(\vec{\theta})^\gamma \, p(\vec{\theta})$. (We can see this more clearly if we write this right-hand side as $\int [\ln \mathcal{L}(\vec{\theta})][\mathcal{L}(\vec{\theta})^\gamma \, p(\vec{\theta})/E(\gamma)] d\vec{\theta}$; the expectation value of $\ln \mathcal{L}$ (first bracket) is with respect to the probability distribution of the second bracket. E must turn up in the denominator to normalize the distribution.) Thus, we have now managed to get

$$\frac{\partial \ln E(\gamma)}{\partial \gamma} = \text{average value of } \ln \mathcal{L} \text{ at a particular } \gamma,$$

and we can find this average by MCMC because we know $\mathcal{L}$ as well as the probability distribution in question, $\mathcal{L}(\vec{\theta})^\gamma \, p(\vec{\theta})$. Writing the right-hand side of (7.23) as $< \ln \mathcal{L} >_\gamma$, our solution for $E = E(\gamma = 1)$ is then

$$\ln E = \int_0^1 < \ln \mathcal{L} >_\gamma \, d\gamma. \qquad (7.24)$$

Finding the average values by MCMC thus consists of sampling from the distribution $\mathcal{L}^\gamma p$, using a Metropolis–Hastings algorithm and a Gibbs sampler, to get expectation values $< \ln \mathcal{L} >_\gamma$. In practice, we would generate chains of random numbers for a set of discrete values of γ, compute the respective values of $< \ln \mathcal{L} >_\gamma$, and perform a numerical integration of a function (say, via Simpson's Rule) fitted to these values. Piecewise integration is also possible. The process may seem elaborate, and indeed it is; however, in many cases it is actually feasible, which is not the case for brute-force integration.

To summarize: to integrate

$$E = \int \mathcal{L}(\vec{\theta}) \, p(\vec{\theta}) \, d\vec{\theta},$$

(i) create chains of random numbers (instances of the variables $\vec{\theta}$) drawn from the pdf

$$\mathcal{L}(\vec{\theta})^\gamma \, p(\vec{\theta}),$$

where γ is a parameter between 0 and 1;

(ii) use these to find the average of

$$\ln\mathcal{L}(\vec{\theta}) \equiv < \ln\mathcal{L} >_\gamma;$$

(iii) and finally do the numerical integration

$$\ln E = \int_0^1 < \ln\mathcal{L} >_\gamma d\gamma.$$

In many dimensions, and with complex likelihood functions, other methods may be needed. Now we are well into research territory, where cosmologists, in particular, are active in developing Skilling's 'nested algorithm' (MacKay, 2003; Liddle *et al.*, 2006; Skilling, 2007).

MCMC methods are elaborate and often computationally expensive. Like most numerical methods, they suffer from the problem that the right answer is not known beforehand. Particularly when used in many dimensions, it is difficult to visualize what is going on and debug code. Great care is needed to run as many checks as possible for problems where the answer is known – typically a multivariate Gaussian, which in any case will often be a good approximation to the likelihood. MC simulations (Section 2.6) can help.

Example To illustrate both the Gibbs sampler and thermodynamic integration, we will generate random numbers following a bivariate Gaussian (Figure 7.4):

$$g(x, y) \propto \exp-\frac{\gamma}{2}\left(x^2 - \frac{9}{5}xy + y^2\right). \tag{7.25}$$

The correlation coefficient should be 9/10. The inverse temperature γ, normally 1, will be used for the thermodynamic integration. The second step is to integrate to get the normalizing factor. Here, as noted, we have to use a proper prior. We used the elliptical prior

$$p(x, y) = 0 \quad \text{if} \quad 20 - \left(x^2 - \frac{9}{5}xy + y^2\right) < 0,$$

$$p(x, y) = 1/\mathcal{N} \text{ otherwise.}$$

Here $\mathcal{N}$ is defined so that

$$\int\int dx\,dy\,p(x, y) = 1. \tag{7.26}$$

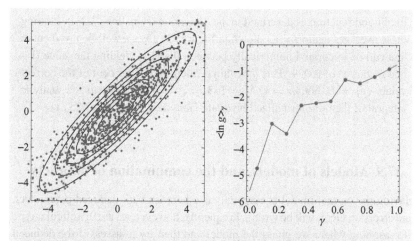

Figure 7.4 (a) The contours of the target bivariate Gaussian, with samples from the Markov chain over-plotted. (b) The values of the expectation $\ln g_\gamma$ are plotted against the 'inverse temperature' γ. Integrating the simple piecewise curve through these points yields a very satisfactory value for the evidence, or normalizing factor of the bivariate.

This prior is non-zero over the whole region where g has any appreciable magnitude; but we have to have it, as otherwise we would be smuggling in an improper prior and the results would be nonsense.

The integral of the product gp is what we will calculate using random numbers, but it can also be calculated by a straightforward numerical integration because only two dimensions are involved. If there were no prior, the integral of a multivariate Gaussian is known analytically; see Section 6.5.

The pairs of random x, y are generated by a straightforward use of the Gibbs sampler, where the required univariate proposal distribution is simply a Gaussian of standard deviation 5. This gives a success rate (at least one of the x or y being new each iteration) of about 40 per cent. A chain 100 000 long, thinned to one in a hundred, gives a correlation coefficient between x and y which is typically within 0.5 per cent of the exact value 0.9. The figure shows the theoretical contours and some of the samples from the chain. The power spectrum also looks reasonable for either x or y, being very nearly white. The variance in y is estimated by the chain to within 5 per cent.

The thermodynamic integration proceeds with a set of 10 values of γ between 0 and 1. For each value, a chain of random numbers is generated and then the average of $\ln g$ is calculated – g is playing the role of the likelihood function as described in the derivation (Equation (7.23)). Plotting these averages against γ as shown in Figure 7.4 gives a well-behaved curve that can be integrated numerically between 0 and 1, yielding the value (for one instance) of 0.099. This is within a fraction of a per cent of the correct value. ($gp = 0.099$; $p = 1/\mathcal{N} = 143 \Rightarrow g = 14.1$, close to the analytic integral of the non-normalized bivariate Gaussian, Equation (7.25).)

7.9 Models of models, and the combination of data sets

Having the correct model is essential, as otherwise both deduced parameters, and errors on them, will be wrong. Frequently, however, we are in a circular type of reasoning where we guess the model and then try to assess if the deduced parameters are reasonable. A useful way of expanding the set of models, as an insurance policy against having the wrong one, is to use *hierarchical models*. These, in turn, make use of the even more impressively named *hyperparameters*. It turns out that, in addition to helping with modelling, these notions are useful in the familiar problem of combining sets of data which have different levels of error. They are also useful in constructing fairly general priors for a Bayesian analysis, avoiding introducing large numbers of parameters. This important aspect of statistical modelling is well covered by Gelman *et al.* (2004).

The idea of the hierarchical model can be illustrated by returning to the example in Section 6.5, where we needed to include some kind of offset in the model for each of our flux measurements. Each term in the likelihood function took the form

$$\frac{1}{\sqrt{2\pi}\epsilon k f_i^{-\gamma}} \exp\left[-\frac{(S_i - (\beta + k f_i^{-\gamma}))^2}{2(\epsilon k f_i^{-\gamma})^2}\right]. \tag{7.27}$$

We are assuming that the offset error β is the same for each measurement. Before, we supposed that the distribution of β was Normal, with a known mean and standard deviation – quite a strong assumption. Suppose that we knew only the standard deviation, but the mean μ was unknown. The likelihood is then

$$\exp\left[\frac{-(\beta - \mu)^2}{2\sigma_\beta^2}\right] \prod_i \frac{1}{\sqrt{2\pi}\epsilon k f_i^{-\gamma}} \exp\left[-\frac{(S_i - (\beta + k f_i^{-\gamma}))^2}{2(\epsilon k f_i^{-\gamma})^2}\right], \tag{7.28}$$

where μ is now a hyperparameter, described (appropriately enough) by a *hyperprior*. So, for hierarchical models, Bayes' theorem takes the form

$$\text{prob}(\alpha, \theta \mid X_i) \propto \mathcal{L}(X_i \mid \alpha)\text{prob}(\alpha \mid \theta)\text{prob}(\theta), \tag{7.29}$$

where, as usual, X_i are the data and θ is the hyperparameter (and may, of course, be a vector). If we integrate out θ, we get a posterior distribution for the parameter α which includes the effect of a range of models.

Example In our radio spectrum example of Section 6.5, we make a simple hierarchical model as described above. Take the standard deviation $\sigma_\beta = \epsilon$ and the prior $\mathrm{prob}(\mu) = \mathrm{constant}$. We compute the likelihood surface by marginalizing over both μ and β; these integrations are not too bad because we have Gaussians, and because we integrate from $-\infty$ to ∞. (More realistic integrations, over finite ranges, get very messy.)

In Figure 7.5 we see the likelihood surface for K and γ, compared to the previous 'strong' model for which we knew μ. There is a tendency, not unexpected, for flatter power laws to be acceptable if we do not know much about μ.

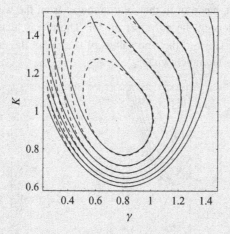

Figure 7.5   The radio spectrum parameter plane, showing the log likelihoods for the two models; the black contours are for the hierarchical model and the dashed contours are for known μ.

In a more elaborate form of a hierarchical model, we can connect each datum to a separate model, with the models being joined by an overarching structural relationship. In symbols, Bayes' theorem then reads

$$\mathrm{prob}(\alpha_i, \theta \mid X_i) \propto \mathcal{L}(X_i \mid \alpha_i)\mathrm{prob}(\alpha_i \mid \theta)\mathrm{prob}(\theta). \qquad (7.30)$$

In a common type of model we may have observations X_i drawn from Gaussians of mean μ_i, with a structural relationship that tells us that the μ_i are in turn drawn from a Gaussian of mean, say, θ. This is a weaker model than the first sort we considered, because we have allowed many more parameters, linked only by a stochastic relationship. In the case of Gaussians there is quite an industry devoted to this type of model; see Lee (2004) for details.

Example Back to our power-law spectrum. If we allow a separate offset β_i at each frequency, then each term in the likelihood product takes the form

$$\exp\left[-\frac{(\beta_i - \mu)^2}{2\sigma_\beta^2}\right] \frac{1}{\sqrt{2\pi}\,\epsilon\,kf_i^{-\gamma}} \exp\left[-\frac{(S_i - (\beta_i + kf_i^{-\gamma}))^2}{2(\epsilon\,kf_i^{-\gamma})^2}\right] \quad (7.31)$$

and we take again the usual (very weak) prior $\text{prob}(\mu) = \text{constant}$. Marginalizing out each β_i by an integration is then exactly the same task for each i, and having done this we can compare the likelihood contours with the very first model of these data (no offsets allowed). The likelihood contours of Figure 7.6 are very instructive. The hierarchical model, by allowing a range of models, has moved the solution away from the well-defined (but wrong) parameters of the no-offset model. The hierarchical likelihood in fact peaks quite close to the true values of (k, γ) but the error bounds on these parameters are much wider.

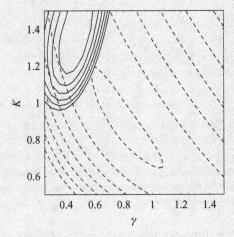

Figure 7.6   The log likelihoods for the two models; the black contours are for the simplest model, with no provision for offsets; the dashed contours are for the weak hierarchical model, allowing separate offsets at each frequency.

This is a general message; allowing uncertainty in our models may make the answers apparently less precise, but it is an insurance against well-defined but wrong answers from modelling. Similarly, the method gives us a way of making fairly general priors without too much 'expense' in parameters.

7.10 Broadening the range of models, and weights

Broadening the range of models is a useful technique in combining data. We saw how to construct a weighted mean in Section 6.4, and that the weights depend strongly on the variance in the data. The optimum weight for an observation of standard deviation σ is just $\frac{1}{\sigma^2}$. This weight turns up naturally in modelling using minimum χ^2.

For example, suppose that we have data X_i, of standard deviation σ_x, and some other data Y_i of standard deviation σ_y. Then, to fit to some model function $\mu(\alpha_1, \alpha_2, \ldots)$ we minimize

$$\chi^2 = \sum_{i=1}^{N} \frac{(X_i - \mu)^2}{\sigma_x^2} + \sum_{i=1}^{M} \frac{(Y_i - \mu)^2}{\sigma_y^2} \tag{7.32}$$

and it is obvious how the different data sets are weighted.

Quite often, the quoted error levels on data are wrong; it is no small task to make accurate error estimates. One simple way of dealing with this is simply to tinker with the σ's in the χ^2 sum, above, so that the minimum value comes out to be about equal to the number of degrees of freedom ($N + M$ less the number of free parameters). This can be a useful technique but of course it is rather arbitrary how we allocate the tinkering between σ_x and σ_y.

Let us broaden our model by allocating weights ξ_x and ξ_y to these data sets. This is a hierarchical model, and the weights are hyperparameters (Hobson *et al.*, 2002). As before, our model depends on some parameters so that $\mu = \mu(\alpha, \beta, \ldots)$. On the assumption of Gaussian residuals, the likelihood function, assuming that σ_x and σ_y are given, is then

$$\mathcal{L}(X_i, Y_i \mid \alpha, \beta, \ldots, \xi_x, \xi_y) \propto \frac{\xi_x^{N/2} \xi_y^{M/2}}{\sigma_x^N \sigma_y^M}$$

$$\times \exp\left[-\sum_{i=1}^{N} \xi_x \frac{(X_i - \mu)^2}{2\sigma_x^2} \right]$$

$$\times \exp\left[-\sum_{i=1}^{M} \xi_y \frac{(Y_i - \mu)^2}{2\sigma_y^2} \right]. \tag{7.33}$$

Bayes' theorem will now tell us the posterior probability distribution for the parameters of our model μ, plus the weights. It would be nice to marginalize out the weights, as in this context they are nuisance parameters.

The tidy aspect of this approach is that it is one of the rare cases in which we have a convincing (uncontroversial?) prior to hand. Hobson *et al.* (2002) show that, on the assumption that the mean value of the weight is unity (perhaps an idealistic assumption), we have simply

$$\text{prob}(\xi) = \exp(-\xi). \tag{7.34}$$

This is derived by the method of maximum entropy, as described in, for example, Jaynes (1968). Carrying out the integration over the ξ's is easy, and we find the posterior probability for our problem to be

$$\text{prob}(\alpha_1, \alpha_2, \ldots \mid X_i, Y_i) \propto \frac{1}{\sigma_x^N} \frac{1}{\sigma_y^M} \times \frac{1}{\left(2 + \sum_{i=1}^{N} \frac{(X_i - \mu)^2)}{2\sigma_x^2}\right)^{N/2+1}}$$

$$\times \frac{1}{\left(2 + \sum_{i=1}^{M} \frac{(Y_i - \mu)^2)}{2\sigma_y^2}\right)^{M/2+1}} \times \text{prob}(\alpha_1, \alpha_2, \ldots). \tag{7.35}$$

Example Here (Figure 7.7) are two noisy spectra of a single line. Both are alleged to have the same noise level, $\sigma = 5$, but one is slightly worse and is not centred at zero, unlike the better one.

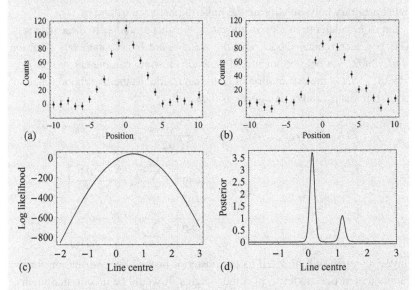

Figure 7.7   (a) and (b) The two synthetic spectra which are our input data. (c) and (d) The log-likelihood function for the combined, unweighted data (left) and the posterior distribution for the line centre, after marginalizing out the weights (right).

For simplicity, let us assume that we know the line to be Gaussian and only its position is unknown. Combining the data, taking the quoted errors at face value, we get a log likelihood for the line centre which peaks some way away from zero. If our prior on the line centre is diffuse, the posterior probability is proportional to the likelihood. Including the data weights as hyperparameters, we get a simple answer after marginalization, shown in Figure 7.7(d): the posterior probability for the line centre shows two clear peaks, the larger at zero (the good data) and the lesser at two units (the poorer data).

Since the weights are an amplification of our model, we may want to know if they ought to be included; this can be calculated in the usual way by computing the odds in favour of or against the more complex model. To do this we need to keep track of all the constants we have elided so far. Here is the full set of equations, for a Gaussian noise model for the data.

Let us index each (homogeneous) set of N_i data by i, and call the covariance matrix C_i, the data vector $\vec{X}_i$ and the model vector $\vec{\mu}_i$. μ_i depends on the parameters of interest. Abbreviating

$$\chi_i^2 = (\vec{X}_i - \vec{\mu}_i)^T C_i^{-1} (\vec{X}_i - \vec{\mu}_i) \tag{7.36}$$

the Gaussian model noise for the ith data set is, as usual,

$$\text{prob}(\vec{X}_i \mid \vec{\mu}_i, \text{no weights}) = \frac{1}{(2\pi)^{N_i/2} \mid C_i \mid^{1/2}} \exp\left(-\frac{1}{2}\chi_i^2\right) \text{prob}(\vec{\mu}_i). \tag{7.37}$$

Introducing a weight ξ_i, and the exponential prior, gives

$$\text{prob}(\vec{X}_i \mid \vec{\mu}_i, \xi) = \frac{\xi_i^{N_i/2}}{(2\pi)^{N_i/2} \mid C_i \mid^{1/2}} \exp\left(-\frac{\xi_i}{2}(\chi_i^2 + 2)\right) \text{prob}(\vec{\mu}_i). \tag{7.38}$$

The Gaussian model for the ith data set, after marginalizing over the weight parameter with respect to the exponential prior, is just

$$\text{prob}(\vec{X}_i \mid \vec{\mu}_i, \text{weights}) = \frac{2\Gamma(\frac{N_i}{2} + 1)}{\pi^{N_i/2} \mid C_i \mid^{1/2}} \left(\frac{1}{2 + \chi_i^2}\right)^{N_i/2+1} \text{prob}(\vec{\mu}_i). \tag{7.39}$$

Each of these distributions depends on the parameters of the model. The odds in favour of weighting the data entail integrating over the parameters (let us abbreviate this by $\int_\alpha$), taking account of any priors prob(α), and then forming the ratio

$$\frac{\int_\alpha \text{prob}(\alpha)\text{prob}(\vec{X}_i \mid \vec{\mu}_i, \text{ weights})}{\int_\alpha \text{prob}(\alpha)\text{prob}(\vec{X}_i \mid \vec{\mu}_i, \text{ no weights})}. \tag{7.40}$$

7.11 Press and Kochanek's method

Press (1997) (together with Kochanek) gives a very nice Bayesian solution to the problem of combination of observations. Its key feature is that it builds in the probability of the ith estimate being 'bad', in the specific sense that it was actually drawn from a distribution with a much bigger standard deviation than the quoted one. Discounting these bad measurements means that we can find a posterior estimate for the weighted mean μ_w that is relatively insensitive to poorly estimated standard deviations. (Notice that the weighted mean of the previous section is a statistic, being only a function of the data; here we are trying to find the posterior distribution of the parameter μ_w.)

The reasoning is as follows (see Exercise 7.8). If the ith estimate is 'good', we suppose it to be drawn from a distribution we denote prob($X_i \mid \mu_w, \sigma_i$); if 'bad' it is drawn from prob($X_i \mid \mu_w, \sigma_i'$). Here we suppose that the two distributions are the same (say Gaussian) but $\sigma_i' > \sigma_i$ according to some prescription we will discuss later.

We also suppose that there is some probability p that any one of the available estimates is good. All estimates are on the same footing in this regard, so a priori we treat them as equally likely to be good.

The status of the entire collection of estimates can be summarized in a vector $\vec{v}$, where $v_i = 1$ if an estimate is good and $v_i = 0$ if an estimate is bad. Given p we can compute the prior probability of the various possible $\vec{v}$'s from the binomial distribution; for example, the prior probability that all the estimates are good is p^N.

Given all this, a neat argument shows that the likelihood function for this set of estimates is

$$\mathcal{L}(X_1, X_2, \ldots, X_N \mid \mu_w, p)$$
$$= \prod_i \big(p \, \text{prob}(X_i \mid \mu_w, \sigma_i) + (1 - p) \, \text{prob}(X_i \mid \mu_w, \sigma_i') \big). \tag{7.41}$$

As usual, the likelihood is a product of terms (and we are assuming that the X_i are independent here). Each term is a weighted combination of the chances of the estimate being drawn from the alleged distribution, or from one which has a bigger standard deviation.

Bayes' theorem then gives us the posterior distribution for μ_w, marginalized over the prior on p in the usual way

$$\text{prob}(\mu_w \mid X_1, X_2, \ldots, X_N)$$
$$\propto \int dp \; \mathcal{L}(X_1, X_2, \ldots, X_N \mid \mu_w, p) \, \text{prob}(\mu_w) \text{prob}(p). \quad (7.42)$$

The normalization follows by integrating over the prior on μ_w. A couple of other interesting posterior distributions can be derived as well. The posterior probability for p is given by a different integration:

$$\text{prob}(p \mid X_1, X_2, \ldots, X_N)$$
$$\propto \int d\mu_w \; \mathcal{L}(X_1, X_2, \ldots, X_N \mid \mu_w, p) \, \text{prob}(\mu_w) \text{prob}(p). \quad (7.43)$$

This is the posterior *distribution* of the probability, given the data and the priors, that any particular estimate will be good. More usefully, we can also calculate the posterior probability (a number, not a distribution) that the kth estimate is good; this is equivalent to the posterior probability that $v_k = 1$ and the other components of $\vec{v}$ are not specified. Defining an ancillary likelihood term as

$$\mathcal{L}_k(X_1, X_2, \ldots, X_N \mid \mu_w, p) = p \, \text{prob}(X_k \mid \mu_w, \sigma_k)$$
$$\times \prod_{\substack{i \\ i \neq k}} \left(p \, \text{prob}(X_i \mid \mu_w, \sigma_i) + (1 - p) \text{prob}(X_i \mid \mu_w, \sigma'_i) \right)$$

we get

$$\text{prob}(i\text{th good}) = \frac{\int d\mu_w \, dp \, \mathcal{L}_k(X_1, X_2, \ldots, X_N \mid \mu_w, p) \, \text{prob}(\mu_w) \text{prob}(p)}{\int d\mu_w \, dp \, \mathcal{L}(X_1, X_2, \ldots, X_N \mid \mu_w, p) \, \text{prob}(\mu_w) \text{prob}(p)}.$$
$$(7.44)$$

Example We now apply this method to some data from astronomers' long struggle with the Hubble constant H_0. Here are 15 estimates, selected at random from John Huchra's (www.cfa.harvard.edu/~dfabricant/huchra/hubble/) compilation. The units are km s^{-1} Mpc^{-1}.

Estimate	Standard deviation	Estimate	Standard deviation
51	5	68	3
55	5	69	7
58	17	69	8
59	8	71	6
60	10	73	4
61	11	81	10
62	8	84	4
62	5		

The straightforward weighted mean of these estimates is 67 with a standard deviation of 8. We now assume (i) that the estimates are independent (actually they might well share systematic errors) and (ii) that they are drawn from either a Gaussian of the stated standard deviation (a good measurement) or from one of twice the quoted standard deviation (a bad measurement). So we have

$$\text{prob}(X_i \mid H_0, \sigma_i) = \frac{1}{\sqrt{(2\pi)}\sigma_i} \exp\left(-\frac{(X_i - H_0)^2}{2\sigma_i^2}\right) \tag{7.45}$$

for good measurements and

$$\text{prob}(X_i \mid H_0, \sigma_i') = \frac{1}{\sqrt{(2\pi)}2\sigma_i} \exp\left(-\frac{(X_i - H_0)^2}{8\sigma_i^2}\right) \tag{7.46}$$

for bad measurements. (We have chosen to make the true standard deviation of bad measurements twice the quoted value, but for a set with more uniform standard deviations we might just choose a single large value.) Finally, we select priors; we have taken the prior for p to be uniform between zero and unity (perhaps a pessimistic view of the community's ability to get its observations right), and the prior for H_0 to be uniform between 40 and 100 km s^{-1} Mpc^{-1}.

Figure 7.8 shows the results. The combination of all the data gives quite a narrow posterior distribution for H_0. It has a very similar mean to the standard weighted mean estimate but is considerably narrower; the method has de-weighted the bad observations more strongly than their error estimates would suggest, and the estimate improves as a result. (In other cases, the bad measurements are not so symmetrically distributed, and the discrepancy between the weighted mean and the Press–Kochanek estimate gets much bigger.)

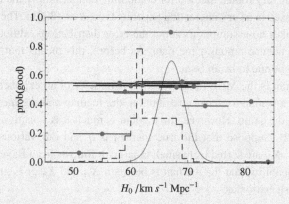

Figure 7.8   Posterior estimates of H_0 from the trial data set. The solid line is the unnormalized posterior distribution of H_0. The filled circles are the estimates of H_0, with quoted error bars indicated; the vertical axis gives the probability that each of these data is good. The dashed line is the distribution for the median. The point at prob(good) = 0.9 is the ordinary weighted mean.

Of interest are the posterior probabilities for the observations to be good. (Recall that this means being consistent with their quoted errors.) Overall the numbers are quite low. This is because a number of estimates are very discordant and this pulls down the posterior p – in other words, overall, astronomers had difficulty getting good error estimates for their determinations of H_0. This is not exactly news but the actual numbers are interesting; also of interest is the evidence that three of the estimates were much too optimistic about their errors than the bulk of the data.

Some other interesting experiments can be tried that confirm the usefulness of this approach. For example, other random selections from the Huchra list give very similar results: the method is robust. We may well suspect that a Gaussian is optimistic in this problem. Replacing it in the likelihoods by an exponential distribution, which has broader wings, indeed considerably boosts the probabilities of most of the observations being good. Finally, for the prior on p we may adopt the considerably more dogmatic Haldane prior (Section 2.3), which tends to insist that answers are either right or wrong. This changes the posterior for H_0 very little, and somewhat diminishes the probabilities of the five outlying estimates being good. All this is in accord with common sense.

7.12 Median statistics

A simple but very robust method for combining data is simply the median. Its use is discussed in a very interesting paper by Gott *et al.* (2001). The advantage is that it makes no assumptions about the error distributions, although it does need independence amongst the data. As before, this means that systematic error will continue to be an issue.

The median is the MLE of the location parameter of an exponential distribution. For a symmetric error distribution, the median and the mean will, on average, be the same. However, the median is much less sensitive to outliers (Section 3.2). Suppose that the true median is q and our various estimates $X_1, X_2, \ldots, X_N$ are sorted from small to large. It follows from Bayes' theorem that the probability that the median is between X_m and X_{m+1} is given by the binomial distribution as

$$\text{prob}(X_m < q < X_{m+1}) = \binom{N}{N-m} \frac{1}{2}^m \frac{1}{2}^{N-m}. \qquad (7.47)$$

The results for the trial data set of Hubble constants are shown in Figure 7.8. The median value for the data is $62 \text{ km s}^{-1} \text{ Mpc}^{-1}$, and the range of likely values for the true median spans about the same range as the posterior distribution from the Press and Kochanek method. The most probable values differ quite appreciably, however. This is because the Press method is making use of extra information; the median method treats all data equally. Interestingly, the median method is sceptical about the same five observations as the Press and Kochanek method.

Exercises

7.1 Goodness of fit. A goodness-of-fit statistic, for example χ^2, has a significance level p defined by

$$p = \int_C^\infty f(\chi^2) \, d\chi^2,$$

where C is the observed value of χ^2 and F is the appropriate distribution of χ^2. Show that p is uniformly distributed if C is indeed drawn from f.

7.2 Monte Carlo integration. The Gaussian or Normal distribution function $(1/\sigma\sqrt{2\pi}) \exp[-x^2/2\sigma^2]$ does not have an analytic integral form. Use Monte Carlo integration to find erf, the so-called error function of

Table B.1. Show that (a) approximately 68 per cent of its area lies between $\pm\sigma$, and that (b) the total area under the curve is unity.

7.3 Metropolis algorithm. Use the Metropolis algorithm to draw samples from a simple distribution, say the Laplacian $\exp - |x|$. Check the match to the target distribution along the chain, using a selection of non-parametric tests (e.g. Kolmogorov–Smirnov).

7.4 Metropolis algorithm continued. Using the chains developed in Exercise 7.3, test the amount of correlation in the chain against various choices of the proposal distribution. Use the standard Fourier methods (Section 9.2) to do this, as outlined in this chapter. Experiment with direct calculation of the correlation coefficient between elements of the chain separated by various amounts.

7.5 Burn-in. Using the chains of Exercise 7.3, make a direct check of the duration of burn-in by comparing with within-chain and between-chain variances.

7.6 Bayes factor. A very common example of the issues around model complexity is given by the simple polynomial fit: adding another term to a polynomial often gives a better fit (smaller chi-square per degree of freedom) but introduces an additional parameter. Use some random data (say, 20 numbers drawn from a Gaussian distribution) and the Bayes factor method to examine the odds in favour of adding one more polynomial term. Test that the Laplace approximation can be used to avoid a multidimensional numerical integration.

7.7 Markov chain Monte Carlo. Extend the previous exercise to compute the actual posterior probabilities for three possible polynomials from a constant through to a quadratic. In this case the evidence will have to be calculated: use an MCMC method to compute the evidence by a thermodynamic integration.

7.8 Consistency of observations. Refer to Press's paper (Press, 1997) to derive the formula quoted in the text for the probability of a given observation being 'good'.

8

Detection and surveys

Watson, you are coming along wonderfully. You have really done very well indeed. It is true that you have missed everything of importance, but you have hit upon the method.

(Sherlock Holmes in 'A Case of Identity', Sir Arthur Conan Doyle)

By a small sample we may judge of the whole piece.

(Don Quixote, Miguel de Cervantes)

'Detection' is one of the commonest words in the practising astronomer's vocabulary. It is the preliminary to much else that happens in astronomy, whether it means locating a spectral line, a faint star or a gamma-ray burst. Indeed, of its wide range of meanings, here we take the location, and confident measurement, of some sort of feature in a fixed region of an image or spectrum.

When a detection is obvious to even the most sceptical referee, statistical questions usually do not arise in the first instance. The parameters that result from such a detection have signal-to-noise ratio so high that the detection finds its way into the literature as fact. However, elusive objects or features at the limit of detectability tend to become the focus of interest in any branch of astronomy. Then, the notion of detection (and non-detection) requires careful examination and definition.

Non-detections are especially important because they define how representative any catalogue of objects may be. This set of non-detections can represent vital information in deducing the properties of a population of objects; if something is never detected, that too is a fact, and can be exploited statistically. Every observation potentially contains information. If we are resurveying a catalogue at some new wavelength, each observation constrains the energy from the object to some level. Likewise, surveying unmapped regions of sky yields information even when there are apparently no detections. In both cases population properties can be extracted, even though individual objects remain obscured in the fog of low signal-to-noise ratio.

This chapter will examine detection, first in the context of the use to which we will put detected objects; it moves on to consider the usefulness of non-detections in deducing properties of populations; and finally it examines notions of detection which say little about individual objects, but which focus instead on population-level properties. In many experiments, we wish to define wide distributions of widely spread parameters: the initial mass function, luminosity function and so on. We may approach these from the point of view of 'detections' and 'non-detections' (the catalogue point of view) or we may attempt to extract the distributions directly from the data, without the notion of detection ever intruding.

8.1 Detection

Detection is a model-fitting process. When we say 'We've got a detection' we generally mean 'We have found what we were looking for'. This is obvious enough at reasonable signal to noise. In examining a digital image, for example, detection of stars (point-like objects) is achieved by comparing model point-spread functions with the data. In the case of extended objects, a wider range of models is required to capture the possibilities.

In all cases, a clear statistical model is required. The noise level (or expected residuals from the model) may be expected in many cases to follow Poisson ($\sqrt{N}$) statistics, or, for large N, Gaussian statistics. The statistics depend on more than the physical and instrumental models. How were the data selected for fitting in the first place? We will see, for example, that picking out the brightest spot in a spectrum (Section 9.5.1) means that we have a special set of data. The peak pixel, in this case, will follow the distribution appropriate to the maximum value of a set of, say, Gaussian variables. Adjacent pixels will follow an altogether less well-defined distribution; Monte Carlo simulation may be the only way forward.

Indeed, much evaluation of detection is done with simulation. 'Model sources' are strewn on the image or spectrum, and the reduction software is given the job of telling us what fraction is detected. These essential large-scale techniques are very necessary for handling the detail of how the observation was made. Evaluating detection level in radio-astronomy synthesis images is an example. The noise level at any point depends at least on gains of all antennae, noise of each receiver, sidelobes from whatever sources happen to be in the field of view, map size, weighting/tapering parameters, the ionosphere, cloud, ... Modelling all this is not just impossible from a computational point of view; vital input data simply are not known. Although complex and varied issues are involved, the basic notions and algorithms of detection remain just as relevant as in apparently simpler cases.

The basic problem from a statistical point of view is the problem of modelling, as discussed in the last chapter. A full Bayesian approach is desirable but computationally intensive and possibly not practical in a large surveying application. An entry point to the Bayesian literature on this subject is Hobson & McLachlan (2003).

We may need a simpler method, and a classical approach is useful. Firstly, we have to ask: what do we really want from the survey we are planning? Are we more concerned with detecting as much as possible (*completeness*) or are we more worried about false detections (*reliability*)? Moreover, we need to know what we want to do with the 'detections' once we have them. Perhaps we should publish, in a catalogue, the complete set of posterior probabilities, at each location, of the observed parameters? Or just the covariance matrix, as an approximation? Or perhaps the marginalized signal-to-noise ratio, integrating away all 'nuisance' parameters? Scientific judgement must be used to answer these questions. The more information we catalogue, the better; and in the Internet age, this is so inexpensive as to be almost mandatory.

From the classical point of view, if we are trying to measure a parameter α, then the likelihood sums up what we have achieved: $\mathcal{L} = \text{prob}(\text{data} \mid \alpha)$. To be specific, suppose that α is a flux density and we wish to set a flux limit for a survey. We are only going to catalogue detections when our data exceed this limit s_{lim}. (Other quantities of astrophysical interest may need a somewhat different formulation, but the essential points remain the same.) Two properties of the survey are useful to know.

(i) The false-alarm rate is the chance that pure noise will produce data above the flux limit:

$$\mathcal{F}(\text{data}, s_{\text{lim}}) = \text{prob}(\text{data} > s_{\text{lim}} \mid \alpha = 0). \quad (8.1)$$

The *reliability* is $1 - \mathcal{F}$, i.e. $\mathcal{F} = 5/100$ gives 95 per cent reliability. This may sound good, but note that it is the infamous 2σ result.

(ii) The *completeness* is the chance that a measurement of a real source will be above the flux limit:

$$\mathcal{C}(\text{data}, s_{\text{lim}}, S) = \text{prob}(\text{data} > s_{\text{lim}} \mid \alpha = S). \quad (8.2)$$

These notions go back as least as far as Dixon & Kraus (1968); an interesting recent treatment is by Saha (1995). There is a close relationship too with the standard Neyman–Pearson method of hypothesis testing (Section 5.1).

We would like to set the flux limit to maximize the completeness, and minimize the false-alarm rate. But higher completeness (or even complete completeness, $s_{\text{lim}} = 0$!) comes at the price of an increasing number of false detections. Moreover, this definition of completeness only takes account of statistical effects. There may be other reasons for missing objects, poor recognition algorithms in particular, which introduce systematic error and bias.

Example Suppose our measurement is of a flux density s and the noise on the measurement is Gaussian, of unit standard deviation. The source we are observing has a 'true' flux density of s_0, measured in units of the standard deviation. We then have

$$\text{prob}(s \mid s_0) = \frac{1}{\sqrt{2\pi}} \exp\left[-\frac{(s - s_0)^2}{2}\right]$$

for the probability density of the data, given the source; and

$$\text{prob}(s \mid s_0 = 0) = \frac{1}{\sqrt{2\pi}} \exp\frac{-s^2}{2}$$

for the probability density of the data when there is no source. Suppose further that our survey limit is s_{lim}. Integrating these two functions from s_{lim} to ∞ (Table B.1) makes it easy to plot up the completeness against the false-alarm rate, taking the survey limit as a parameter. The methodology is shown in Figure 8.1, and the results appear in Figure 8.2. High completeness does indeed go hand in hand with a high false-alarm rate.

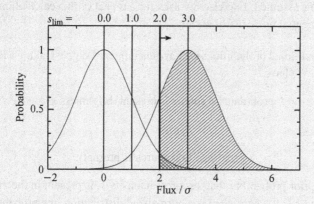

Figure 8.1 Completeness versus false-alarm rate: a sample calculation. Here we take the true flux $s_0 = 3$ (in units of the noise standard deviation), and we focus on a survey with completeness limit of 2 units. The Gaussian centred on 0 units is pure noise, and it pokes its tail beyond the survey limit of 2; this cross-hatched area is the false-source region, totalling 2.3 per cent. The Gaussian centred on 3 units is the true flux modified by noise of unit standard deviation. Its low-end tail falls below our survey limit of 2 units and such objects are lost; the hatched + cross-hatched area shows the proportion of objects obtained at or above the survey limit – the completeness – as 84.1 per cent. The result is the point on the $s_0 = 3$ curve of Figure 8.2 at $s_{\text{lim}} = 2$ at $(0.023, 0.841)$.

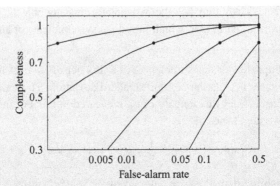

Figure 8.2   Completeness versus false-alarm rate: the calculation plotted for source flux densities in terms of σ_{noise} ranging from 1 unit (right) to four units (left). The survey limits s_{lim} are indicated by the dots, starting at zero on the right and increasing by one unit at a time. For example, a 4σ source and a 2σ flux limit give a false-alarm rate of 2 per cent and a completeness of 98 per cent – with the Gaussian noise model.

However, there seem to be quite satisfactory combinations for flux limits and source intensities of just a few standard deviations. In real life no one would believe this, mainly because of outliers not described by the Gaussians assumed. Exercise 8.4 asks for a repeat of this calculation using an exponential noise distribution.

The conditional probabilities we have encountered suggest taking a Bayesian approach. We have

$$\text{prob}(\text{data} \mid \text{a source is present, brightness } s)$$

and

$$\text{prob}(\text{data} \mid \text{no source is present}).$$

Take the prior probability that a source, intensity s, is present in the measured area to be $\epsilon N(s)$, where $N(s)$ is a normalized distribution, the probability that a single source will have a flux density s. The prior probability of no source is $(1 - \epsilon)\delta(s)$; δ is a Dirac delta function. Then the posterior probability

$$\text{prob}(\text{a source is present, brightness } s \mid \text{data})$$

is given by

$$\frac{\epsilon\,\text{prob}(\text{data} \mid s)N(s)}{\epsilon \int \text{prob}(\text{data} \mid s)N(s)\,ds + (1 - \epsilon) \int \text{prob}(\text{data} \mid s = 0)}.$$

Integrating this expression over s gives the probability that a source is present, for given data.

Example Pursuing the previous example, take the noise distribution to be Gaussian and take the prior $N(s)$ to be a simple uniform distribution from zero to some large flux density – a very uninformative prior! The value of ϵ reflects our initial confidence that a source is present at all, and so in many cases will be small. Figure 8.3 shows that the posterior distribution of flux density s peaks at the value of the data, as expected; the role of ϵ is to suppress our confidence of a detection in low signal-to-noise cases. Again, we see that for Gaussian noise, 4σ data points mean detection with high probability. Real life is more complicated.

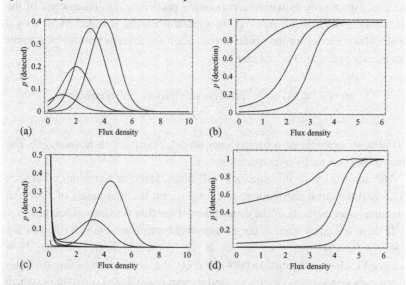

Figure 8.3 Plot (a) shows the probability of a detected source of flux density s; the curves correspond to measurements of 1 to 4 units (as before, a unit is one noise standard deviation). A prior $\epsilon = 0.05$ was used. On (b) these curves are integrated to give the probability of detection at any positive flux density, as a function of the data values; the curves are for $\epsilon = 0.5, 0.05$ and 0.005. Plots (c) and (d) show the results of the calculation for the power-law prior, truncated at 0.1 unit.

Using a power-law prior $N(s) \propto s^{-5/2}$ gives results rather similar to the example of Figure 2.8, which ignored the possibility that no source might be present. As we noted in the previous example, the rarity of bright sources in this prior now means that we need a much better signal to noise to achieve the same confidence that we have a detection.

A Bayesian treatment of detection gives a direct result; from the figures in the previous example, we may read off a suitable flux limit that will give the desired probability of detection. This is affected by the prior on the flux densities, but often we will have a robust idea of what this should be from previous survey parameters such as source counts or number–magnitude relations.

In many cases, however, the notion of detection of individual objects is poorly defined. Images or spectral lines crowd together, even overlap as we reach fainter and fainter. Within the region we measure, several different objects may contribute to the total flux. Even if only one object is present, if the source count $N(S)$ or number–magnitude $N(m)$ relation is steep, it will be more likely that the flux we measure results from a faint source plus a large upward noise excursion, rather than vice versa – see the following section. In these cases we can expect only to measure population properties, i.e. parameters of the flux-density distribution $N(s)$. If these parameters are denoted by α, then a probabilistic model for the observations, when the average number of sources per measurement area is less than 1, is

$$\text{prob}(\alpha \mid \text{data}) \propto \sum_s \text{prob}(\text{data} \mid s)\text{prob}(s \mid N, \alpha)\text{prob}(\alpha).$$

(This is an example of a hierarchical model, discussed in Section 7.9. The quantities α are really hyperparameters.)

The summation in this equation will often denote a convolution between $N(s)$ and the error distribution; given a prior on the parameters of s we can obtain a better estimate of the distribution of the flux densities of sources.

If there are many sources per measurement area (and this will often be the case for faint sources) then we are in the 'confusion-limited' regime. Now we need to draw a distinction between $N(s)$, the distribution of flux densities when only one source contributes, and a more complicated distribution which takes account of the possibility that several sources may add up to give s. This complicated situation is considered in Section 8.6; the details for the simpler case are left to Exercise 8.2, and they are very similar to the previous examples.

In summary, detection is a modelling process; it depends on what we are looking for, and how the answer is expressed depends on what we want to do with it next. The simple idea of a detection, making a measurement of something that is really there, only applies when signal to noise is high and individual objects can be isolated from the general distribution of properties. At low signal to noise, measurements can constrain population properties, with the notion of 'detection' of individual objects disappearing.

8.2 Catalogues and selection effects

Typically, a body of astronomical detections is published in a catalogue. On the basis of some clear criteria, objects will either be listed in the catalogue, or not. If they are not, usually we know nothing more about them; they are simply 'below the survey limit'.

Most astronomical measurements are affected by the distance to the object. In Euclidean space, a proper motion, for a fixed velocity of the star, becomes a smaller angle inversely as the distance to that star. Apparent intensity drops off as the square of the distance. Other effects may be more subtle; the ellipticity of a galaxy becomes harder to detect, depending on distance, the blurring effect of seeing, and the detailed luminosity profile of the galaxy. The common factor in all these examples is that we measure a so-called *apparent* quantity X and infer an *intrinsic* quantity by a relationship $Y = f(X, R)$ where R is the distance to the object in question. The function f may be complicated, for observational reasons and also because it may depend on a distance involving redshift and details of space-time geometry.

We take a simple and definite case (remembering that the principles will apply to the whole range of functions f). We observe a flux density S and infer a luminosity L given by

$$L = SR^2;$$

we are considering a flat-space problem. The smallest value of S we are prepared to believe is S_{lim}; if a measurement is below this limit, the corresponding object does not appear in our catalogue. (As usual, we use upper-case letters to denote measured values of the variable written in lower case.)

Our objects (call them 'galaxies') are assumed to be drawn from a *luminosity function* $\rho(l)$, the average number of objects near l per unit volume. Using only our catalogue set of measurements $\{L_1, L_2, \ldots\}$, however, we will not be able to reproduce ρ at all. Instead, we will get the *luminosity distribution* η, where

$$\eta(l) \propto \rho(l)V(l). \tag{8.3}$$

Crucially, $V(l)$ is the volume within which sources of intrinsic brightness l will be near enough to find their way into our catalogue. We get

$$\eta(l) \propto \rho(l)\left(\frac{l}{S_{\text{lim}}}\right)^{3/2}. \tag{8.4}$$

Obviously η will be biased to higher values of luminosity than ρ. This sort of bias occurs in a multitude of cases in astronomy, and is often called *Malmquist bias*.

Example We return to our toy universe. To each of the 10^6 objects in this universe we assign a luminosity according to a power-law luminosity function of slope -3. The minimum luminosity is unity. We retain our 10^6 'distances' to the objects, the r_i chosen so that they are uniformly distributed out to $R_{max} = 1.0$. It is readily shown that the normalized luminosity function in this universe is $\rho(l)dl = [2 \cdot 10^6/(4\pi/3)]l^{-3}dl$. We now conduct a survey, calculating a flux from each object to be $s_i = L_i/r_i^2$, and setting a survey limit of $s_{lim} = 200$ units. We find some 1400 objects above the survey limit, and the plot of luminosity versus distance is shown in Figure 8.4(a). In Figure 8.4(b) we show the distribution of the luminosities of the objects found in the survey. The law shown is $l^{-3/2}$, as indicated by Equation (8.4); however, in one of power-law's tricks, it appears as $l^{-1/2}$, because the plot is a histogram binned in $\Delta\log_{10} l$, and $d(\log_{10} l) = (1/2.3026)\,d(\ln(l)) = (1/2.3026)\,l^{-1}dl$.

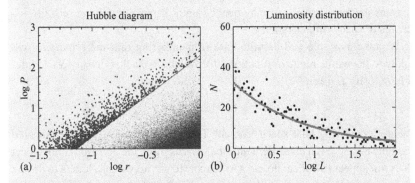

Figure 8.4   (a) The luminosity distance diagram for our toy universe with luminosities, distances and a survey limit as described above. All 10^6 sources are plotted as very faint dots; the majority are, of course, crowded into the bottom right corner, relatively low luminosities predominating from our chosen steep luminosity function, and larger volumes increase numbers at larger distances. Filled circles are used for objects which are 'detected' – some 1400 out of the 10^6 objects in the universe. (b) The luminosity distribution of these objects, the histogram of luminosities of the objects detected in the survey.

If we choose a different and perhaps more realistic form of the luminosity function, we get the correspondingly different luminosity distribution, but still in accord with Equation (8.4), as shown in the following example. (We note that other factors, most notably cosmic evolution, modify observed luminosity distributions, particularly in the radio regime where the change from anticipated forms such as shown in these two examples is extreme.)

Example The luminosity function of field galaxies is well approximated by the Schechter function

$$\rho(l) \propto \left(\frac{l}{l_*}\right)^{\gamma} \exp\left(-\frac{l}{l_*}\right),$$

in which we take $\gamma = 1$ and $l_* = 10$ for illustration. To obtain the form of the luminosity distribution in a flux-limited survey, we multiply the Schechter function by $l^{3/2}$. The differences between the luminosity function and luminosity distribution are shown in Figure 8.5.

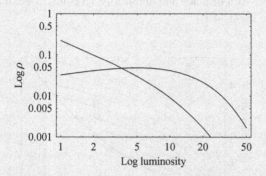

Figure 8.5 The luminosity function ρ (steep curve) and the (flat-space) luminosity distribution are plotted for the Schechter form of the luminosity function.

Malmquist bias is a serious problem in survey astronomy. The extent of the bias depends on the shape of the luminosity function, which may not be well known. More seriously, the bias will also be present for objects whose properties correlate with something that is biased. For example, the luminosity of giant HII regions is correlated with the luminosity of the host galaxy, so that any attempt to use the HII regions as standard candles will have to consider the bias in luminosity of the hosts.

Malmquist bias arises because intrinsically bright objects can be seen within proportionately much greater volumes than small ones. Because most of the volume of a sphere is at its periphery, it follows that in a flux-limited sample the bright objects will tend to be further away than the faint ones – there is an in-built distance–luminosity correlation. See Figure 8.7 and discussion.

Example We adopt a Schechter function with $\gamma = 1$ and $l_* = 10$ for the purposes of illustration. The probability of a galaxy being at distance R is proportional to R^2, in flat space. The probability of it being of brightness *l* is proportional to the Schechter function. The probability of a galaxy of luminosity L, located at distance R being in our sample is

$$\text{prob(in sample)} = 1 \quad L < S_{\text{lim}} R^2$$
$$= 0 \quad \text{otherwise.}$$

The product of these three probability terms is the bivariate distribution prob(l, r), the probability of a galaxy of brightness l and distance r being in our sample. This distribution is shown in Figure 8.6; there is a clear correlation between distance and luminosity. (It is this effect that produces diagrams like Figure 4.1.)

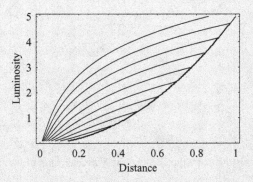

Figure 8.6   Contour plots of the bivariate prob(l, r). The contours are at logarithmic intervals; galaxies tend to bunch up against the selection line, leading to a bogus correlation between luminosity and distance.

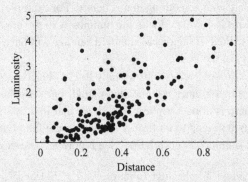

Figure 8.7   Results of a simulation of a flux-limited survey of galaxies drawn from a Schechter function.

A direct check of this is to simulate a large spherical region filled with galaxies whose luminosities are drawn from a Schechter function, and then select a flux-limited sample. (The Schechter function has to be truncated at $l > 0$ as it otherwise cannot be normalized.) Figure 8.7 shows the effect indicated by the contours of Figure 8.6.

The luminosity–distance correlation is widespread, insidious and very difficult to unravel. It means that for flux-limited samples, intrinsic properties correlate with distance; thus two unrelated intrinsic properties will appear to correlate because of their mutual correlation with distance. Plotting intrinsic properties – say, X-ray and radio luminosity – against each other will be very misleading. Much further analysis is necessary to establish the reality of correlations, or (more generally) statistical dependence. Such analyses may require detailed modelling of the detection process. Take the case of measuring the ellipticity of galaxies – distant ones may well look rounder because of the effects of seeing. As more distant galaxies seem to be more luminous as well, we are on course for deducing that round galaxies are more luminous or vice versa. A detailed model will be necessary to establish the relationship between true ellipticity, measured ellipticity and the size of the galaxy relative to the seeing disc.

Example We take the same simulation as before, but attribute two luminosities to each galaxy, drawn from different Schechter functions. These might be luminosities in different colour bands, for example, and by definition are statistically independent. If we construct a flux-limited survey in which a galaxy enters the final sample only if it falls above the flux limit in both bands, we see in Figure 8.8 that a bogus but convincing correlation emerges between the two luminosities.

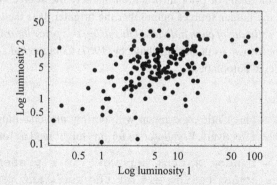

Figure 8.8   Results of a simulation of a flux-limited survey of galaxies, where each galaxy has two statistically independent luminosities associated with it.

An effect that competes with Malmquist bias is caused by observational error. The number of objects as a function of apparent intensity $N(s)$, the number counts or source counts, usually rises steeply to smaller values of s – there are many more faint objects than bright ones. In compiling a catalogue, we in effect draw samples from the number count distribution, forget those below s_{lim}, and convert the retained fluxes to luminosities. The effect of observational error is to convolve the number counts with the noise distribution. Because of the steep rise in the number counts at the faint end, the result is to contaminate the final sample with an excess of faint objects. (The generally steep counts ensure that an object of observed apparent flux density is much more likely to be a faint source with a positive noise excursion than a bright source with a negative excursion.) This can severely bias the deduced luminosity function towards less-luminous objects. This is *Eddington bias* (Eddington, 1913), a bias (Figure 8.9) rediscovered at irregular intervals to this day in different survey wavebands, either accidentally or analytically (Jauncey, 1967; Mihalas & Binney, 1981; Jensen *et al.*, 1998; Coppin *et al.*, 2006). The effect is present to a greater or lesser extent in every sensitivity limited survey. (It does not occur if the observational error is a constant fraction of the flux density, and the source counts are close to a power law.)

There are two approaches to 'correct' for the effect, depending on the aim of the experiment: the count itself may be corrected, or each individual flux may be corrected. The choice goes to the heart of the frequentist versus Bayesian debate. The frequentist would say that the best and objective measure of the flux comes from repeated measurements and appropriate combination of these; this is the information you have in the experiment. When you have measured all the fluxes, you can form a count, a secondary process, and *it is this count to which the bias correction can be applied*. The Bayesian approach is to correct the fluxes on the basis of prior information, namely the source count telling you just how the fainter sources outnumber the brighter. It is then possible to make *a better estimate of each individual flux using this prior information*. This has come to be known as debiasing (Jauncey, 1967; Coppin *et al.*, 2006) – the count is corrected automatically.

Example A Monte Carlo experiment with our toy universe illustrates the Eddington bias effectively. Pseudo-code for this might read as follows:

```
(i) Draw a random x from a power law representing a
    source count (using either the cumulant method or
    the inverse-function method) and record in a list X.
```

(ii) Draw a random y from a single Gaussian representing
the measurement error.

(iii) Find the observation $z = x + y$ and record in a list Z.

(iv) Repeat steps 1 to 3 as often as you like, say 10^6 times.

(v) Sort the data lists X and Z into a histogram with bins
of a convenient width.

(vi) Plot both up in log-log form and estimate the power-
law slopes. In the case of X, check that you get out
the power law you put in.

Repeat, varying the input power law and the noise (the σ of the Gaussian),
to get a feel for the bias that the noise introduces. The results of one such
experiment are shown in Figure 8.9.

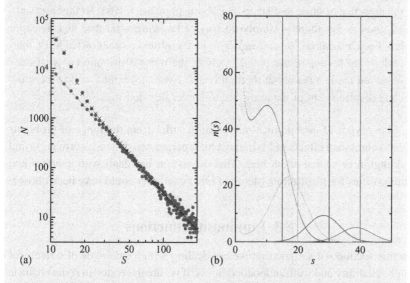

(a)   (b)

Figure 8.9   (a) The dots on the straight line show a source count (number of
sources on the 'sky' per interval of 2 units in flux), from error-free measurements, with each object in our toy universe ascribed a luminosity $L = 1$; thus
from our central position each source offers a flux $S_i = 1/R_i^2$. The straight line
is the theoretical power law of slope $-5/2$ for our Euclidean universe. The dots
and crosses following the upwardly curving line represent the distorted source
count resulting from Eddington bias assuming a Gaussian flux measurement
error of $\sigma = 5$ units, the dots calculated via the MC pseudo-code of the text
and the crosses calculated from Equation (8.5). (b) Plots of $[f(S, s)n(S)]$ from
Equation (8.5), for single apparent flux densities of 15, 20, 30 and 40 units
($3\sigma, 4\sigma, 6\sigma, 8\sigma$) given an underlying law of slope $-5/2$ and Gaussian errors
of $\sigma = 5$ units.

If the measurement error distribution is known accurately as well as the underlying population law $n(S)\mathrm{d}S$, we may calculate the observed law point by point:

$$n'(S)\mathrm{d}S = \int_0^\infty f(S, s)n(s)\mathrm{d}s, \qquad (8.5)$$

where f is the (unit area) error distribution. In the example, we have assumed f to be a Gaussian of $\sigma = 5$ units, i.e. $\exp[0.5(S - s)^2/5^2]$ and a power law for $n(s)$ of slope $-5/2$; the crosses represent the calculation. (Eddington bias is the steepening of the source/magnitude count rather than the apparent overestimate of individual flux densities.)

The right panel of Figure 8.9, shows the integrands of Equation (8.5) for our examples of the calculations of individual distorted flux measurements. This demonstrates graphically what happens to fluxes and survey data in the presence of noise and steep underlying population laws. At an observed 15 units $= 3\sigma$, there is simply no way of knowing what true flux density has been measured, or conversely – in the absence of knowledge of the underlying law or of the precise tails of the noise distribution – what the observed count says about the real count. Even at 40 units $= 8\sigma$, there is clear displacement of the integrand to fainter flux densities.

Many types of astronomical observation suffer from the range of problems due to combined effects of Malmquist bias, parameter–distance correlation and Eddington or source-count bias. This discussion has dealt with galaxies and luminosities for illustration; plenty of other examples could have been chosen.

8.3 Luminosity functions

In this section we assume that we are dealing with a catalogue of objects, of high reliability and well-understood limits. If we are interested in some intrinsic variable l (say a luminosity), then of frequent importance is the luminosity function $\rho(l)$, the space density of objects of luminosity between l and $l + \mathrm{d}l$. In principle we could get an approximation to ρ by measuring L_i for all of the objects in some (large) volume. In practice we need another way, because high luminosities are greatly over-represented in flux/magnitude-limited surveys, as we have seen.

8.3.1 Luminosity functions via the $V_{\max}$ method

One of the best methods to estimate $\rho(l)$ is the intuitive $V_{\max}$ method (Rowan-Robinson, 1968; Schmidt, 1968). The quantities $V_{\max}(L_i)$ are the maximum volumes within which the ith object in the catalogue could lie, and still be in the catalogue. V_{max} thus depends on the survey limits, the distribution of the objects in space, and the way in which detectability depends on distance. In the simplest case, assume a uniform distribution in space. Then, given the $V_{max}(L_i)$, an estimate of the luminosity function is

$$\hat{\rho}(B_{j-1} < l \leq B_j) = \sum_{B_{j-1} < i \leq B_j} \frac{1}{V_{max}(L_i)} \qquad (8.6)$$

in which its value is computed in bins of luminosity, bounded by the B_j.

The V_{max} method has much to recommend it. As an MLE (Marshall *et al.*, 1983), it has minimum variance for any estimate based on its statistical model. The errors are uncorrelated from bin to bin and can easily be estimated – the fractional error in each bin is close to $1/\sqrt{N_j}$, where N_j is the number of objects in each bin. More accurate error estimates can be obtained by a bootstrap, thereby taking account of the differing $1/V_{max}$ contributions from individual objects in the sample. Like any method involving bins, the estimate is biased because it can only return the average value over the width of the bin. This bias may be significant in steep regions of the luminosity function.

The main practical issue is simply the determination of V_{max}; as we have seen, choosing the flux limit of a survey affects the number of sources that are missed, the number of bogus ones that are included, and the extent to which faint sources are over-represented. In general these complicated effects are best examined with Monte Carlo simulations, as even a rough idea of the thing we want to know (the luminosity function) suffices to check these biases. In practice the processes of survey evaluation and calculating the luminosity function are iterative.

With V the volume defined by the distance to the source as its radius, the distribution of V/V_{max} is very useful in estimating the actual limit of a survey. If the correct flux limit has been used in the calculation of V_{max} for each object, then we would expect V/V_{max} to be uniformly distributed between zero and one. This can easily be checked by, for example, a Kolmogorov–Smirnov test. In fact this test can be regarded as a model-fitting procedure to estimate the effective flux limit of a survey. For large cosmological distances (say those corresponding to $z > 0.2$), this technique is upset by cosmological evolution, the derivation of which was a driving force behind development of the technique (Schmidt, 1968).

The literature on the V_{max} method is, justifiably, vast; Willmer (1997) provides a summary.

Example Taking the previous simulation based on the Schechter function, a flux limit of 20 units gives a sample of about 200 objects. The distribution in luminosity (Figure 8.10) shows a strong peak at about 5 units, related to the characteristic luminosity $l_* = 10$ for the simulation. Faint sources are greatly under-represented, because they are only above the flux limit for small distances. Applying the $V_{\max}$ method and bootstrapping to derive error bars gives Figure 8.11. Because $V_{\max}$ is so small for the faint sources, the few faint sources in the sample give a large contribution to $\hat{\rho}$ – although the errors are correspondingly large. For simplicity the luminosity functions have been normalized, so giving luminosity probability distributions; the two are related by a number density.

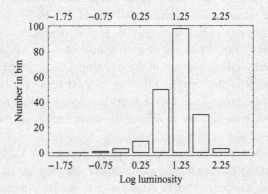

Figure 8.10   The luminosity distribution for the simulation, in bins 0.5 dex wide, derived from the Schechter function of the previous example with a flux limit of 20 units.

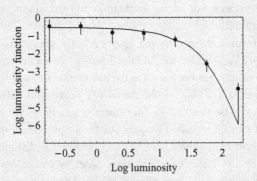

Figure 8.11   The input luminosity function for the simulation (solid line) and the estimate via $V_{\max}$ (points). The error bars are the interquartile range, estimated from a bootstrap.

A key assumption of the simple form of the V_{max} method is that the objects of interest are uniformly distributed in space. If this is not a good assumption (and it is not in most cosmological investigations), then there are three ways of making a better estimate.

One simple improvement is to bin the data into narrow ranges of distance, and estimate the luminosity function within each bin. As we can see from Figure 8.7, at large distances we will know nothing about low luminosities, a consequence of the agnosticism of this approach. The approach is further limited by the decreasing numbers of objects as the number of distance bins is increased.

We can further consider spatial dependence by making somewhat stronger assumptions. Our data is the set of pairs (R_i, L_i) and our task is to compute the *bivariate* function. Obviously we are not going to be able to do this without some constraints on the form of the bivariate distribution. The usual assumption is that the distribution factorizes, so that

$$\rho(r, l) = \rho_0(l)\phi(r). \tag{8.7}$$

This just means that the *form* of the luminosity function does not change with distance, but the normalization can. Now ρ_0 refers to the local luminosity function $\rho(r = 0, l)$. This method extrapolates information from small distances and low luminosities into the bottom-right portion of Figure 8.7.

In this case, the standard estimator of ρ is the C^- method, due originally to Lynden-Bell (1971) and redeveloped by Chołoniewski (1987). This method is not nearly as intuitive as the V_{max} method, but it is important because it is also an MLE. The likelihood function, from which both methods are derived, is given by Marshall *et al.* (1983).

The C^- estimator is best described by a piece of pseudo-code.

```
Arrange the data (R_i, L_i) in decreasing order of l
set C_i = 0
for each L_j < L_i : add 1 to C_i if this source is within
V_max(L_i) otherwise, go to the next L_j until finished.
```

Remarkably, the C numbers suffice to determine the cumulative luminosity function:

$$\int_0^{L_i} \rho(l)\,\mathrm{d}l \propto \prod_{k \leq i} \left(\frac{C_k + 1}{C_k} \right) \tag{8.8}$$

with

$$\frac{C_1 + 1}{C_1} = 1$$

as the starting point. The constant of proportionality is the inverse of the largest $V_{max}(L_i)$ in the sample; it can also be obtained by requiring that the estimated distribution $\hat{\rho}$ yields a total number of detections that matches the observed luminosity distribution. If there are ties in the sample, the simplest remedy is to shuffle the data by small amounts (say, a tenth of the observational error) so that the C^- algorithm can be applied straightforwardly.

Obtaining the result as a cumulative distribution is slightly inconvenient, but a conversion to binned form is easy enough. This yields errors that are more independent from bin to bin, and as usual can easily be computed by bootstrap. The distance distribution ϕ (or *evolution function*) can also be be extracted by similar methods, if required; see Chołoniewski (1987) and the cited references.

If parametric forms are known for ρ and ϕ, then a normal modelling method can be used; see Sections 8.3.2 and 8.3.3. In fact, the C^- method obtains an analytic solution for the form

$$\rho(l, r) = \sum_i a_i \delta(l - L_i) \sum_j b_j \delta(r - R_j), \tag{8.9}$$

where distances are denoted by r and R, and the a_i and b_j are the parameters of the luminosity distribution and evolution function, respectively. The V_{max} method is obtained from a similar model via maximum likelihood, except that the distribution with distance is assumed to be uniform. Models may be available with far fewer parameters (the Schechter function only has two, for example) and then a model fit will usually give lower random errors. As usual, however, care is needed to be sure that the model represents reality.

The two approaches we have described are at opposite ends of the spectrum of assumptions: fitting factorizable functions $\rho(r, l)$, or simply counting objects in bins of distance. Intermediate between these two are the 'free-form' methods (Peacock, 1985; Dunlop & Peacock, 1990), which attempt to fit fairly general functions to the data populating the r–l plane. Examples of these functions are

$$\sum_{i,j} a_i (\log r)^i (\log l)^j,$$

where the cross-terms break the factorizability. However, the use of a relatively small number of terms in the expansion does permit extrapolation into areas of the r–l plane unpopulated by observations.

8.3.2 Luminosity functions via maximum likelihood; the SOS method

The V / V_{max} method has as its main shortcoming the need to bin data. From intensive investigation over the years, we have much prior information about the forms of luminosity function; and it is frequently possible to adopt a form in which we can have reasonable expectation that it will fit the data. We can test to see if it does; and modify the form accordingly. In this way it is possible to map luminosity functions in a model-dependent way, and with care, obtain substantially more accuracy of representation than available via the V/V_{max} method. Here ML methods come into their own.

Marshall *et al.* (1983) provided an early demonstration of the power of this technique. Their analysis of a sample of 32 X-ray-loud quasars produced a form for the luminosity function, and a quantitative measure of how much cosmic evolution the luminosity function underwent, all from a sample of 32 X-ray-selected quasars.

The single-object-survey (SOS) technique described here is a variant, introduced to consider how to combine (a) several complete samples of sources with differing survey-selection properties, and complete samples in which objects have spectra so dissimilar that the selection (survey) lines in the $L-z$ plane (luminosity versus redshift or distance; see Figure 8.4) differ dramatically. This approach of individual $L-z$ planes was first developed for a sample of flat-spectrum radio quasars, whose spectra show wildly different forms and hence different K-corrections (Wall *et al.*, 2005). It was generalized to use the Marshall ML method by Wall *et al.* (2008) for a sample of 35 sub-mm galaxies (SMGs).

Consider the sample as a single homogeneous set of N objects, for which $\rho(z, L)(\partial V/\partial z)dz \, dL$ is the number in volume element $(\partial V/\partial z)dz$ in luminosity element dL. The first procedure is to chop up the $L-z$ plane into array elements or cells small enough so that each is occupied by either one or zero objects. We set up individual $L-z$ planes for each object, because in general the sky fraction $\Omega_i(z, L)$ accessible to each object i is unique – we can in this way absorb the issues that each of our objects is: (a) observable over an area of different physical size; and (b) has its own flux-density limit line in the $L-z$ diagram. This individual plane $\Omega_i(z, L)$ is thus essential in introducing the feature of the single-object-survey, by which each object is treated as having unique access to the $L-z$ plane (Figure 8.4). The treatment is analogous to the final survey having been done as N individual surveys finding a single source each. The unique area accessible to each object on the $L-z$ plane is determined by the individual survey cut-off line for each object; below this line the omega-plane cells take the value of zero, while above it the numbers are the sky fraction covered by the survey for the object. Having calculated the individual $\Omega_i(z, L)$ planes, we proceed as follows.

The $\mathcal{L}$(ikelihood) function for the ith object is the probability of observing *one* object in its (dz, dl) element times the probability of observing *zero* objects in all other (dz, dl) elements accessible to it. The Poisson model is the obvious one for the likelihood:

$$f(x \mid \mu) = \frac{e^{-\mu} \mu^x}{x!},$$

(8.10)

where μ is the expected number.

With $\rho(z, L)$ as the full description of space density,

$$\mu = \lambda(z, L)\, dz\, dL, \quad \text{for } \lambda = \rho(z, L)\Omega(z, L)(\partial V / \partial z).$$

(8.11)

Taking the probabilities of one or zero objects from (8.10),

$$\mathcal{L} = \prod_i^N \lambda(z_i, L_i)\, dz\, dL\, e^{-\lambda(z_i, L_i)\, dz\, dL} \prod_{j \neq i}^N e^{-\lambda(z_j, L_j)\, dz\, dL},$$

(8.12)

where i denotes the elements of the $(L-z)$ plane in which objects are present and j denotes all others. From this, defining $S = -2 \ln \mathcal{L}$, then

$$S = -2 \sum_{i=1}^N \ln \rho(z_i, L_i)$$

$$+ \sum_{i=1}^N \int_z \int_L \rho(z, L)\Omega_i(z, L)\frac{\partial V}{\partial z}\, dz\, dL + \text{constant}.$$

(8.13)

By way of example, consider simple factorizable density evolution of the form $\rho(L, z) = \rho(z{=}0, L)\phi(z)$; consider further the adoption of a single power-law luminosity function,

$$\frac{dN}{dL} = \rho(L, z) = \frac{\rho_0}{L_*} \phi(z) \left(\frac{L}{L_*}\right)^{-\alpha}.$$

(8.14)

With $l \equiv L/L_*$, we have the local luminosity function as $\rho(z{=}0, L) = (\rho_0/L_*)l^{-\alpha}$. For the evolution function we again adopt a power law, $\phi(z) = (1 + z)^k$.

If we substitute these assumptions into Equation (8.4) and set the derivative with respect to ρ_0 to zero, we get an MLE estimate for ρ_0:

$$\rho_0 = \frac{N}{\sum_{i=0}^N \int_z \int_l (1 + z)^k\, l^{-\alpha}\, \Omega_i(z, l)\, (\partial V / \partial z)\, dz\, L_* dl}.$$

(8.15)

Putting this back into Equation (8.4) gives the algorithm for the likelihood:

$$S = -2 \sum_i^N \ln[(1+z_i)^k \, l_i^{-\alpha}]$$

$$+ 2N \ln \sum_i^N \int_z \int_l (1+z)^k \, l^{-\alpha} \, \Omega_i(z,l) \left(\frac{\partial V}{\partial z} \right) dz \, dl$$

$$+ (2N - 2N \ln N). \tag{8.16}$$

Of course the assumptions on the luminosity function and its evolution may be modified ad infinitum. We might wish to use as the basic form the Schecter luminosity function or the broken power law, very common and versatile:

$$\rho(l) = \rho_0 \left[\left(\frac{l}{l_b} \right)^\beta + \left(\frac{l}{l_b} \right)^\gamma \right]^{-1} \tag{8.17}$$

with l_b as the break luminosity between the two power laws.

There is again a range as large as imagination in attempting to fit data on space density at different redshifts via density evolution, luminosity evolution, density-dependent luminosity evolution, power-law evolution (e.g. of the form $(1+z)^\alpha$), exponential evolution (say $\exp(M\tau_b)$, where M is a constant and τ_b is look-back time). All of these may be modified by some form of tapering off to the highest redshifts (e.g. Wall *et al.*, 2008) to mimic a redshift cut-off or epoch of creation of our objects.

Example Back to the toy universe. We know the answer for our luminosity function here: we put in a power law l^{-3}. Can we get it back directly with the SOS approach? In fact, because we have not ascribed spectral forms to our objects, they do all have the same $\Omega_i(l, z)$ planes; except for dividing these into tiny cells, these planes look similar to the (l, r) plane of Figure 8.4, with $\Omega_i = 1$ above the survey-limit line and 0 below the line. When these planes have been calculated and the cell coordinates of all 1400 objects determined in the Ω planes, we calculate the ML algorithm and find the results shown in Figure 8.12. We have deduced bulk properties of our million-object universe very accurately from a survey which detected only 1400 (1.4 per cent) of them. The power of such techniques is substantial. Marshall's original sample of 32 quasars found a luminosity function and evolution for X-ray quasars which has been shown to be accurate, although, of course, the size of modern-day samples adds much detail, generally using procedures which are similar to the above.

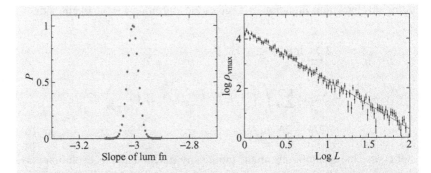

Figure 8.12   (a) The likelihood function computed from Equation (8.16) for a range of slopes of the luminosity function and converted to probabilities; the known slope is -3.0. (b) The luminosity function computed via the $1/V_{max}$ method and shown in a log–log plane. Note that the anticipated slope is now -2.0 (see Figure 8.4(b)) because of binning choice. The line through the points comes directly from the ML solution, which found a slope of -3.004 and a value for ρ_0 of 474 000, indistinguishable from the known values of (-3.0, 477 465).

A key reason that the example works so well is that this universe is homogeneous and the power law a simple one: the fitted model is the same as the input model. We have also assumed that ϕ is a constant, $k = 0$, no evolution. *Model error may matter just as much as $\sqrt{N}$-type error* in these applications. Your favourite model may fit quite well but there may be others, of various functional forms, equally satisfactory. Model comparison (à la Bayes perhaps) is important, and the point holds for the next example as well.

8.3.3 Luminosity functions via source counts and redshift distributions

As we saw in Section 8.2, a sky survey to a fixed intensity limit results in a sample of sources whose redshift distribution enables us to calculate a *luminosity distribution $\eta(l)$*, the probability distribution of luminosities for objects with flux densities or apparent magnitudes above a survey limit. In practice, determining luminosity distributions, even for samples of relatively bright objects, is long and arduous, as complete or near-complete redshift data for sample members are required. Despite the rapid advance towards deep large-area sky surveys at many wavelengths, then, the following situation generally prevails: at some frequency/passband we have redshift information which may be complete for a sample at a relatively bright flux/magnitude limit, and we have source-count/number–magnitude information extending to a much lower apparent intensity limit. This latter, the $N(S)$, is put together from wide-angle surveys at relatively bright intensities and small area surveys extending far deeper; the steep form of counts in all passbands ensures that we can examine the sky surface area of our objects from very small-area surveys at very faint intensities. How best can we use relatively complete redshift information at one or more relatively bright survey levels together with extensive number counts to much fainter levels in this respect? We wish to interrogate these data to find, e.g., the form of the luminosity function ρ, how this changes with cosmological epoch (redshift), and/or how it compares to a theoretical or prior model of evolution.

With an observed luminosity distribution $\eta(L)$ at survey sensitivity limit $S_{\lim}$, one simple procedure to do so is as follows (Wall *et al.*, 1980). (a) For the luminosity function $\rho(L, z)$, assume a factorization of the form $\rho(L, z) = \rho_0 \phi(L, z)$, where ρ_0 is the local luminosity function, namely $\rho_0(L, z = 0)$. (We have met this factorization in Section 8.3.1 and the previous section; it looks like *density evolution*, a change in the number-density of objects with epoch, but it may be shown to be the equivalent of *luminosity evolution*, a change in luminosity of objects with epoch.) (b) From the luminosity distribution and $\phi(L, z)$, calculate the local luminosity function via

$$\rho_0(L)\mathrm{d}L = \frac{\eta(L, S > S_{\lim})\mathrm{d}L}{\int_{z=0}^{z_{L,S_{\lim}}} \phi(L, z)\mathrm{d}V},$$ (8.18)

where DV is the co-moving volume element and $z_{L,S_{\lim}}$ is the redshift out to which an object of luminosity L may be seen at $S = S_{\lim}$. (c) Compute the source-count/number–magnitude relation via

$$N(> S) = \int_0^\infty \int_{z=0}^{z_{L,S_{\lim}}} \phi(L, z)\rho_0 \mathrm{d}V$$ (8.19)

(d) Compare this with the observed $N(S)$; return to step (b) to modify the evolution function; proceed to the next comparison. The process may be automated, using a minimization procedure on the comparison statistic, e.g. χ^2, to find a set of optimum parameters for an evolution model. The comparison may be set up as a likelihood, and Bayesian methodology used to insert prior knowledge; it may be generalized to use more than one luminosity distribution, partially defined luminosity distributions or incomplete redshift information, and count data at different frequencies/passbands.

Example A toy flat universe with $R_{max} = 1.0$ has 10^6 sources selected randomly from a density law $\phi(r) = (1 + r)^{(k+\gamma r)}$ with $k = 10.0$ and $\gamma = -10$, and a luminosity law L^{-3}, with a lower limit to the luminosities of 1.0. The survey limit from which the luminosity distribution is obtained has a sensitivity limit of $S_{lim} = 200$ units (where $S = L/R^2$); the source count is assumed to extend 10 times deeper, to a sensitivity limit of 20 units.

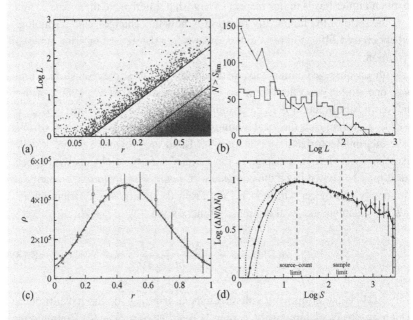

Figure 8.13   A toy universe with 'evolution' as prescribed in the text. (a) Luminosity–redshift plane (Hubble diagram); the 10^6 sources are plotted as pin-points, with the 1117 sources above the survey cut-off as bright dots. The survey cut-off at $S = 200$ units is the upper diagonal line, while the source-count comparison is made down to $S = 20$ units, the lower line. (b) Histogram of the luminosity distribution for the 1117 sources in the sample. For comparison, the dots + solid lines show this distribution for 10^6 objects obeying L^{-3} but with uniform ($\propto r^3$) space density. (c) The density distribution; dots give the densities per Δr, while the model from which the dots were selected is the solid line. The open circles with error bars show densities from the 'observed' sample of 1117 objects, calculated in 10 radius slices with the $1/V_{max}$ method. (d) Dots plus error bars show the 'observed' source count for the 10^6 objects in relative differential form – see text. The 'survey' limit at $S_{lim} = 200$ units is shown together with a source–count comparison limit a factor of 10 lower at 20 units. The solid black line is the best-fit model, which from minimum χ^2 finds values of (k, γ) close to the input values (10, −10). The dashed lines represent models which fit the counts down to 20 units almost as well; see text.

We wish to try to recover the input parameters (k, γ) with an application of the technique just described. The panels of Figure 8.13 tell the story, via the 'Hubble diagram', the luminosity distribution, the density distribution and the source count. Our 'knowledge' is officially confined to the brightest 1117 sources which appear above the survey limit; we have to infer the properties of the 10^6 sources from this ~ 1 per cent. Plots (c) and (d) hence contain some 'privileged information' which we know because we did the simulation. The density plot mainly checks our success in randomly selecting our sample according to the given density law, showing that we can recover the general form of the density law via the sample using the $1/V_{max}$ method (Section 8.3.1). The (privileged) source-count plot extends down to intensities far below our self-imposed 'observational' limit for the count of 10 times fainter than the sample limit.

The count for the 10^6 objects is in relative differential form, ΔN_0 representing the number of objects predicted in each narrow radius slice for a uniformly filled universe. The initial rise is due to the 'evolution' imposed by the power $k = 10.0$ in our assumed form for the function ϕ; the fall to faint intensities is largely due to the cut-off at larger radii imposed by the power γr with $\gamma = -10.0$. The contours of χ^2 in the $k = \gamma$ plane show a well-defined $k-\gamma$-plane minimum at $(k, \gamma) = (9.2, -8.9)$ close to the input values of $(10, -10)$. The best-fit model gives the source count shown by the solid line in the lower-right diagram; the dotted count lines are from models which fit the counts down to 20 units almost as well. However, the parameters k and γ are tightly related because, e.g., increasing k produces a steeper rise in the counts, which may be counteracted by a decrease in γ to lower the counts. The χ^2 contours in the (k, γ) plane are thus a long narrow trough. Following the line of the trough, (k, γ) pairs of $(6.0, -4.0)$ and $(12.0, -13.6)$ give values of χ^2 which exceed the minimum value at $(9.2, -8.9)$ by only ~ 0.2, nowhere near the $\Delta \chi^2$ needed to reach even the 68 per cent (1σ) significance level. The source counts for these two models are shown as the dotted lines in Figure 8.13(d). This serves to demonstrate the power of the source counts. We have used privileged data to plot these to extremely faint intensities. If we had adopted a defined source count to, say, 5 units instead of 20, *both these models would have been rejected at a high level of significance* – and the ends of the trough would have been shrunk to define the parameters very tightly.

A similar procedure was used by Shanks *et al.* (1984) to demonstrate strong cosmological evolution for galaxies selected in the blue passband. The procedure is a basis for the landmark study of radio AGN by Dunlop & Peacock (1990). In this, a likelihood analysis was used to examine the epoch dependence of space density for radio AGNs, and minimal assumptions were made about the form of the evolution. The study gave the first clear indication of a diminution in co-moving space density of active galaxies at redshifts above 2.

8.4 Tests on luminosity functions

8.4.1 Error propagation

A luminosity function may be used to derive some other parameter – an estimate of a contribution to background light, for example. Propagating the errors, whether from a binned estimate or a model fit, is straightforward enough, as long as we have a simple analytic relationship between the desired parameter, and the model parameters. If not, a simulation may be the easiest solution.

If the luminosity function has been derived by ML, then an asymptotic error estimate is available (Section 6.1). Suppose we have a model, with parameters $\vec{\alpha}$, and we are interested in the error bars on some function $e = f(\vec{\alpha})$. The 'unconstrained' ML is $\mathcal{L}(\vec{\alpha} = \hat{\vec{\alpha}})$ and the constrained ML is $\mathcal{L}(\vec{\alpha} = \hat{\hat{\vec{\alpha}}}, e = f(\hat{\hat{\vec{\alpha}}}))$. The classical theory of the likelihood ratio tells us that

$$-2 \log \mathcal{L}\left(\vec{\alpha} = \hat{\hat{\vec{\alpha}}}, e = f(\hat{\hat{\vec{\alpha}}})\right) + 2 \log \mathcal{L}\left(\vec{\alpha} = \hat{\vec{\alpha}}\right) = \Delta \qquad (8.20)$$

is asymptotically distributed as χ^2, with one degree of freedom. The first term may be calculated with numerical routines for constrained maximization, and so an error bar for e can be obtained. For instance, a value of $\Delta = 4$ corresponds to a confidence level of 95 per cent. Avni (1978) discusses the technique in an astronomical context.

8.4.2 Luminosity-function comparison

We may, however, have two estimates for different types of object, and we may want to know if the luminosity functions are different. Here the range of possible tests is very wide. Considered as probability distributions (so normalized to unity) it is possible to adapt many of the tests described in Chapter 5. The chi-square test can be applied directly to the differences between luminosity functions derived in binned form. The methodology of other tests may also be applied. For example, the Kolmogorov–Smirnov statistic would be a natural one to use for a cumulant derived by the C^- method. In general, the distribution of the t statistic (under the null hypothesis) would have to be derived by a Monte

Carlo simulation of the experiment, because the fluctuations in the numbers in the bins are not simply related to the number of objects in each bin.

Another type of test is based on the likelihood ratio (Jenkins, 1989), and is applicable to cases where the luminosity functions have been derived in parametric (or binned) form from an ML analysis. This idea is discussed further below.

It is important to emphasize yet again that these tests can reject the null hypothesis (the distributions are the same), but without alternative hypotheses we have no idea of their statistical power. If we do have well-defined alternative hypotheses, then Bayesian methods of model choice could be used. The methods that are based on the likelihood function could be adapted for this purpose, because the likelihood function is half of what is needed to calculate the evidence, the other half being the prior.

8.4.3 Correlation: multivariate luminosity functions

A further sort of test is correlation, leading on to the subject of multivariate luminosity functions. If we generate a sample (say from X-ray observations) we obtain a catalogue which we may then resurvey at, say, radio wavelengths. Retaining the objects which are detected at both wavelengths, we can construct a *bivariate luminosity function* $\rho(l_X, l_R)$. The most straightforward way of doing this is by a generalization of the V_{max} method. To obtain the V_{max} for each object, compute its V_{max} for each of the variables for a particular object, and take the minimum. The justification is simply that an object will drop out of the catalogue if it is below the detection limit in either band (Schmidt, 1968).

Multivariate luminosity functions take much effort to construct. However, they do provide a solution to the problem of bogus luminosity–luminosity correlations, mentioned earlier in this chapter. The main problem is the increase in the number of bins; four times as many for a bivariate function, nine times as many for a trivariate. These bins become sparsely populated with objects.

If we have an estimator of (say) a bivariate luminosity function of X-ray and radio luminosity, three possibilities are available to see if l_X and l_R are correlated. The easiest is by simple inspection of $\rho(l_X, l_R)$ which may show an obvious 'ridge line'. Another possibility is that some statistic, say the median l_X, computed from the luminosity function in narrow slices of l_R, will correlate with l_R. Here we could use end-to-end Monte Carlo simulations of a correlation coefficient to establish the significance of any result.

Example Phillips *et al.* (1986) reported an emission-line survey of an optical-magnitude-limited sample of nearby galaxies. They derived an emission-line luminosity function, binned into one-magnitude ranges of absolute magnitude. Dividing by the optical luminosity function gives an estimate of the fraction of galaxies that are emitting at a given emission-line power. Moreover, integrating these normalized luminosity functions gives an estimate of the fraction of galaxies that have emission-line power anywhere in the range sampled by the survey. As seen in Figure 8.14, the emission-line luminosity function shifts to higher powers at brighter absolute magnitudes. Clearly, the Malmquist bias of the original sample would make it impossible to make an unbiased estimate of the emission-line luminosity function, which is why the data were binned into magnitude ranges.

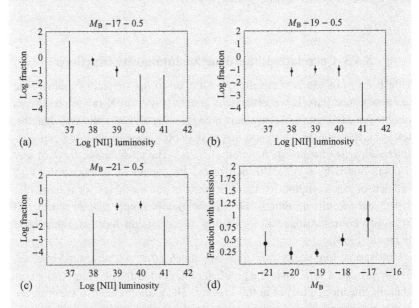

Figure 8.14   (a), (b) and (c) Estimates of the normalized emission-line luminosity function, derived from the V_{max} method. (d) The integral of the normalized functions, plotted against absolute optical magnitude. The fraction of galaxies producing emission lines appears to fall with increasing optical luminosity.

These data have been corrected, following an erratum to the original paper which illustrates the pitfalls of this type of analysis. For the sample on which the emission-line survey was based, the normalization of the optical luminosity function depends somewhat on distance. But, being a flux-limited sample, the emission-line luminosity correlates with distance too. This means that normalizing the emission-line luminosity function must take into account *which* optical luminosity function to use, depending on the spread of distances in each magnitude bin.

Finally, we need to remember that correlation is just one case of statistical dependence. Variables may be related non-linearly – for example, $\rho(l_X, l_R)$ may be steeper (as a function of l_X) at larger l_R, without changing its median. Quite generally, what we want to know is whether

$$\rho(l_X, l_R) = \rho_X(l_X)\rho_R(l_R)$$

is statistically plausible.

Probably the best tests to use in this case are those based on the likelihood ratio. Suppose we fit both factorizable and unfactorizable models of the luminosity function, using ML. Call the ML in each case $\mathcal{L}_f$ and $\mathcal{L}_{uf}$. The log likelihood ratio

$$\mathcal{R} = \log \frac{\mathcal{L}_f}{\mathcal{L}_{uf}} \qquad (8.21)$$

will give an indication of which model is better – we have encountered this general idea in a Bayesian context in Chapter 6. If we are fitting many parameters (and more than one poses difficulties), it is easier to use the *maximum* of the likelihood to derive the ratio. Evidently the ratio will depend on how many free parameters we have in the competing models; classical results tell us that $\mathcal{R}$ is distributed asymptotically as χ^2, with a number of degrees of freedom that depends on the number of free parameters in each model. As usual, a pragmatic conclusion is not to reach for the tables of χ^2, but rather to regard $\mathcal{R}$ as a potentially useful test statistic, and derive its distribution by Monte Carlo for the problem to hand. This approach is described by Schmitt (1985) in the context of survival analysis, but is applicable whenever a likelihood technique is used.

8.5 Survival analysis

When we produce a primary sample of objects in astronomy, we do so by making a series of measurements and picking out the ones we regard as detections. The results often find their way into catalogues, of which venerable examples are the New General Catalogue or the 3CR catalogue. Objects which are not in the catalogue – usually because they are below the flux limit – are simply unknown. Since in general we do not know if there is anything there at all, quoting an upper limit for every position or wavelength surveyed is not a useful thing to do.

However, frequently an established primary sample is then resurveyed in some other way; we may investigate the Hα luminosity of galaxies in the NGC, for instance. In this case, it is very useful to quote upper limits for the undetected galaxies, because we know that such limits refer to real objects. Sometimes a resurvey may yield lower limits as well. If we were to measure X-ray and radio flux densities for the NGC galaxies we would probably obtain both upper limits and lower limits for the radio to X-ray spectral index.

The branch of statistics that deals with limits is called *survival analysis*. The term arises in medical statistics, where at the conclusion of a study some of the subjects may have survived and some died. For presumably unrelated reasons, measurements which are only limits are called 'censored'. The methods of survival analysis were introduced into astronomy by Avni *et al.* (1980), Feigelson & Nelson (1985), Schmitt (1985), and Isobe *et al.* (1986). Other astronomers had independently discovered aspects of the technique, but these papers offer the best introductions. A useful text is Kalbfleish & Prentice (2002). Bayesian approaches to missing data are discussed in Gelman *et al.* (2004).

Survival analysis offers (i) estimation of intrinsic distributions (like luminosity functions), (ii) modelling and parameter estimation, (iii) hypothesis testing, and (iv) tests for correlation and statistical independence, for cases in which some of the available measurements are limits. The key assumption is that the censoring is random; this means that the chance of only an upper limit being available for some property is independent of the true value of that property. This assumption is often met for flux-limited samples. For an object of true luminosity L and distance R, the condition for censoring is that

$$\frac{L}{R^2} < S_{\text{lim}},$$

the flux limit for the survey. If R is a random variable, independent of L, and S_{lim} is fixed, then the chance of censoring is independent of L. Evidently a careful examination of the way in which a sample was selected is necessary to determine that survival analysis is applicable.

8.5.1 The normalized luminosity function

To be definite, suppose we select a sample of objects at wavelength A and then resurvey the sample at wavelength B. For some objects, we will achieve a detection and so have a measurement of luminosity L_{B}; for others, we will only have an upper limit L_B^U. The methods of survival analysis use the detections, and upper limits, to reconstruct the distribution of L_B. This will be proportional to the luminosity function ρ_B. However, it is vital to remember that the censoring has to be random. Also, the luminosities L_A will have Malmquist bias; if L_A and L_B are correlated, then of course the estimate of the distribution of L_B will also be biased. In general it is safest to calculate the estimate in narrow bins of L_A. Indeed, this is one way of checking for a relationship between L_A and L_B in the sample, as we shall see.

Two equivalent algorithms are available for computing the normalized luminosity function. If we are happy to bin the data (both the detections and the upper limits) into intervals of L_B, the estimated probability per bin $\hat{p}_k$ can be derived by a recursive relation due to Avni *et al.* (1980):

$$\hat{p}_k = \frac{n_k}{M - \sum_{j=1}^{k} \frac{u_j}{1 - \sum_{i=1}^{j-1} \hat{p}_i}}. \tag{8.22}$$

This intimidating formula in fact results from a straightforward ML argument. (Avni *et al.* give an expression for the likelihood function; it can be useful in various tests.) In the formula, n_k is the number of detected objects in bin k; u_k is the number of upper (or lower) limits allocated to bin k; and M is the total number of observations (detections plus limits). To use the formula with upper limits, number the bins from large to small values of the observed quantity; conversely for lower limits. In either case, undetected objects must lie in higher-numbered bins than the bin where their limit is allocated. Calculation begins with bin 1, for which the solution is

$$\hat{p}_1 = \frac{n_1}{M - u_1}.$$

Allocating limits to bins takes a little care. For narrow bins the scheme used should not matter, but for wider bins a little experimentation may be instructive; the problem here is bias, as is usual with wide bins. This method will produce a normalized distribution as long as the highest-numbered bin contains detections, not limits. This makes sense; in the case of upper limits, the highest-numbered bin is the faintest, and if there are limits in the faintest bin we have no way of using them.

Example Here are some data from Avni *et al.* (1980), giving the distribution of X-ray to optical luminosity spectral index for quasars. In this case the u_k are lower limits corresponding to no detections in the X-ray band and the k are the indices of Equation (8.22). ($\hat{K}$ is the Kaplan–Meier estimator of the next example.)

k	n_k	u_k	$\hat{p}_k$	$\hat{\mathcal{K}}_k$
1	2	0	0.057	0.057
2	1	1	0.029	0.086
3	4	1	0.122	0.204
4	4	0	0.122	0.326
5	3	1	0.096	0.418
6	6	0	0.191	0.612
7	3	3	0.128	0.709
8	1	1	0.051	0.758
9	2	0	0.102	0.879
10	1	1	0.102	0.939

We see from Figure 8.15 that inclusion of the upper limits does give a little more information. Since the method should extract the distribution that was subject to the censoring, the reconstructed distribution will be proportional to the number of quasars per unit volume with each spectral index. As Avni *et al.* discuss in detail, much depends on the selection of the sample in the first place. Here an optically selected sample was subsequently surveyed at X-ray wavelengths, and only at X-rays were upper limits available; the original selection will therefore have biased the sample to optically luminous quasars.

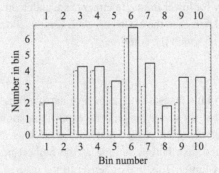

Figure 8.15   Distribution of spectral indices (optical to X-ray) for a sample of optically selected quasars, showing the observed distribution (dashed boxes) and the estimated true distribution (solid boxes) after including the lower limits.

Distances were available for all the objects in this sample, so that in fact a $V_{\max}$ method could have been used to reconstruct the X-ray luminosity function. The retention of upper limits in the analysis means that *no distances would be necessary* to reconstruct the luminosity function, to within a constant, using Avni's method.

An alternative estimate of the cumulative distribution is provided by the Kaplan–Meier estimator. The Kaplan–Meier estimator is better known in the wider statistical world, and it is an MLE just like the Avni estimator. It has the advantage of not relying on any binning scheme. However, being cumulative, errors are highly correlated from one point on the estimate to the next.

$$\hat{\mathcal{K}}(L_k) = 1 - \prod_{i=1}^{k-1}(1 - d_i/n_i)^{\delta_i} \tag{8.23}$$

is the Kaplan–Meier estimator of the cumulative probability distribution, at the kth observation. As with the Avni estimator, this formula will work for either upper or lower limits.

For lower limits, arrange the observations in increasing order. Then d_i is the number of observations of L_i and n_i is the number of observations equal to or larger than L_i. By 'observations', we here mean either detections or non-detections. δ_i is 1 for a detection and zero for an upper limit. For upper limits, arrange the observations in decreasing order. Then d_i is the number of observations of L_i and n_i is the number of observations equal to or smaller than L_i. In both analyses, ties in the detections can be removed by shuffling the data by amounts small compared to observational error.

Example Using the data from the previous example, we can calculate the Kaplan–Meier estimator for the spectral indices. The results are in the '$\hat{\mathcal{K}}_k$' column in the data table of the previous example. If we form a cumulant from the Avni estimator, we find that the two results are very similar (Figure 8.16), as expected since both are MLEs. The treatment of the upper limit in the last bin follows a slightly different convention in the two methods.

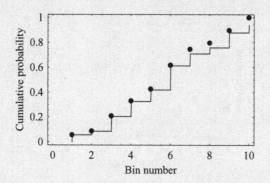

Figure 8.16 The Kaplan–Meier estimator (solid line), compared to the cumulant derived from the Avni estimator (dots).

Since these estimators are derived from a likelihood function, we might expect that there would be a formula for the variance on the estimate. This is indeed the case; it is called Greenwood's formula, and we refer you to Feigelson & Nelson (1985) for details. Being an asymptotic formula, it is not terribly useful in practice. Fortunately, we can estimate errors in other ways – either by a direct Monte Carlo simulation, or by a bootstrap on the sample we have. Bootstrapping censored data is not well investigated (Feigelson & Nelson, 1985) but we have found it to be satisfactory. In their review paper on the bootstrap, Efron & Tibshirani (1986) work through an example of bootstrapping censored data.

Example Returning to our simulated field-galaxy sample selected at one wavelength, we find on 'resurveying' at another wavelength that we have 67 detections and 317 upper limits. The simulation allocated luminosities from independent Schechter functions at both wavelengths. Binning the data gives the histogram shown in Figure 8.17; note that upper limits are counted in one bin lower down than would be the case for equivalent detections. Applying the Avni estimator, we find a luminosity probability distribution (Figure 8.18) that agrees well with the input theoretical distribution. The error bars are derived from a bootstrap, and the errors are reasonably uncorrelated from one bin to the next. The luminosity function itself is estimated by finding the constant of proportionality (galaxies per unit volume) which gives the correct total number of observed galaxies. This means matching the estimated luminosity probability distribution to the luminosity distribution η. (Do not confuse the luminosity probability distribution with a luminosity distribution; the terminology is unhelpful, but is, unfortunately, established.)

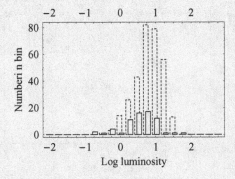

Figure 8.17   The luminosity distribution, and upper limits, for the field-galaxy simulation; there are 67 detections and 317 upper limits. The bins (dashed) for the upper limits are slightly displaced for clarity.

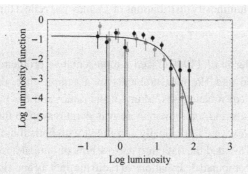

Figure 8.18 The luminosity probability distribution (black dots), and theoretical distribution (solid line), together with bootstrapped error estimates (the interquartile range is shown). The lighter dots are a V_{max} estimate, displaced slightly in luminosity for clarity.

Figure 8.18 also demonstrates that V_{max} and survival analysis results are in close agreement. This is what we should expect; both are MLEs, based on rather similar models, and the MLE for a given model is unique. The advantage of survival analysis is not that it gives better estimates of luminosity functions, but rather, that it will help in correlation analysis, or the reconstruction of distributions which do not need distances (like the spectral index distribution of Figures 8.15 and 8.16).

8.5.2 Modelling and parameter estimation

Once we have obtained a luminosity probability distribution ($\hat{\rho}(L_B)$ in our example) we may well want to estimate some other quantity from it, or decide if it differs from some other distribution. The same remarks apply as in the case of an ordinary luminosity function, except that we must never forget the Malmquist bias of the primary sample.

One useful technique, given enough data, is to divide the data into bins of L_A and compute a distribution of L_B for each bin; call these $\hat{\rho}(L_B \mid L_A)$. With luck (and enough data) we may be able to estimate a location parameter, say a median, at each slice of L_A. This sort of analysis may well answer the question of whether L_A and L_B are correlated. Error analysis, as usual, can be via bootstrap or direct Monte Carlo. We may also need to compare estimates, say $\hat{\rho}_1(L_B)$ and $\hat{\rho}_2(L_B)$. Perhaps sample 1 consists of one morphological type, and sample 2 of another. Again, we have to be extremely careful of Malmquist bias; the samples may have different distributions of L_A, and any difference in the luminosity distributions of L_B may just reflect this, plus L_A-L_B correlation.

> **Example** Sadler *et al.* (1989) faced a representative problem in this area. They had radio and Hα measurements of a sample that was originally selected at optical wavelengths. Many of the radio- and Hα measurements were upper limits and, moreover, there was good reason to think that both of these variables were intrinsically correlated with optical luminosity.
>
> Sadler *et al.* divided the data into narrow bins of optical absolute magnitude, and then computed distributions of radio luminosity and Hα luminosity, using survival analysis.

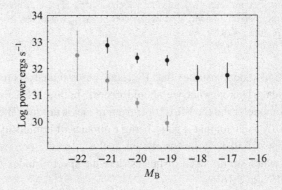

Figure 8.19   The sample of Sadler *et al.* (1989): the 30th-percentile radio power (light symbols) and emission-line power (dark symbols), as a function of absolute magnitude.

As can be seen in Figure 8.19, the 30th percentile of these distributions correlates well with absolute magnitude in each case. However, it is clearly not easy to establish whether radio and Hα luminosity are correlated, given this mutual correlation. Sadler *et al.* (1989) used Schmitt's (1985) factorizability test to show that radio and Hα emission did not correlate.

Error estimates on parameters derived from distributions can be calculated analytically, by likelihood ratios, or by simulation, as discussed for luminosity functions (Section 8.4.1).

8.5.3 Hypothesis testing

If we wish to test two distributions of observations against each other, using detections as well as limits, we have a number of choices. In all cases, however, we have to be aware of how our samples were selected in the first place, in case this forces differences to exist. (And, as ever, we stress that one can only reject the null hypothesis by these tests; see Section 5.1). In general we will expect a problem whenever the variable of interest is correlated with the variable used to define the sample. The Malmquist bias of the defining variable will then be manifest in the other variable. If the bias is not the same for the two samples (and it depends on observational method), then a bogus difference will be detected. Feigelson & Nelson (1985) give a useful introduction to the test statistics available. Distributions for these are known under the null hypothesis, in the asymptotic limit; it is probably best to derive small-number distributions by Monte Carlo or bootstrap simulations.

Some ideas for the test statistic are familiar from the Wilcoxon–Mann–Whitney test (Section 5.4.3). Suppose our two samples are drawn from the same probability distribution. If we combine the two samples we want to test and order them in size – intuitively we expect the observations from the two samples to be randomly intermingled. If the 'rank' (position in the sorted list) of observations from one of the samples were to be, say, systematically low, we would suspect a difference. Evidently a similar procedure could be used for data containing limits, as we would expect limits to be randomly intermingled in just the same way. Constructing a test statistic depends on the penalty we assign for non-random intermingling, and how we distribute this penalty between detections and limits. Feigelson & Nelson (1985) described two variations on this idea, the Gehan and log-rank tests. A major concern in these tests must be the distribution of the limits, as these are affected both by observational technique and by intrinsic differences between the samples. As always, the result of the test will be to give the probability that the differences between the distributions of the data is due to chance. Asymptotic distributions are known for the statistics, but simulation will be more reliable for small samples.

The Gehan test is probably the simplest to use. We describe the procedure for the case of no ties, which can always be arranged for experimental data; the test is somewhat simpler in this case.

Suppose we have two samples of data, labelled A and B, including both detections and limits.

Arrange the *detections* in order; ascending order for data with lower limits, descending order for data with upper limits. Number the observations; this gives each datum a rank. Call the ith rank for data from sample A r_{iA}.

For the ith detection from sample A, calculate n_{iA}, the number of observations of A which are to the right. By 'right' we mean data that are greater than or equal to, the ith observation (in the case of lower limits) or less than or equal to, the ith observation (in the case of upper limits). Thus, this part of the calculation uses the limits.

The number of limits from sample A between detection i and detection $i + 1$ is m_{iA}.

The Gehan statistic is then

$$\Gamma = \sum_{\text{detections in A}} (n_{iA} - r_{iA}) - r_{iA}m_{iA}. \qquad (8.24)$$

This is asymptotically distributed as a Gaussian of mean zero and variance

$$\sigma^2 = \sum_{\text{detections}} n_{iA}n_{iB} \qquad (8.25)$$

in which the assumption of 'no ties' has simplified the formula given in Feigelson & Nelson (1985).

Example We simulated two samples of objects, one drawn from the field-galaxy Schechter function with a characteristic luminosity $L_* = 10$, and the other with $L_* = 30$. In one sample there were 23 detections and 149 limits; in the other, 45 detections and 167 limits. The estimated luminosity functions are in Figure 8.20, and show an appreciable difference. The Gehan test gives $\Gamma/\sigma = 3.3$, significant at the 0.1 per cent level (if the asymptotic approximation holds for these small numbers, this far out in the wings). Simulation could check.

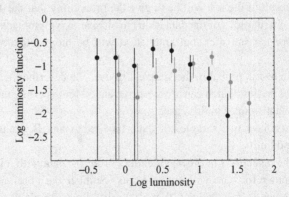

Figure 8.20 Luminosity functions for two simulated samples drawn from field-galaxy Schechter functions (see text), estimated by the Avni method with bootstrap errors. The error bars are the interquartile ranges.

Other possibilities (Feigelson & Nelson, 1985) include a species of Kolmogorov–Smirnov test on the estimated cumulative distributions. The theoretical (and notational, and naming) situation is very complicated; different tests are sensitive to different things, and the best advice is to try each on Monte Carlo simulations of the problem to hand.

One quite simple test (Jenkins, 1989) is based on the likelihood function for the Avni estimator. This test compares the likelihood of two possibilities. In one, the pooled data from both samples are used to estimate a single distribution. In the other, the separate sets of data are used to derive two distributions. The test takes into account the larger number of parameters that are available when dealing with the data separately, and has the same theoretical basis as the methods described for estimating confidence limits on parameters of luminosity functions (Avni, 1978). In detail, if we have Avni estimates $\hat{p}_k^A$ and $\hat{p}_k^B$ for two samples A and B, plus an estimate $\hat{p}_k$ for the pooled samples, we can then compare the log likelihoods with the statistic

$$
-2 \log \Lambda = -2 \left(\sum_{j=1}^{k} n_j \log \hat{p}_j + \sum_{j=1}^{k} u_j \log \sum_{i=j}^{k} \hat{p}_i \right)
$$
$$
+ 2 \left(\sum_{j=1}^{k} n_j^A \log \hat{p}_j^A + \sum_{j=1}^{k} u_j^A \log \sum_{i=j}^{k} \hat{p}_i^A \right)
$$
$$
+ 2 \left(\sum_{j=1}^{k} n_j^B \log \hat{p}_j^B + \sum_{j=1}^{k} u_j^B \log \sum_{i=j}^{k} \hat{p}_i^B \right) \tag{8.26}
$$

in the same notation as before. The test works only if the separate and pooled data are binned into the same k cells, each with at least one detection; in this case the distribution of $-2 \log \Lambda$ is asymptotically chi-square with $k - 1$ degrees of freedom. Experimentally, it is found that it can be quite a long way from chi square with typical amounts of data, and it is best to simulate the distribution. The test is simple to use if the Avni estimators have already been computed; the main nuisance is the need to ensure that the rightmost, or highest-numbered bin, does not contain only a limit. This can be achieved by making this bin arbitrarily large, but it is best to alter the binning scheme in the same way for all the distribution function estimates that are used in the test.

8.5.4 Testing for correlation or statistical independence

Testing for correlation or statistical independence is an area in which survival analysis has something very useful to offer. This is because it deals automatically with the pernicious luminosity–distance correlation that appears in flux-limited samples. Recall that to test for correlation using survival analysis, we need a primary sample, followed by observations of two further parameters. As noted, the Malmquist bias of the primary sample may well affect any conclusions based on resurveying the sample. If we can safely focus on correlations of the two new parameters only, thus assuming that mutual correlations with the primary selection parameter do not matter, then we may use various survival-analysis regression techniques. Because these incorporate limits, they deal automatically with mutual correlations with distance – the bane of any correlation analysis of intrinsic parameters. It remains crucial that the two sets of data are censored in the same way as the distribution of the limits amongst the data can affect the results of tests.

Isobe *et al.* (1986) gave a detailed review, essential reading for application of these types of test. Broadly, we may test for correlation or we may fit regression lines. Isobe *et al.* carried out tests with simulations of flux-limited samples and found several methods which do avoid the trap of the correlation with distance.

The generalized Kendall rank correlation test is fairly simple to use. We start with $n + m$ observations of pairs (X_i, Y_i). In m of these, both variables are detected and the pair is completely known; in the remaining n, either or both of the variables may be censored. Each variable is then ranked. We give a procedure for data with upper limits, but an obvious alternative will work for lower limits. In pseudo-code:

```
create a square matrix a of size (n + m) × (n + m)
initialize it to zero
for each Xi
if Xj > Xi and Xj is detected, set aij = 1
if Xj < Xi and Xi is detected, set aij = −1
```

Repeat this procedure to create a matrix b for the y-variable. This method is assigning a very simple rank, depending on whether a variable is *definitely* known to be bigger than, or less than, the one with which it is being compared.

The Kendall statistic is just

$$\kappa = \sum_{i=1}^{n+m} \sum_{j=1}^{n+m} a_{ij} b_{ij} \tag{8.27}$$

and is asymptotically Gaussian, of variance

$$\sigma^2 = \frac{4}{(n+m)(n+m-1)(n+m-2)}$$

$$\times \left(\sum_{i=1}^{n+m} \sum_{j=1}^{n+m} \sum_{k=1}^{n+m} a_{ij} a_{jk} - \sum_{i=1}^{n+m} \sum_{j=1}^{n+m} a_{ij}^2 \right)$$

$$\times \left(\sum_{i=1}^{n+m} \sum_{j=1}^{n+m} \sum_{k=1}^{n+m} b_{ij} b_{jk} - \sum_{i=1}^{n+m} \sum_{j=1}^{n+m} b_{ij}^2 \right)$$

$$+ \frac{2}{(n+m)(n+m-1)} \sum_{i=1}^{n+m} \sum_{j=1}^{n+m} b_{ij}^2 \sum_{i=1}^{n+m} \sum_{j=1}^{n+m} a_{ij}^2. \tag{8.28}$$

A useful generalization, to cope with partial correlation, is given by Akritas & Siebert (1996).

Example In our usual simulated galaxy sample, we select at one wavelength and then observe at two more. Each of the assigned luminosities is drawn from a Schechter function, and is independent of the others. Retaining the upper limits only, we obtain the convincing 'correlation' between data at the two new wavelengths shown in Figure 8.21. (There are 87 detections of both variables, out of a primary sample of 349.) However, the Kendall rank correlation calculation yields $\kappa/\sigma = 0.56$, showing that the use of upper limits has automatically retrieved the true non-correlation. A correlation was never detected in repeated runs of the simulation. The distribution of κ, for sample sizes of around 30, was quite markedly non-Gaussian. For samples of typical astronomical size (= small), it would be worth estimating the distribution of the test statistic by Monte Carlo.

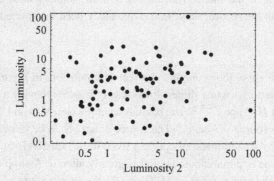

Figure 8.21 The apparent correlation between two new luminosities in the doubly resurveyed sample.

Quite often in astronomy, intrinsic parameters are so widely scattered that it is unrealistic to look for a correlation of X with Y, in the sense of trying to identify some linear relationship plus scatter. It may make more sense to ask if the variables are statistically independent, a more agnostic question. This amounts to asking if the probability distribution $\rho(x, y)$ can be factorized into $\rho_x(x)\rho_y(y)$. Schmitt (1985) developed a useful test for this based on the Avni estimator; it is rather fiddly to use, and there is a detailed discussion of the practical issues by Sadler *et al.* (1989).

8.6 The confusion limit

In many cases of astronomical interest, we find that faint objects are much more numerous than bright ones. Faint objects therefore crowd together; ultimately they start to be unresolved from each other and our signal becomes a mixture of objects of various intensities, blended together by the point-spread function of our instrument. Examples include radio sources, spectral lines in the Lyman-α forest, and faint galaxies observed in the optical and IR wavebands.

The notion of the confusion limit was first developed during a memorable controversy amongst radio astronomers and cosmologists in the 1950s, the source-count/Big Bang/steady state controversy – see Scheuer (1991) for a historical perspective. The root of the problem was instrumental, wildly different source counts being obtained at Sydney (Mills Cross; essentially filled aperture) and Cambridge (interferometer). In an enviable paper written at the heart of the storm, Scheuer (1957) analysed the statistics of the source counts and showed that the Cambridge results were seriously affected by *confusion*. Because of the wide beam of the interferometer, many radio sources were contributing to each peak in the record; these had erroneously been interpreted as discrete sources.

Example To show the pronounced effect of confusion, in Figure 8.22 we show a simulation of a one-dimensional scan of sources obeying a Euclidean source count $N(f) \propto f^{-5/2}$. The beam is a simple Gaussian and there is, on average, one source per beam. (The source count has to be truncated at the faint end to avoid infinities, of course.) Even in this relatively benign case, we see that the apparent source count is greatly altered. A simple count of the peaks in the record gives an ML slope for the source count of -1.7 with standard deviation 0.3 (Section 6.1), very different from the true value. In the case of an interferometer, the presence of sidelobes biases the faint counts to much steeper than true values; the apparent cosmological evolution this implies was the subject of the original controversy.

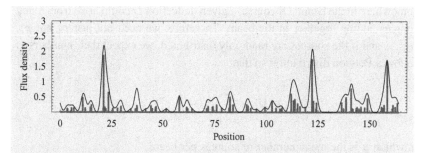

Figure 8.22 A confusion simulation at a level of one source per beam area. The input sources are shown as vertical lines, with the solid line representing the response when observed (convolved) with a Gaussian beam.

The technique developed by Scheuer is known to astronomers as '$p(D)$', or 'probability of Deflection'. The word 'deflection' refers to the deflections of the needle of a chart recorder and is now hallowed by long usage. However, the $p(D)$ technique has been used at the following wavelengths at least: radio (Wall & Cooke, 1975; Windhorst *et al.*, 1993), sub-mm (Hughes *et al.*, 1998), far IR (Glenn *et al.*, 2010), near IR (Jenkins & Reid, 1991), optical (Webb *et al.*, 1992, Lyman-α-line), and X-ray (Scheuer, 1974; Barcons *et al.*, 1994). The method derives the probability distribution of measurements in terms of the underlying source count, which may be recovered by a model-fitting process. Its benefit is that information is obtained from sources which are much too faint to be 'detected' as individuals.

Full details of $p(D)$ are given in the papers of Scheuer (1957), Scheuer (1974) and Condon (1974). However, the steps in the derivation of the distribution are interesting and we outline them here.

Consider a one-dimensional case, for simplicity. A source of brightness f is observed with a beam, or point-spread function, denoted $\Omega(x)$. Here x is the distance (in angle, wavelength or whatever) from where our instrument is pointed. We measure an intensity

$$s(x) = f\Omega(x).$$

If the sources have a source count $N(f)$, so that the number of sources of intensity near f per beam is

$$\int N(f)\Omega(x)\,dx$$

then the observed source count, for just one source in the beam at a time, is the result of a calculation involving conditional probability. From this we obtain $p_1(s)$, the probability of an intensity s resulting from just one source somewhere in the beam. Of course, a given deflection D could arise from many sources adding together in the beam. Therefore, we need not just p_1 but p_2, $p_3, \ldots$ and if the sources are randomly distributed, we expect their numbers to follow a Poisson distribution so that

$$p(D) = \sum_{k=1}^{\infty} p_k(s) \frac{\mu^k}{k!} e^{-\mu}$$

in which μ is the mean number of sources per beam.

To do this summation we need $p_k(s)$; this is the probability distribution of an intensity s which is the sum of k intensities drawn from the distribution p_1. In Section 3.3.3 we showed that the probability of a sum was given by the autocorrelation of the distribution of the terms of the sum, assuming them to be identically distributed. This means that there is a simple relationship between the Fourier transforms:

$$P_k(\omega) = P_1(\omega)^k.$$

Here, upper case denotes a Fourier transform and ω is the Fourier variable.

Putting all this together, we get Scheuer's result for the Fourier transform of the $p(D)$ distribution

$$P(\omega) = \exp(R(\omega) - R(0)) \tag{8.29}$$

in which

$$r(s) = \int N\left(\frac{s}{\Omega(x)}\right) \frac{dx}{\Omega(x)} \tag{8.30}$$

contains the source count N. R is the Fourier transform of r.

Analytic solutions are available for $\tilde{r}(\omega)$ when $N(f)$ is a power law (Condon, 1974), but the inverse transform to get $p(D)$ has to be done numerically. In real life we often need to take account of differential measurement techniques in which measurements from two positions are subtracted to avoid baseline errors (Wall & Cooke, 1975; Wall et al., 1982). In addition the ideal $p(D)$ is always convolved with a noise distribution. All of this needs to be included in the modelling process which recovers the parameters of $N(f)$. The derivation of source counts from $p(D)$ is another technique in which population characteristics are derived from observations of discrete objects or features without forming an object list or catalogue.

Example Wall & Cooke (1975) applied the $p(D)$ technique for filled aperture telescopes to extend the 2.7-GHz radio source counts to much fainter levels than could be achieved by identifying individual sources; their results are shown in Figure 8.23.

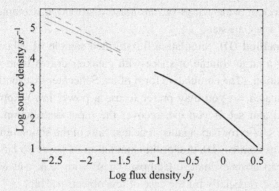

Figure 8.23 The 2.7-GHz counts from Wall & Cooke (1975): the darker line is derived from ordinary source counts with error bars not much wider than the line, while the $p(D)$ results are shown in grey, the dashed lines representing one standard deviation of the fitted parameters.

A more sophisticated version of the technique was subsequently used at 5 GHz (Wall *et al.*, 1982), and data from this experiment are shown in the minimum-χ^2 model-fitting example of Figure 6.3. The technique continues to be used to extend source counts, e.g. Glenn *et al.* (2010), and the counts from deeper survey observations carried through subsequently have invariably shown agreement with the $p(D)$ estimates.

Exercises

8.1 Source counts and luminosity function. Derive the relationship between the number count and the luminosity function for a general luminosity function; show that the result takes a simple form for a power-law luminosity function.

8.2 Noise and source-count slope. Generate data from a power-law source count and add noise; by an ML fit, investigate the effect of the noise level on the inferred source-count slope. Use the results from Exercise 8.1 to show the effect of the noise on the luminosity function.

8.3 Survey completeness and noise. Make a 1D Gaussian signal plus noise plus baseline, fit a profile, verify completeness versus signal-to-noise ratio. Do the same for an empty field.

8.4 Reliability and completeness. Calculate the relationship between reliability and completeness for an exponential noise distribution. This shows the effect of wide wings on the noise distribution. Compare with the result for a Gaussian.

8.5 V_{max} method (D). Simulate a flux-limited sample of galaxies by populating a large volume of space with galaxies drawn from a Schechter distribution. (The cumulative form of the Schechter distribution is rather complicated, so you may prefer to use a power law.) Apply the V_{max} method and see if you can recover the input distribution. Check the simple $\sqrt{N}$ error bars against repeated runs of the simulation.

8.6 Error estimates (D). Adapt the simulation of Exercise 8.5 to produce bootstrap error estimates. Compare these with $\sqrt{N}$ and Monte Carlo estimates, especially for the case of few objects per bin.

8.7 Luminosity–distance 'correlation' (D). Adapt the simulation of Exercise 8.5 to the case for which each galaxy has two independent luminosities assigned to it (at different wavelengths, say). Check that these luminosities show a bogus correlation unless upper limits are included in the analysis. Adapt the simulation to produce intrinsically correlated luminosities and show that the Kendall test can detect these correlations.

8.8 Parameter error estimates. Use the X-ray and radio data from Avni *et al.* (1980), as given in the example in the text, to work out the mean spectral index in their survey. Using their likelihood function as a starting point, work out error bounds on the mean, using a likelihood ratio. You will need to use a Lagrange multiplier in the maximization of the likelihood.

8.9 Source counts from confusion (D). In a confusion-limited survey where there are potentially several sources per beam, the apparent source count can be very different from the true one. On the assumption that sources can lie anywhere in the beam and are not clustered, derive the result for the source count

$$r(s) = \int N\left(\frac{s}{\Omega(x)}\right) \frac{\mathrm{d}x}{\Omega(x)}$$

as given in Section 8.6.

8.10 Monte Carlo and confusion. Design your own confusion-limit experiment. Use a version of the toy universe to derive a source count: produce a patch of sky like that of Figure 2.10; pick a beam size such that your sky survey becomes totally confusion limited; convolve the points in the patch of sky with your beam; compile a $p(D)$ distribution of at least 1000 samples more than one beam area apart; analyse this distribution to find parameters which describe the count below the confusion limit; assess errors; compare with the source count originally prescribed for the source population.

In the exercises denoted by (D), data sets are provided on the book's website; or create your own.

9

Sequential data – 1D statistics

The stock market is an excellent economic forecaster. It has predicted six of the last three recessions.

(Paul Samuelson)

The only function of economic forecasting is to make astrology look respectable.

(John Kenneth Galbraith)

In contrast to previous chapters, we now consider *data transformation*, how to transform data in order to produce improved outcomes in either extracting or enhancing signal.

There are many observations consisting of sequential data, such as intensity as a function of position as a radio telescope is scanned across the sky or as signal varies across a row on a CCD detector, single-slit spectra, time-measurements of intensity (or any other property). What sort of issues might concern us?

(i) trend-finding; can we predict the future behaviour of data?
(ii) baseline detection and/or assessment, so that signal on this baseline can be analysed;
(iii) signal detection, identification, for example, of a spectral line or source in sequential data for which the noise may be comparable in magnitude to the signal;
(iv) filtering to improve signal-to-noise ratio;
(v) quantifying the noise;
(vi) period-finding; searching the data for periodicities;
(vii) correlation of time series to find correlated signal between antenna pairs or to find spectral lines;

(viii) modelling; many astronomical systems give us our data convolved with some more or less known instrumental function, and we need to take this into account to get back to the true data.

The distinctive aspect of these types of analysis is that the feature of interest only emerges after a transformation. Take filtering as a simple example; after smoothing, we are able easily to see the feature of interest in a previously noisy spectrum. But what now? Further modelling is suggested after examining the cleaned-up data, and ideally this will be done following the Bayesian methods of Chapters 6 and 7. In this case, the smoothing may only be used in the exploratory stage of the analysis.

Alternatively, the transformation may be an integral part of the final analysis. If we were looking for periodicity in a data set, the Fourier transform would be an obvious first step, followed by model-fitting to the peaks so revealed. In this case, the statistical properties of the transform are very important for the modelling step.

In this chapter, we discuss by means of examples the statistical and computational techniques employed. We refer to sequential data as 'scans' – they are in many cases, but the sampling may be in the frequency/wavelength domain (spectra), in the time domain (time series) or in the spatial domain (true scans).

The computational aspects alone would justify a large textbook, and we will only give the briefest of outlines; correspondingly, statistical detail here is thinner than in other chapters. Instead, we concentrate on general advice on the statistical issues involved. Excellent detailed guides at graduate level are Bendat & Piersol (1971) and Papoulis & Unnikrishna Pillai (2002).

9.1 Data transformations, the Karhunen–Loeve transform, and others

We are concerned here with expansions in orthogonal functions, a method most familiar from the Fourier series. Moving from one presentation of the data to another may have advantages; noise may be isolated, or features of importance emphasized. Such transformations have a close affinity with principal component analysis (PCA, Section 4.5); the main features can be extracted from a baffling jumble of data. However, what we extract depends entirely on the *basis set* we use. How to use data transformations is a craft, with experience playing a large part as guide.

We start with a scan $f(t)$; t is some kind of sequential or ordered index, time, space or wavelength perhaps. Invariably, f is sampled at discrete intervals and so our data are a finite set $\{f(t_1), f(t_2), \ldots\}$. From a statistical point of view, this set will be described by some sort of multivariate distribution function; to make much progress, we hope it will be Gaussian, in which case the covariance matrix of the f's will be a sufficient description.

We start out by ignoring the (vital) differences between finite-length scans, sampled at discrete intervals, and introduce the ideas with an idealized case. In certain (mathematical) circumstances, a long scan $f(t)$ may be represented by

$$f(t) = \int_{-\infty}^{\infty} F(\omega) B(t, \omega) \, d\omega \qquad (9.1)$$

in which the basis functions are B and the expansion coefficients are F. If our scan $f(t_j) \equiv f_j$ is sampled at discrete times t_j, then the coefficients $F(\omega_i) \equiv F_i$ will have different values for each of the (discrete) values of ω, labelled ω_1, $\omega_2, \ldots$ Our expansion then reads

$$f(t_j) = \sum_i F_{\omega_i} B(t_j, \omega_i). \qquad (9.2)$$

To be useful, we need transformations which can be reversed; in these cases we get equations of the form

$$F(\omega_j) = \sum_i f_{t_i} B'(t_i, \omega_j) \qquad (9.3)$$

with sampling at discrete values of t, and with some known relationship between B and B'. If B is the complex exponential, we have the familiar Fourier transforms and series.

Before we specialize to the Fourier case, we indicate a way of constructing transformations other than the dominant Fourier transforms.

The covariance matrix of the coefficients

$$C_F = \begin{bmatrix} E[F(\omega_1)F(\omega_1)] & E[F(\omega_1)F(\omega_2)] & \cdots \\ E[F(\omega_2)F(\omega_1)] & E[F(\omega_2)F(\omega_2)] & \cdots \\ \vdots & \vdots & \ddots \end{bmatrix} \qquad (9.4)$$

tells us everything we need to know about F, provided that the statistics are Gaussian; the components of F are then described by a multivariate Gaussian. It turns out that a basis set which gives a diagonal C is very efficient at capturing the variance in the data, and then the data variation is compressed as much as possible into the smallest number of coefficients F_ω. Clearly this has advantages for reducing the volume of the data, and may also be useful in isolating noise in some of the coefficients. Requiring that C_F be diagonal leads quite directly to the Karhunen–Loeve equation (Papoulis & Unnikrishna Pillai, 2002) which, for our discrete case, is an eigenvalue problem:

$$R\vec{B} = \lambda\vec{B}. \tag{9.5}$$

The matrix R is closely related to the autocorrelation function which we will encounter soon, and is

$$R_{ij} = E[\,f_n f_{n+(i-j)}\,]. \tag{9.6}$$

(In this equation, we are assuming for simplicity that f has been reduced to zero mean value; we are also assuming that f is *stationary*, which means that R_{ij} does not depend on the index n.) When the basis functions form an orthogonal set, R is evidently just the covariance matrix of the original data components $f(t_i)$. If we have a reasonable model for the statistics of our data, we can construct R and solve the Karhunen–Loeve equations. Now the reason for introducing this approach: the eigenvectors $\vec{B}$ will be discretized basis functions, the B_ω introduced earlier. Depending on the structure of the data, they may indeed be the familiar sines and cosines of Fourier analysis; but other, quite ordinary looking assumptions will yield different basis functions. Fourier analysis is therefore not unique, and if we are interested in data compression we may well want to construct tailor-made functions that do a better job.

Apart from the systematic Karhunen–Loeve method, we may try basis functions from the abundant menagerie of mathematical physics. The many special functions arising in the solution of standard partial differential equations generally have suitable orthogonality conditions, and in some problems may happen to provide just the behaviour needed. An example is the set of Chebyshev polynomials, which, when used as a finite series to approximate a function, will minimize the maximum error. See Andrews (1985) for further information on special functions.

Wavelets represent another possible source of suitable basis functions. We introduce these briefly later in this chapter; at this stage we need only note that this is yet another way of transforming data, which may work well for the particular problem to hand.

By now it will be apparent that choosing a basis set, and using it, is a modelling problem. The more we know about our data, the better the choice we will make, although we may also need specific mathematical properties that go with certain choices. There is therefore the possibility of a thoroughgoing Bayesian approach; we assert Equation (9.2), and with knowledge of the statistics of f, deduce the posterior multivariate distribution of the components F. This, as always, is The Answer; we may propagate uncertainties in a rigorous way through to the final results we infer from the transformation. The limitations, again, are the usual ones: the computational burden, which may be prohibitive, and the lack of useful prior information. If it is possible to be Bayesian, then a conceptually simpler analysis is permitted than the classical approaches we now describe.

9.2 Fourier analysis

The Fourier transform, however, remains king amongst the data transforms, and there are numerous reasons for this. Perhaps the most weighty is a simple practical one – the existence of the fast Fourier transform (FFT), perhaps the most-used algorithm on the planet. We will encounter the FFT later in this chapter.

Many, if not most, physical processes at both macro and micro levels involve oscillation and frequency: orbits of galaxies, stars or planets, atomic transitions at particular frequencies, spatial frequencies on the sky as measured by correlated output from pairs of telescopes. It is natural to examine the frequencies composing data streams. The amplitudes of these frequencies may be the answer (as in the case of detection of a spectral line); they may be adjusted to find the answer (as in digital filtering); or they may reveal unwanted and damaging features in the data stream imposed instrumentally.

In astronomy, as in many physical sciences, there is frequent need to measure signal from a data series. In measuring a specific attribute of this signal such as redshift, the power of Fourier analysis has long been recognized (e.g. Sargent *et al.*, 1977; Tonry & Davis, 1979). Solutions to many questions posed of the data lie in taking the 1D scan to pieces in a Fourier analysis.

Fourier theory (e.g. Bracewell, 1999, and note the simple treatment in the monograph by James, 1995) indicates that any continuous function may be represented as the sum of sines and cosines, i.e.

$$f(t) = \int_{-\infty}^{+\infty} F(\omega) \exp^{-i\omega t} dt, \qquad (9.7)$$

where the function F representing the phased amplitudes of the sinusoidal components of f is known as the Fourier transform (FT).

FTs have a number of enormously important mathematical properties, and these carry over (with some caveats, described in Bracewell, 1999 and Bendat & Piersol, 1971) to the real life of finite-length, discrete transforms, the province of the FFT.

- The FT of a sine wave is a delta function in the frequency domain. This is why we use the FT, or its relatives, to look for periodicities in data.

- The FT of $f \otimes g$, the cross-correlation or convolution of functions f and g, is $F \times G$. Many instruments produce data which result from a convolution with a stable instrumental function: for example, linewidths in spectra are convolutions of intrinsic line shapes with velocity dispersion functions.
- The transform of $f(t + \tau)$ is just the transform of f, times a simple exponential $\exp^{-i\omega\tau}$. Use of this *shift theorem* has measured many redshifts.
- The Wiener–Khinchine theorem states that the *power spectrum* $| F(\omega) |^2$ and the autocorrelation function $\int f(\tau)f(t + \tau)\,d\tau$ are Fourier pairs. The autocorrelation function, as noted earlier, is very closely related to the covariance matrix and hence is a fundamental statistical quantity. Its relationship to the power spectrum is the basis of every digital spectrometer.
- Closely related is Parseval's theorem; this relates the variance of f, and the variance in the mean of f, to the power spectrum. We give the details later; this theorem is very useful in cases where we have correlated noise, especially the prevalent and pernicious '$1/f$' noise discussed below.
- The FT of a Gaussian is another Gaussian. Given the prevalence of Gaussians in every walk of astronomical and statistical life, this is a very convenient result.

Much astronomy deals with uniformly sampled functions, spectra at wavelength intervals, the output of a receiver/bolometer sampled at fixed time intervals, for example. In contrast, time-varying phenomena such as observations of variable stars or quasars require techniques for dealing with irregular sampling and gappy data.

The discrete Fourier transform (DFT) has a number of special features. If the function is sampled N times at uniform intervals Δt in the spatial (observed) frame, the total length in the t-direction is $L = \Delta t \times (N - 1)$, and the result is the continuous function multiplied by the 'comb' function, producing a function $f'(t)$ which (with the interval in spatial frequency as $\Delta v = 2\pi/\Delta t$) may be represented (e.g. Gaskill, 1978) either as a sum of sines and cosines

$$f'(t) = A_n \Sigma \sin(n\Delta v) + B_n \Sigma \cos(n\Delta v); \qquad (9.8)$$

or as a cosine series

$$f'(t) = A'_n \Sigma \cos(n\Delta v + \Phi'_n), \qquad (9.9)$$

where amplitudes A'_n and phases ϕ'_n are given by

$$A'_n = \sqrt{A^2_n + B^2_n}, \quad \phi'_n = \arctan\left(\frac{A_n}{B_n}\right). \qquad (9.10)$$

In the latter formulation, obtaining the DFT produces – by virtue of the 2π cyclic nature of sine and cosine – a 'Fourier-transform plane' for $f'(t)$ which shows the amplitudes mirror-imaged about zero frequency, with a sampling in spatial frequency at intervals of $2\pi/\Delta t(N-1)$ and a repetition of the pattern at intervals of $2\pi/\Delta t$.

There are five criteria for successful discrete-sampling.

1. The *Nyquist criterion* or *Nyquist limit* guarantees that there is no information at spatial frequencies above $\pi/\Delta t$. (Consider the silly case of a signal which is a spatial sine wave of wavelength $2\Delta t$: sampling at intervals of Δt finds points of identical amplitude and thus does not carry information on amplitude or phase of this spatial frequency.) Thus, the sampling interval Δt sets the highest spatial frequency $2\pi/\Delta t$ which can be present; if higher frequencies are present in the data, this sampling rate loses them.

2. At the same time, the *sampling theorem* (Wittaker, 1915; Shannon, 1949) indicates that any bandwidth-limited function can be specified *exactly* by regularly sampled values, provided that the sample interval does not exceed a critical length (which corresponds approximately to half the FWHM resolution, i.e. for an instrumental half-width B, $f'(t) \to f(t)$ if $\Delta t \leq B/2$). In practice any physical system is indeed band-pass limited (although noise added by the subsequent detector is not necessarily so), and therefore with adequate sampling interval, the signal may be fully recovered.

3. To avoid any ambiguity – *aliasing* – in the reconstruction of the scan from its DFT, the sampling interval must be small enough for the amplitude coefficients of components at frequencies as high as $\pi/\Delta t$ to be effectively zero. If $A'_n \geq 0$ for components of frequency this high, the positive high-frequency tail of the repeating $A'(\nu)$ tangles up with the negative tail of the symmetric function repeating about $\nu = 2\pi/\Delta t$ to produce an indeterminate transform.

4. The sampling span or scan length must be long enough. The lowest frequencies which harmonic analysis can delineate are at $2\pi/N\Delta t$. Such low-frequency spatial components may be real as in the case of a stellar spectrum, or may be instrumental in origin as for sky scans with a single-beam radio telescope. In either case, to have any chance of distinguishing the signal from these low-frequency features, the scan length must exceed the width of single resolved features by a factor preferably ≥ 10. This issue of the 'contaminant' low frequencies is considered below.

5. A related point is *windowing*. The DFT simply approximates an infinite integral by a finite one, thus in effect assuming that the function being transformed is zero outside the finite range of integration. Often this is not

the case. From the convolution theorem, it follows that the DFT is convolved with the transform of the *window function*. The simple form of the DFT has, in effect, a window function that is a top hat, zero outside the region of integration and unity within it. Better window functions involve removing trends from the data before transforming (sometimes called *prewhitening*) and tapering the ends of the scan to zero with functions such as the Hanning window. See Press *et al.* (2007) for details.
6. The integration time per sample must be long enough so that the signal is not lost in the noise.

In practice many data sets satisfy these properties; if yours does not, then either take steps – if possible – to ensure that it does, or be aware of consequences. By design, sampling is usually frequent enough to maintain resolution, to obtain spatial frequencies beyond those present in signal, and to avoid aliasing. By design we take spectra or scans over ranges substantially greater than the width of the features. But despite experiment design, the Universe may not oblige with enough photons to satisfy (5), while our instruments or sky + object circumstance may require some analysis to eliminate (4).

9.2.1 The fast Fourier transform

The FFT, discovered by Cooley & Tukey (1965) is a clever algorithm which does the transform of N points in a time proportional to $N \log N$, rather than the N^2 timing of a brute-force implementation. It has a number of quirks, amongst which are its typical arrangement of its output data, and its normalization – see, for example, Bendat & Piersol (1971), Bracewell (1999), or Press *et al.* (2007) for details. Although its discovery defined a generation of signal processing, the algorithm was apparently known to Gauss – even before Fourier had discovered his series in 1808.

9.2.2 Statistical properties of Fourier transforms

For data assessment or model-fitting in the Fourier domain, we need to know the probability distribution of the Fourier components and their derived properties. There are detailed discussions of these matters in Bendat & Piersol (1971).

Most scans will contain both signal and noise. Because 'signal' can mean so many things, the theory typically deals with scans which are entirely noise. To deal with signal, we need a specific model for its form, and an idea that, for example, scan = signal + noise. We 'add in' the noise properties with appropriate analysis, because noise can be characterized more easily and generally than signal. For this comparatively simple case in which the 'data' f are pure Gaussian noise, of known covariance C_f, there are analytical results for the Fourier components, as well as for the power spectrum and the autocorrelation and cross-correlation functions. We will focus on the practical case of the uniformly spaced DFT, as implemented by a standard fast transform. As usual, we assume that f is of zero mean; also that we have just one set of data. Our best estimate $\hat{F}$ is just provided by Equation (9.3), applied to the single set of data we have.

The real and imaginary components of each component $\hat{F}_{\omega_i}$ are then independent Gaussian random variables, and each component is uncorrelated with the others; so the covariance matrix C_F is diagonal. However, this is a very specific result, and depends on doing the discrete transform on data sampled at uniform intervals, returning exactly as many components as there are measured data points. Non-uniform sampling or embedding the data in zeroes to sample the transform more finely, will result in correlations between the components. This is a manifestation of the windowing problem; complicated window functions give complicated correlations between components of the DFT and may move noise from one part of it to another.

A useful result is the following, a version of Parseval's theorem: an estimate of the variance σ^2 in our data $f(t_1)$, $f(t_2)$, ... is just the integral of the estimated power spectrum. For a DFT estimate,

$$\hat{\sigma}^2 = \sum_i \mid \hat{F}_{\omega_i} \mid^2 . \tag{9.11}$$

A related and equally useful result is the variance in the estimated mean of f:

$$\mathrm{Var}[\hat{\mu}] = \mid \hat{F}_0 \mid^2 . \tag{9.12}$$

For both relations, one of several possible scaling factors has to be divided out of the answer, depending on the FFT implementation. Invariably we do not know the value of the power spectrum at zero – it will have been artificially set to zero by prior removal of a mean value to avoid ringing problems with DFT – but we can extrapolate from values of ω where it is known. This is a very useful check in cases where we have correlated noise in the data, and the simple '$1/\sqrt{N}$' rule for the error in a mean will not apply. We discuss this point later in the context of '$1/f$' noise.

The components of the estimated power spectrum $\mid \hat{F}_{\omega_i} \mid^2$, in the simple case, will be distributed like

$$\chi^2 \mid F_{\omega_i} \mid^2$$

with two degrees of freedom in χ^2.

This leads to a surprising and important result. Since this distribution does not depend on the number of observations $f(t_1)$, $f(t_2)$, ... it follows that the DFT method of estimating the power spectrum is *inconsistent*; the signal to noise on the components is unity and does not improve, no matter how much data we have. The reason is simple: longer scans give finer sampling of the transform, degrading the signal to noise at the same rate as the greater quantity of data tries to improve things for us. To improve the signal to noise, we have to average components together, effectively smoothing the power spectrum. This leads to bias errors, for example where sharp peaks in the spectrum will be reduced in amplitude by smoothing. We might try to split up the data into shorter sections, and average the estimated spectra from the short sections, but then the same bias problem will resurface because of the reduced spectral resolution that we will get from shorter scans. In fact the only satisfactory solution is to take more data, and average power spectra at full resolution.

We also notice from this analysis that the distribution function, for the power spectrum components, depends on the 'true' spectrum $\mid F_{\omega_i} \mid^2$, usually what we are seeking.

The same problem surfaces in the case of the autocorrelation or cross-correlation functions; before we can do the error analysis, we need the answer we are looking for, because the true value of these functions enters into the distribution function of the estimates. Worse, the discrete values of the correlations, such as we might obtain via a DFT, are highly correlated amongst themselves. This means that the error level in a correlation function is difficult to represent and we certainly cannot use simple techniques like χ^2 to assign confidence levels to fitted parameters. A typical example of a parameter derived from a correlation function might be the relative velocity between two objects, as determined from the peak in the cross-correlation of their spectra. Simulation is really the only practical way to derive the probability distribution of the measured position of the peak.

Why do we have these perverse difficulties in estimating power spectra? The Wiener–Khinchine theorem tells us that if we know the power spectrum, we know the autocorrelation function, and that means we know the covariance matrix which defines our data f. In other words, estimating a power spectrum is closely allied to estimating a probability distribution function; and here it is familiar that we have a trade-off between signal to noise and bias. Regarding the distribution as a histogram, we can have either large bins (good signal to noise but bias) or narrow bins (the opposite).

A common use of power spectra is the estimation of some instrumental response function. If we have input test data $f(t)$ and output data g, in many cases

$$g(t) = f(t) \otimes h(t) \qquad (9.13)$$

meaning that the instrument introduces a convolution with some response function h. In the Fourier domain we then have

$$G(\omega) = F(\omega)H(\omega). \qquad (9.14)$$

A prominent example of this technique in astronomy is the 'Fourier Quotient', a method which measures velocity dispersion and redshift simultaneously in galaxy spectra (Sargent *et al.*, 1977). Here the input g is a suitable template stellar spectrum, and the output f is the target galaxy spectrum; a model function, containing a velocity dispersion parameter and an overall redshift, is then fitted to G/F. While successful, it turns out that high signal to noise is required in the template spectrum, essentially because of the appearance of its transform in the numerator of the expression for the response function:

$$\hat{H}(\omega) = \frac{\hat{G}(\omega)}{\hat{F}(\omega)}. \qquad (9.15)$$

At values of ω where $F(\omega) \simeq 0$, very wide noise excursions occur in $\hat{H}$. The method has to be used with some care. The quotient may have a non-Gaussian distribution so that goodness-of-fit tests with χ^2 could be very misleading.

To estimate the error distribution of $\hat{H}$ we need the coherence function

$$\gamma^2(\omega) = \frac{\mid F(\omega)G(\omega)^* \mid^2}{\mid F(\omega) \mid^2 \mid G(\omega)^* \mid^2} \qquad (9.16)$$

and we will usually have to insert estimates of all the transforms. Evidently, if there is no noise in the system, we will have $H = G/F$ and $\gamma = 1$. The estimate of H follows an F-distribution

$$\mid \hat{H}(\omega) - H(\omega) \mid^2 \leq \frac{2}{n-2} \mathcal{F}_{2,n-2}(1 - \hat{\gamma}^2(\omega)) \frac{\mid \hat{G}(\omega) \mid^2}{\mid \hat{F}(\omega) \mid^2} \qquad (9.17)$$

in which $\mathcal{F}_{2,n-2}$ is an F-distributed random variable with 2 and $n - 2$ degrees of freedom, and n is twice the number of separate spectral components that are averaged to yield a single estimate; a single component has two degrees of freedom, because of its independent real and imaginary parts. We can see that this is an approximate result for a confidence interval on $\hat{H}$, as the right-hand side contains estimates (and may even contain guesses).

The occurrence of the term $n - 2$ is of importance. It is telling us that we must smooth the power spectrum before we can use it, since the F-distribution is not defined if one of its degrees of freedom is zero. We can also see that errors will be very large at frequencies where the spectrum of the template approaches zero.

The discussion so far has only considered the data $f(t_1)$, $f(t_2)$, ... to be noise; we have not yet added in the effects of systematic signal. In many astronomical problems, matters are a great deal more complicated. The input distribution functions are unlikely to be Gaussian. The central limit theorem will quickly give a Gaussian core to a distribution in many cases, but the wings can dominate the results of statistical testing and will converge slowly to Gaussian form (Newman *et al.*, 1992). Most astronomical data are not just noise; there is a signal, usually also poorly known. Incorporating this into paper-and-pencil statistical analysis is very involved. An example is in Jenkins (1987).

In many cases, the only method for obtaining reliable errors on derived parameters is a detailed Monte Carlo simulation, which can build in all the messy aspects of a real observation. The analytical results we have sketched do, however, provide valuable guidance; they tell us that power spectra will have problems of consistency and bias, that correlation functions will contain highly correlated errors, and that we will probably have to sacrifice detail in estimates of response functions. These are pointers to the behaviour of Fourier analysis in real cases and indicate that we do need a reasonable idea of basic statistical properties – the power spectrum or correlation function – to make much progress in understanding our data when it is in the form of scans.

9.3 Filtering

Filtering is an area in which analysis by Fourier or other techniques can play a significant role. Before we begin filtering data, however, as usual we need to ask what we want to achieve.

Filtering always has two related aims: to reduce noise; and to compress data. Suppose for concreteness that we have a noisy spectrum, containing an emission line. Using a suitable filter (a simple running mean can help) will usually reduce noise and make the line more prominent. What does this achieve? If we want to measure some parameter of the line, say the height, filtering the data may make it possible to make a measurement 'off the screen' with a cursor. This kind of real-time, fast data assessment is a very common application of filtering; it also provides attractive data for publication. However, any more detailed analysis will involve fitting a model, perhaps a Gaussian, to the line. The usefulness of filtering is less obvious here. A fitting procedure requires starting estimates (line location, width, baseline level) to converge to the correct answer; the filtered data will provide these. Also, fitting algorithms will be stabilized, and prevented from converging to wrong answers, if they operate on less noisy data. Unfortunately, since filtering alters the statistical properties of data, the analysis of the fitting procedure will probably be more complicated. However, it is worth remembering that any instrument will filter data to some extent and this effect may have to be modelled anyway.

9.3.1 Low-pass filters

Fourier filtering to improve signal-to-noise ratio can be highly effective. The reason is simple: if noise is shot noise or photon noise, it is 'white', and its spectrum extends flat to the limit given by the sampling theorem. Provided that the signal is not governed by high-frequency components, tapering off the amplitudes of high frequencies is a winning strategy. (Recall that the FT of a Gaussian is another Gaussian, so that if instrumental response or line shape is anything like Gaussian, there should be little high-frequency information.)

It is simple to manipulate the transform of the data to cut out the higher frequencies. An example is shown in Figure 9.1. Whatever we do by chopping out or reducing the amplitudes at high frequencies is bound to decrease the noise, but it must decrease some signal as well, particularly if signal is on small scales in the spatial domain. Chopping is generally a poor idea, however; square filters produce ringing in the signal, so that a tapering to high frequencies is desirable. There are many techniques for assessing how to taper. In fact, assuming that the noise is additive, it is readily shown both by minimizing the variances and by conditional probabilities that an estimate of the optimum filter is given by

$$F(f) = \frac{|\,S(f)\,|^2}{|\,S(f)\,|^2 + |\,N(f)\,|^2},\tag{9.18}$$

where S is the Fourier transform of the signal and N is the Fourier transform of the noise.

This is *Wiener filtering*. It requires us to assess or model the FT of both noise and signal. This is difficult, of course, if signal and noise have similar power spectra, but then, no filter can cope under these circumstances.

Example The example of Figure 9.1 shows a Gaussian sitting on a flat baseline, with Gaussian random noise added, as in a photon-starved observation. It is possible to guess at models in the Fourier plane for the noise and signal components. Here we knew how to model this; we knew the FT of both signal and noise, and as a result, drawing in the separate components in the Fourier plane, making our Wiener filter according to Equation (9.18), is straightforward as the diagram shows; and the result indicates its efficacy. However, we generally do know properties of the signal, from, e.g., instrumental response, so that the signal FT model can usually be constructed. Even without this, the robustness of the procedure is impressive. Take away the eye-guiding lines from panel (b) of Figure 9.1 and approximate the signal with a triangle, say, and the noise with a straight line at some level, and discover that very similar results to panel (d) are obtainable with minimum a-priori knowledge.

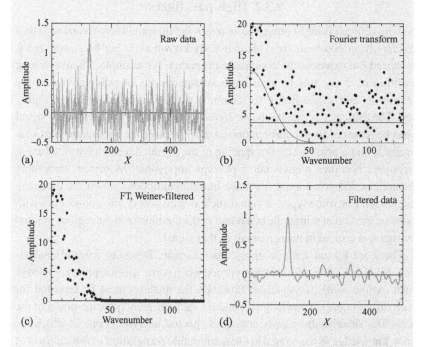

Figure 9.1   A Wiener filter in action. The raw data of (a) is a Gaussian sitting on a flat baseline, with random Gaussian noise added. The DFT in (b) shows the signal and noise components, modelled by the Gaussian and horizontal curves, respectively. The Wiener filter, applied in the frequency domain, produces the DFT of (c), and the reverse transform produces the greatly improved signal/noise of (d).

There are many types of numerical filter available, most of them having been developed for real-time applications. Such filters are *causal*, that is, they only use 'past' data. One of the most famous causal filters is the *Kalman filter*. Most astronomical applications are not so restricted and the main problem is probably the range of choice. Lyons (1997) is a good source of ideas for filtering.

The *Savitsky–Golay filters* are worth a separate mention. These operate by fitting low-order polynomials to a sliding window on the data. Unlike the filters which operate via the Fourier domain, a Savitsky–Golay filter need not inevitably broaden features in order to reduce noise. On the other hand, their effect on data and signal to noise is not as simply visualized. Press *et al.* (2007) show a nice set of examples of Savitsky–Golay filters in action, together with code for an algorithm.

9.3.2 High-pass filters

A more difficult issue is presented in removing unwanted low-frequency components from observations. This is usually known as *fitting baselines*, and it is carried out to assess the continuum in spectra, for example. There is a long tradition of doing this by eye; but least-squares fits of polynomials, heavy smoothing and spline-fitting are common ways of proceeding in the digital age. The difficulty is inevitably *the signal*. Those parts of the scan with signal must be removed from consideration in order to place the continuum; and with irregular and a-priori unknown spacing of the signal, development of a formal technique becomes prejudicial or perhaps impossible. Moreover, smoothing techniques and polynomial fits make initial assumptions which the data may not justify. For some types of signal such as emission or absorption lines with extreme breadth of wings, the behaviour of the continuum in the regions masked by signal is critical in measurement of that signal.

There are formal tools to apply. For example, Bayesian spectral analysis (e.g. Sivia & Carlile, 1992) is appropriate when some specific prior knowledge such as line-width is available. However, the analysis must be repeated for each different prior-knowledge set and for each different question posed of the data. The situation frequently arising in spectral analysis is one in which the prior knowledge is the somewhat unquantifiable recognition of which parts of the spectrum are signal-free, while very general parameter sets (e.g. line-shape, line-width, line-flux, equivalent width, centroid position) may be required from the measurements.

Unlike low-pass filtering in which separation of signal and noise in the Fourier plane is relatively straightforward, the problem here is the tangle between the two. Gaussians helped us in the former because their transforms drain away so fast at high frequencies. But at low frequencies? This is where Gaussians have most of their harmonic signal. We have to be cleverer than just staring at the transform.

A simple technique, *minimum-component filtering*, based on DFT and harmonic analysis is described by Wall (1997). The key to success is forming a baseline array by patching across regions in which signal is clearly present. The sequence to follow is this:

(i) *patching*: forming a 'baseline array' from the original data series by patching across regions of the scan where signal is evident;

(ii) *end matching*: subtracting from this baseline array a first approximation to the patched scan, obtained with a linear fit, a very low order polynomial or a heavy-smoothing estimate;

(iii) *Fourier transforming* the resultant baseline array;

(iv) *removal of the high frequencies* by applying a heavy-taper multiplicative filter in the Fourier plane to taper off the higher-frequency Fourier amplitudes;

(v) *reverse transforming* using these minimum remaining components; and

(vi) *gradient restoration*, by adding back in the first approximation (step (ii)) to the baseline.

We are concerned with data sets of the type represented by the optical spectra in Figure 9.2. The appearance is dominated by substantial emission or absorption lines covering more than 30 per cent of the length of the spectrum; moreover in the second case, the continuum slope is severe.

In addition to its objectivity and its ease of application, a further advantage of the technique is that an analysis of the error introduced by the baseline assessment can be carried out (Wall, 1997), an aspect seriously lacking in most baseline assessment (and therefore line-intensity estimation) in the literature. The following points emerge:

- Except for very narrow signal, the baseline difference, i.e. the error in continuum assessment due to minimum-component fitting, will not dominate errors.

- The patch width is critical. For noisy situations in which the signal is relatively weak, it is imperative to choose a patch width as narrow as possible. Gaussians cut off quickly and for the weaker signals, patch and flux measurement should be confined to $\pm 3\sigma_s$, as determined from an accurate estimate of σ_s (the instrumental standard deviation). For strong signals, the patch should be significantly broader.

- For even the weakest apparent signals it is crucial to patch over the region in which signal measurement is carried out. This may seem self-evident, but when signal is weak it is tempting to fit a minimum-component

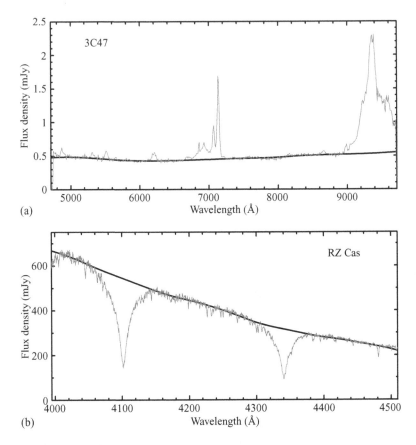

Figure 9.2   (a) A spectrum of 3C47 obtained by Laing *et al.* (1994) with the faint
object spectrograph of the William Herschel Telescope, La Palma. The redshift
is 0.345; broad lines of the hydrogen Balmer series can be seen, together with
narrow lines of [OIII]. (b) A spectrum of the A star RZ Cas (Maxted *et al.*, 1994).
The continuum obtained with the minimum-component technique is shown as the
black line superposed on the original data.

continuum with minimal patching because the fit *looks* satisfactory. This is
not so; the error introduced in the flux measurement may far exceed the noise
uncertainty.

The analysis indicates how rapidly errors in equivalent widths can esca-
late with continua which are curved, i.e. which have low-frequency components present, even when the procedures for continuum assessment and signal measurement are well defined. When yet broader wings are involved, the
errors produced will be substantially greater. The analysis goes some way to

explaining why estimates of line fluxes in the literature can differ by a factor of two, even with reasonable signal to noise.

It should be noted that 'removal' of continuum is *high-pass filtering*, removal of the lowest frequencies. In conjunction with low-pass filtering, a *band-pass* filter has been generated, one which cuts off towards low frequencies *and* towards high frequencies.

9.3.3 An integrated approach

We see that analysis of a scan will often involve some kind of baseline-fitting procedure, plus a localized fitting procedure to derive, say, line widths and positions. The baseline parameters are only required as a step on the way to some final answer and so are classical nuisance parameters.

From a Bayesian point of view, we may be able to formulate the whole problem as a standard model-fitting procedure. From this we will derive joint posterior distributions for line parameters (the interesting ones) $\vec{\lambda}$ and for baseline parameters $\vec{\beta}$. Since the baseline parameters are not required, we can marginalize them out with an integration over $\vec{\beta}$ and its prior. This gives us the distribution of the interesting parameters, with the effects of the uncertainty in the baseline included.

It may be difficult to formulate the baseline problem in such a conceptually clear way; as we have seen, a good deal of judgement may be involved. However, in principle, each of the human decisions involves a set of parameters. We should be able to formulate a Bayesian estimation procedure for these, and this will have benefits in making the procedure more objective, and reproducible.

Quite often, baselines will be the result of '$1/f$' noise (Section 9.7). This results in the sort of large, aimless wanderings that are quite difficult to fit with harmonics. The usual remedy is to fit polynomials to the lowest frequencies. Another possibility, if the baseline statistics (power spectrum or correlation function) are reasonably well known, is to construct the associated Karhunen–Loeve functions. By definition, using these to approximate the baseline will do the best possible job with the smallest number of coefficients.

9.4 Correlating

9.4.1 Redshifts by correlation

We have seen how the shift theorem can be used, in the Fourier quotient method, to measure a redshift. Redshifts are a common and important example of an offset between two scans. As we have seen, there are some disadvantages to the quotient method. Direct *cross-correlation* between a template and target spectrum will generally yield a peak in the cross-correlation function; a modelling procedure can give redshifts and velocity dispersions (Sargent *et al.*, 1977; Tonry & Davis, 1979). This is a successful and widely used technique. The best method for error analysis in this case is direct simulation, because of the highly correlated nature of the errors in a correlation function.

9.4.2 The coherence function

We have met the coherence function between scans f and g briefly before; it is estimated by

$$\hat{\gamma}^2(\omega) = \frac{\mid \hat{F}(\omega)\hat{G}(\omega)^* \mid^2}{\mid \hat{F}(\omega) \mid^2 \mid \hat{G}(\omega)^* \mid^2}. \tag{9.19}$$

The estimation is done in the usual way for power spectra: either by smoothing the power spectrum, or by averaging several power spectra derived from separate scans. The coherence function is just the correlation coefficient between f and g in frequency space.

The coherence is extremely useful in cases where we have an input f and output g and we want to find out more about the 'black box' that changes f into g. If the box has a purely linear effect (like many simple physical systems), then $g = f \otimes h$ for some h, and $\gamma = 1$. More usually, of course, we have noise ϵ, not present in the input, so that $g = f \otimes h + \epsilon$. Now, depending on the frequency content of the noise and the input, we will have structure to γ, which will generally be less than one. Other interesting reasons for $\gamma < 1$ will be that the causal relationship between f and g is non-linear, or that there are extra causal factors in play. These will not be present in the input we know about, and, depending on their frequency content, will lower the coherence.

Example Here is a simple example with some simulated data. We have a relationship for our synthetic data

$$g(t) = f \otimes h + \epsilon(t) + b(t)$$

in which f is white Gaussian noise, h is a Gaussian filter, ϵ is noise added at the output side of the box, and b is an unrelated low-frequency effect (obtained in this case by vigorous recursive filtering of Gaussian white noise).

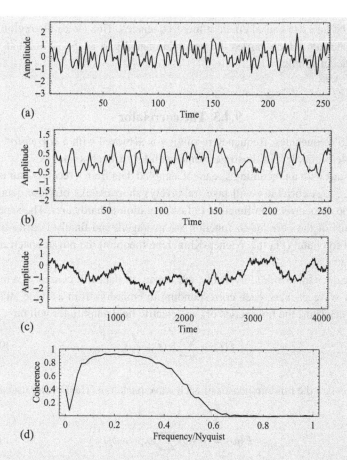

Figure 9.3   (a)–(c) The input to the system f; the input, convolved, with a small amount of noise added $f \otimes h + \epsilon(t)$, the extraneous effect $b(t)$, and (d), the coherence between f and g.

In Figure 9.3 we see the input data, the output (somewhat filtered, apparently) and the extraneous low-frequency effect. Finally the coherence function between f and g shows a loss of coherence at low frequencies (because of the extraneous effect, which is not present in the input) and the loss at high frequencies (due to noise, which is likewise not present in the input). At intermediate frequencies there is a region relatively unaffected by noise, in which our box must be a linear system, where only the input f affects the output. This means that we can model our box as a simple convolution of input data with an instrumental function and we also suspect that there must be an extra causal effect at low frequencies. This yields the region of the spectrum which carries the uncontaminated part of the signal that we can model simply.

9.4.3 The correlator

At radio frequencies, frequency resolution is achieved with a *correlator*. The principle is simple, but illustrates some useful statistical points.

We start with an incoming stream of sampled data from a receiver, our usual $f_{t_1}, f_{t_2}, \ldots$ A correlator will take (relatively) short chunks of these data and form the autocorrelation function (a fast operation in hardware). The separate estimates of the correlation function are averaged, and finally Fourier transformed to obtain (via the Wiener–Khinchine theorem) the power spectrum of the data.

Why does this work? Physically, our stream of data will consist of a multitude of wave packets, each corresponding to emission from a single atom or molecule. Thus, the time series of, say, electric field amplitudes will be

$$f(t) = \sum_i w(t + \phi_i), \qquad (9.20)$$

where ϕ_i are the random phases of each wave packet w. The Fourier transform is

$$F(\omega) = W(\omega) \sum_i \exp(\iota \omega \phi_i) \qquad (9.21)$$

and the average power spectrum will be

$$\mid W(\omega) \mid^2 E[\mid \sum_i \exp(\iota \omega \phi_i) \mid^2]. \qquad (9.22)$$

The exponential term, being an average of positive quantities, will converge to some positive value as more and more chunks of data are averaged – even although the phases are random. By contrast, the average Fourier transform will contain the term

$$E\left[\sum_i \exp(\iota \omega \phi_i) \right]$$

which will converge to zero.

Example A simulation of this procedure is shown in Figure 9.4.

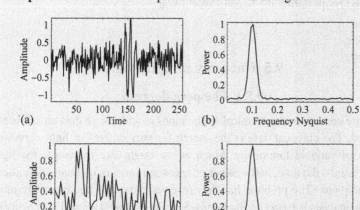

Figure 9.4 (a) Part of the input data stream for the correlator, consisting of 64 wave packets, randomly located, with on average one per 128 units of time. (b) The derived power spectrum from forming the autocorrelation function over 128 time units. (c) The average Fourier transform of 1-bit quantized data, again averaged in 128-long chunks. (d) The power spectrum derived from the quantized data with the same averaging.

A further key feature of the digital correlator is the quantization – astonishingly little sensitivity is lost by digitizing at the 1-bit level, simply recording whether f is positive or negative. This speeds up data rates and reduces numbers of operations; for a given processing speed far higher resolution is possible, this dependent on the number of channels in the shift-and-add of the correlator, rather than the sampling speed of the data (as long as this is high enough to exceed the Nyquist criterion).

Given the correlation coefficient ρ between the data values f_{t_i} and f_{t_j}, we know that the joint distribution function is a bivariate Gaussian. It is then a fairly simple marginalization calculation (see Chapter 8 of Thompson *et al.*, 2001) to compute the probabilities of quantized values like

$$\mathrm{prob}(f_{t_i} > 0 \text{ and } f_{t_j} > 0)$$

and so on; from this, the quantized correlation coefficient ρ_q can be calculated. The result is beautifully simple:

$$\rho_q = \frac{2}{\pi} \sin^{-1} \rho \qquad (9.23)$$

and is called the van Vleck equation. It has the distinction of having been a classified result during the Second World War.

9.5 Unevenly sampled data

9.5.1 The periodogram

There are numerous astronomical applications in which scan data are unevenly sampled. The classical case is the search for periodicities in light curves of objects of variable luminosity. Much as we might like to sample the light output evenly, daytime, bad weather or telescope-time-assignment committees may intervene. The problem has thus received extensive treatment and most modern analysis is based on the Lomb–Scargle method (Lomb, 1976; Scargle, 1982). A concise description of the Lomb normalized periodogram is given by Press *et al.* (2007). The key feature is that the method weights the data on a 'per point' basis rather than on a 'per time interval' basis as does the FFT; an even better feature is that the null hypothesis can be tested rigorously. Scargle (1982) showed that in testing for a peak at a predefined frequency ω, the height of the peak at this point, $Y(\omega)$, has an exponential probability distribution with unit mean; the probability that $Y(\omega)$ lies between $Y(> 0)$ and $Y + dY$ is $\exp(-Y)dY$. (Note that this is not the same as the probability of a certain peak height when the peak is selected from the spectrum as being the biggest; there order statistics apply – see Section 3.4.) If n independent frequencies are considered, then the probability that none gives a value $> Y$ is $(1 - e^{-Y})^n$. Thus

$$P(> Y) = 1 - (1 - e^{-Y})^n \qquad (9.24)$$

represents the significance level of any peak $Y(\omega)$. But this raises the embarrassing question of what is n – how many independent frequencies have we looked at? In the limit of interest, when significance levels are $\ll 1$, $P(> Y) = ne^{-Y}$, scaling linearly with the estimate of n, so that n need not be estimated precisely. Monte Carlo experiments (Horne & Baliunas, 1986) show that if N is the number of scattered but approximately evenly spaced data points which oversample the range up to the Nyquist frequency, then $n \sim N$, and there is little difference for n between random spacing and equal spacing. When a larger frequency range is sampled, n increases proportionally.

These points raise two further questions. Firstly, how can we sample frequencies beyond the Nyquist? Recall that the Nyquist frequency refers to *equally spaced data* with sampling interval Δt; it is $2\pi/\Delta t$. With randomly spaced data evenly distributed through the sampling series, an equivalent (but non-physical) Nyquist frequency can be obtained from the mean time-interval. However, the fundamental limitation of equally spaced data is avoided by unequally spaced data. It is possible to sample well above the equivalent Nyquist frequency without significant aliasing; see the example of Figure 9.5. A similar situation arises with 2D and 3D statistics of space distribution, as discussed in Chapter 10; clustering on scales much smaller than the mean separation between objects can be sampled if the objects are randomly sampled.

Example Figure 9.5 demonstrates the operation of the Lomb–Scargle periodogram, continuous and gappy data, as implemented with the *Numerical Recipes* routines of Press *et al.* (2007).

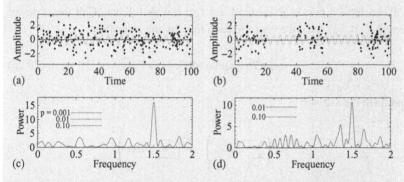

Figure 9.5 The Lomb–Scargle periodogram method. (a) and (c) Randomly spaced data generated by a sine wave of amplitude 0.7 and period 1.5 with noise of unit variance superposed. (b) and (d) Data taken at the same sampling rate but with serious gaps to approximate night-to-night sampling of optical astronomy.

For the continuous data, even with the sine wave shown as a guide, the eye cannot pick out the periodicity. The Lomb–Scargle periodogram has no doubts whatsoever. For the gappy data, note the reduced significance of the peak, as well as the serious aliasing resulting from windowing the data as shown.

Secondly, what about the more usual situation for astronomers with data seriously clumped, e.g. into nighttime observations? Monte Carlo to the rescue again: generate synthetic data sets by holding fixed the number of data points and their sampled locations, generate synthetic sets of Gaussian noise using these, find the largest values of $Y(\omega)$, and find the best fit of the distribution in

Equation (9.24) to determine n. The example of Figure 9.5 shows what happens with gappy data: aliasing becomes serious. With data of even poorer quality than that shown in Figure 9.5 (no problem for astronomers), choosing the right peak amongst these is the issue; and here, folding techniques come into play. In the simplest instance, observing a similar data stream some time later will enable a choice to be made; only one of the frequencies will have the right phase to give anything like a reasonable fit.

9.5.2 Times of arrival

A rather different kind of unevenly sampled data arises in pulsar timing or gamma-ray astronomy. Here we sometimes have rather small numbers of events, times of arrival of pulses or photons. Do these times betray a period?

If we have a period P in mind, we can test as follows. Call the times of arrival $t_1, t_2, \ldots$ Assign a phase to each time by the algorithm

$$\phi_i = 2\pi \ [\text{remainder of } (t_i/P)] \tag{9.25}$$

and form the statistic

$$R^2 = \left(\sum_{i=1}^{n} \cos \phi_i \right)^2 + \left(\sum_{i=1}^{n} \sin \phi_i \right)^2 \tag{9.26}$$

and for $n > 10$, $2R^2/n$ is distributed as χ^2 (Table B.6) with two degrees of freedom.

This is a classical test (the *Rayleigh test*). If R is large, it is unlikely that the phases are random. This will happen if we have guessed the correct period, so we would then infer (illegally, of course) that the period is indeed P.

We may also wish to determine P, which we would do simply by searching for a value of P that maximizes R. Having determined a parameter from the data, we will lose one degree of freedom from χ^2 in the significance test.

Details of this, and more elaborate tests, are in De Jager *et al.* (1989).

9.6 Wavelets

One disadvantage of Fourier analysis, and its relatives, is a loss of information about where in a scan things may be happening. Take the spectrum of Figure 9.1

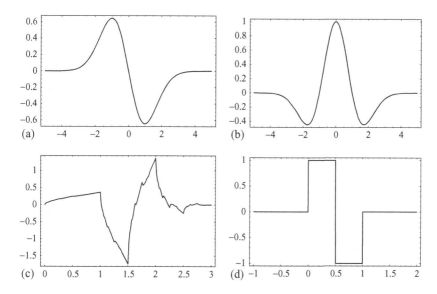

Figure 9.6   Four possible wavelets; a wavelet decomposition will be in terms of scaled and translated versions of each of these. (a) Asymmetrical; (b) Mexican hat; (c) Daubechies (this wavelet is actually a fractal); (d) Haar.

as an example; the noise level might well be different in the spectral line, but a Fourier filter applies the same degree of smoothing everywhere. This feature is a result of the basis functions, the sines and cosines, being infinite in extent. In fact their infinite extent is the cause of many of the difficulties associated with transforms of finite-length data streams.

In many cases we would like a transform which picks out details of frequency content while preserving information about where in the scan those particular frequencies are prominent. There are approximate ways of doing this with Fourier transforms, by taking transforms in short windows which slide along the data, but this has obvious disadvantages. *Wavelets* offer a better way.

A wavelet is a short function which, being convolved with the data, gives some frequency (or scale) information at a particular location in the scan. By placing the wavelets at different places in the scan ('translating') and changing their widths ('scaling') it is possible to obtain a frequency decomposition which preserves some location information. Figure 9.6 shows some examples of wavelets in current use; it is worth noting that there are particular mathematical restrictions on what kind of function can be a wavelet. As can be seen from the figure, different wavelets are likely to be sensitive to different things; the asymmetrical wavelet, for example, will be sensitive to local gradients, while the Mexican hat will be good at picking out oscillations.

Wavelet analysis is a huge and growing field; useful references include Daubechies (1992), Koornwinder (1993), Strang (1994) and Bruce *et al.* (1996). An implementation of a discrete wavelet transform is given in Press *et al.* (2007), along with the usual wise advice. Walker (1999) is an excellent textbook.

Much of the attraction of wavelets is that they can give very effective filtering and data compression. A well-known triumph for wavelets was the decision by the FBI to use a wavelet-based technique for the digitization and compression of their fingerprint database. From an astronomical point of view, we often deal with scans where important properties change from place to place; there are noisy regions in a spectrum, for instance, or times when a light curve seems to show quasi-periodicity. Wavelets offer new possibilities in data assessment, and a whole new armoury of filtering techniques, especially those requiring different filtering in different parts of a scan.

Example In Figure 9.7 we compare wavelet filtering to the Wiener filtering of the example spectrum in Figure 9.1. We chose to filter with Haar wavelets; others were not as satisfactory.

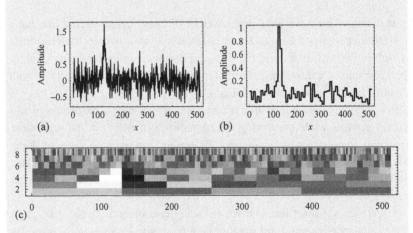

Figure 9.7   (a) The original spectrum of Figure 9.1; (b) the filtered spectrum; (c) the wavelet coefficients as a function of location (horizontal axis) and scale (vertical axis, finest scales at the top).

The greyscale plot shows the strength of the various wavelets as a function of position in the spectrum. The finer-scale wavelets are at the top of the plot. From this plot, dropping the three finest scales of wavelet coefficients is suggested as a suitable simple filter. The result is quite pleasing as noise is markedly reduced without much loss of resolution in the spectral line.

One useful aspect of the wavelet transform is that it gives some information about when (say, in a time series) particular frequencies occur. Similar information can be provided, in a pragmatic way, using the FT. If a time series be divided into sub-sections, and the power spectrum formed for each sub-section, these can be stacked in a diagram rather like Figure 9.7. The limitation is the short duration of each sub-section, with corresponding loss of resolution in frequency. Pattern recognition applied to such diagrams, looking for signatures in frequency and time, is the basis of the tune-recognition application Shazam, for example.

9.7 Detection difficulties: $1/f$ noise

The bane of the experimenter's life is so-called $1/f$ noise; see the excellent review by Press (1978). It is a major reason why filtering theory, which looks so good in contrived examples (Figure 9.1) fails to live up to its promise; why 2σ results are not results; and why increased integration time fails to produce the expected improvements in signal-to-noise ratio, the simple $1/\sqrt{N}$ improvements we naively expect from averaging N samples.

$1/f$ noise is so called because it has a power spectrum which is inversely proportional to the Fourier variable – frequency, if we are dealing with a time series. Hence the name. It is sometimes called flicker noise, and is a particular case of 'pink' noise of various kinds, in which low frequencies dominate. An even more extreme example is Brownian or random-walk noise. As the name suggests, this arises when successive values of the noise are obtained by adding a random number to the previous value. Random-walk noise arises when we integrate a scan; for example, we may integrate a time series of accelerations to deduce the velocity time series.

Example Figure 9.8 shows two simulations of low-frequency noise, obtained by starting with white noise, multiplying the Fourier transforms by $1/\sqrt{f}$ or $1/f$, and taking the inverse transform of the result. The first case gives $1/f$ noise (recall that the $1/f$ dependence is in the power spectrum, not the transform) and the second gives a random walk or Brownian motion, which has a steep $1/f^2$ power spectrum.

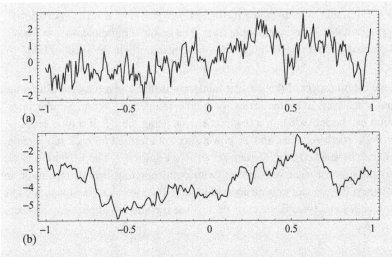

Figure 9.8   (a) Flicker $(1/f)$ noise of unit variance. The data are implicitly band-limited by the finite sampling rate. (b) A random walk of unit variance.

Despite much theoretical work (it crops up in everything from Beethoven symphonies to traffic flow), it is not known why $1/f$ noise is so common. However, its presence (or the presence of one of its near relatives) is usually the reason why averaging large amounts of data does not produce the improvement expected.

$1/f$ noise has the remarkable property of having infinite variance: the longer you watch it, the larger its excursions become. We can immediately see that this behaviour follows from the fact that the variance on a scan is the integral of its power spectrum. For sampled data of finite length, the variance will depend on the integral of the power spectrum between the Nyquist frequency and the first frequency above zero – this will be $1/L$ if L is the scan length. Thus, the variance of sampled $1/f$ noise will be proportional to

$$\int_{1/L}^{f_{Nyq}} \frac{1}{f}\,\mathrm{d}f = \ln(Lf_{Nyq}) \tag{9.27}$$

and so will grow logarithmically with scan length. Qualitatively, we see from the simulations in the example that the noise is highly correlated from one sample to the next; we expect that averaging will not work well. In fact it does not work at all for a noise spectrum of $1/f$ or steeper.

Recall from an earlier result (Section 9.2.2) that the variance on a mean $\hat{\mu}$, derived from a scan f of length L, is

$$\mathrm{var}[\hat{\mu}] =\mid F(0)\mid^2 /L. \tag{9.28}$$

For white noise, which is uncorrelated for adjacent or successive samples of f, we have $\mid F(0) \mid^2 = \sigma^2$, the variance on the scan, and the expected $1/\sqrt{L}$ behaviour follows. If, however, we have $1/f$ noise dominating the power spectrum at low frequencies, the best idea we have of the power spectrum at zero is its value at a frequency $1/L$. Now we see that the variance on the mean is independent of L!

Usually we will have white noise dominating the power spectrum for frequencies greater than some value ω_0, in other words for scans shorter than $1/\omega_0$. As the scans lengthen, however, we will start to uncover the $1/f$ noise below ω_0. Therefore, a general model for the variance on the mean level of a scan of length L will be

$$\text{var}[\hat{\mu}] \simeq \frac{a}{L} + b, \tag{9.29}$$

where a and b are parameters which describe the noise levels in white noise and $1/f$ noise. Note the analogy to the discussions of low-pass and high-pass filtering (Section 9.3): dealing with the slowly varying component may be considered as a baseline issue. Of course we must hope that our signal comes at higher frequencies, or it (and our experiment) is lost.

Because the variance diverges for $1/f$ noise, another measure of variation – the Allan variance – is useful. For our scan f it is defined by

$$\sigma_A^2 = \frac{1}{2} E[\, (f_{t_i} - f_{t_{i-1}})^2 \,] \tag{9.30}$$

and the differencing of successive samples will help to remove the enormous variance which is carried in long-term drifts. Generalizations can be made by changing the distance between the two samples involved; this is the scale of the Allan variance. There is a fascinating connection between the Allan variance and wavelets: the Allan variance is directly related to the variance in the Haar wavelet coefficient, at the same scale.

Exercises

9.1 Fourier transform and FFT. Use a direct numerical integration to do a numerical Fourier transform of an oscillatory function, say a sine wave or a Bessel function. Compare the timings with an off-the-shelf FFT routine, checking how many oscillations you can fit in your region of integration before the FFT accelerates away from the direct method.

9.2 Wiener filtering and $1/f$ noise (D). Make some synthetic data along the lines of the example in Figure 9.1, and make it work with a Wiener filter for uncorrelated Gaussian noise. Now generate some $1/f$ noise. Add this into the input spectrum, and perform the filtering again, without taking account of the extra low-frequency noise in the form of the Wiener filter. Does the $1/f$ noise affect (a) the line profile parameters, (b) the baseline parameters?

9.3 **Periodogram (D).** Consider the Lomb–Scargle periodogram method as formulated by Press *et al.* (2007); use the *Numerical Recipes* routines to test the following issues.

(a) If we can sample at much above the pseudo-Nyquist rate, how much? Where does this run out? Why in practice can we not realize the sampling at these high frequencies provided by scattered time measurement?

(b) The lines of probability in Figure 9.5 are roughly correct for the random uniform coverage of the left set of data. For the data on the right, uniformity has been assumed and the probabilities in the diagram are incorrect. Use the *Numerical Recipes* routine and the Monte Carlo technique outlined to determine how they should be adjusted.

9.4 **Properties of the power spectrum of periodic data.** From the max/min statistics analysis of Section 3.4:

(a) Find the probability density function equivalent to Equation (9.24) for *minimum* values.

(b) Show that the peak of the distribution function of the maximum in the power spectrum of data N long is $ln N$.

9.5 **Power spectrum of signal + noise.** For a signal containing a deterministic signal S and Gaussian noise x, show that the noise distribution in each component of the power spectrum is in general a non-trivial combination of χ^2 and Gaussian noise.

9.6 $1/f$ **noise.** Harmonic analysis (sampling, Fourier transforming) of Beethoven's symphonies indicate that their power spectra follow the $1/f$ law to a good approximation. Consider why this should be so. See Press (1978) for a few hints.

9.7 **Filtering and mean values.** Take your favourite implementation of the FFT, and form the power spectrum of a scan consisting entirely of uncorrelated Gaussian noise. Integrate the power spectrum; is the answer the variance of the input data? If not, why not? Now convolve the data with your favourite (normalized) filter. From the zero frequency of the power spectrum, what is the variance in the mean? Does it change if you change the width of the filter? Explain.

9.8 **Baselines (D).** Fit a Fourier baseline interactively to a spectrum containing a moderately obvious but contaminated line. Now, separately, fit a Gaussian to the line and give your best estimate of the uncertainty in the total flux in the line. Compare this with a complete Bayesian analysis, fitting the same number of harmonics plus Gaussian *ab initio*, and then marginalizing out the baseline parameters.

In the exercises denoted by (D), data sets are provided on the book's website; or create your own.

10

Statistics of large-scale structure

An examination of the distribution of the numbers of galaxies recorded
on photographic plates shows that it does not conform to the Poisson law
and indicates the presence of a factor causing 'contagion'.

(Neyman et al. 1953)

God not only plays dice. He also sometimes throws the dice where they
cannot be seen.

(Stephen Hawking)

The distribution of objects on the celestial sphere, or on an imaged patch of
this sphere, has ever been a major preoccupation of astronomers. Avoiding here
the science of image processing, province of thousands of books and papers,
we consider some of the common statistical approaches used to quantify sky
distributions in order to permit contact with theory. Before we turn to the
adopted statistical weaponry of galaxy distribution, we discuss some general
statistics applicable to the spherical surface.

10.1 Statistics on a spherical surface

The distribution of objects on the celestial sphere is the distribution of directions
of a set of unit vectors. Many other 3D spaces face similar issues of distribution,
such as the Poincaré sphere with unit vectors indicating the state of polarization
of radiation. Geophysical topics (orientation of paeleomagnetism, for instance)
motivate much analysis.

Thus, this is a thriving sub-field of statistics and there is an excellent handbook (Fisher *et al.*, 1987). The emphasis is on statistical modelling and a variety
of distributions is available. The Fisher distribution, one of the most popular,
plays a similar role in spherical statistics to that played by the Gaussian in
ordinary statistics.

In astronomy we usually need different distributions, often those resulting from well-defined physical processes. The distribution of galaxies within clusters is an example. These distributions remain poorly understood and so the emphasis is on non-parametric methods. Here, spherical statistics does have some useful techniques to offer.

If we have a set of directions, defined by n unit vectors $\{X_i, Y_i, Z_i\}$ in a Cartesian system, we can ask if they could have been drawn from a uniform distribution over a sphere. *Rayleigh's test* forms the statistic

$$\mathcal{R}^2 = \left(\sum_{i=1}^{n} X_i\right)^2 + \left(\sum_{i=1}^{n} Y_i\right)^2 + \left(\sum_{i=1}^{n} Z_i\right)^2 \tag{10.1}$$

and for $n > 10$, $3\mathcal{R}^2/n$ is distributed as χ^2 (Table B.6) with three degrees of freedom. For $n < 25$, use the tables of critical values of $\mathcal{R}$ in Table B.13.

If the directions are not uniformly distributed, a useful estimate of their direction is the spherical median. This statistic (call it $\vec{M}_s = \{\lambda, \mu, \nu\}$) minimizes the sum of the arc lengths from each datum. The sum

$$\sum \arccos(X_i\lambda + Y_i\mu + Z_i\nu)$$

usually has to be minimized numerically to solve for the parameters λ, μ and ν. There is an asymptotic distribution available for $\vec{M}_s$, but its calculation is rather complicated. A bootstrap (Section 6.6) will give prob(M_s) directly.

Example The 10^6 sight lines to objects in our toy universe of Section 2.6, Figure 2.10, were chosen at random. We used Rayleigh's test to check this statement, and we found a value for $3\mathcal{R}^2/n = 1.59$, a highly acceptable value of χ^2 for 3 degrees of freedom, indicating no reason to suspect that our chosen directions were not random.

Next we considered just the first 50 sight lines and changed 10 of these from random to coincide with the z-axis. The result for $3\mathcal{R}^2/n$ is now 12.0, a value of χ^2 rejecting uniformity at the 0.01 level of significance. This seems too decisive for such a small sample, so we repeated the exercise for nine further selections of 50 directions, changing 10 each time to coincide with the z-axis. Only one of these realizations had a higher value of $3\mathcal{R}^2/n$. In fact we found only 5/10 to show 'rejection' of uniformity at a significance level better than 0.1, while the rest could be deemed to show no strong evidence of non-uniformity. This demonstrates the need to determine the distribution of the test statistic using, e.g., the bootstrap method before leaping to a (in this case known) conclusion.

In each of the 10 realizations we considered the spherical median, and calculated a set of values of $\sum \arccos(X_i\lambda + Y_i\mu + Z_i\nu)$, expecting of course that the answer would be $(\lambda, \mu, \nu) = (0, 0, 1.0)$. In each realization we did indeed find the sum to be minimized at our expected direction.

There are many further issues that such analysis can illuminate. We enumerate some here; applicable methodology for practical usage is on the book's website, or see Fisher *et al.* (1987).

1. The next question might well be **'is the true median some particular direction?'** Constructing a suitable test statistic in this case requires the use of rotational matrices as described in textbooks on spherical trigonometry, e.g. Murray (1983); see the website. The test statistic is asymptotically distributed as χ^2 (with two degrees of freedom). As usual, for astronomically sized samples, a bootstrap is a good way of deriving the distribution and doing the test.

2. Next consider **'undirected lines' or axes.** These are familiar in astronomy; the normals to orbital planes are an example. Simple, useful analyses can be made that test against the null hypothesis of a uniform distribution on the sphere, in favour of the bipolar hypothesis of clustering of axis-orientation. (Tests (1) and (2) are variants of PCA, Section 4.5).

3. Test (2) depends on the principal axis being unspecified. **If the principal axis is specified**, with direction cosines $\{\lambda, \mu, \nu\}$, then a different (but known) test statistic is needed.

4. Tests are also available **to check if distributions are rotationally symmetric**, and to test for **a specific value of the principal axis**; see Fisher *et al.* (1987) for details.

5. If we have r **distinct samples**, we may ask if they all have the same median direction. To answer this, we compute the medians for the pooled sample and for each of the r samples. The test statistic derived from this provides a non-parametric test, but large samples are required, as is the case for other non-comparison tests discussed by Fisher *et al.* (1987).

6. There are useful methods available for **correlation and regression**. If we have sets of measurements of directions (or unit vectors) in pairs, of the form $(\vec{X}_i, \vec{X}'_i)$, we may wonder if the directions $\vec{X}_i$ and $\vec{X}'_i$ are correlated. The components of these data are $\{X_i, Y_i, Z_i\}$ and $\{X'_i, Y'_i, Z'_i\}$. To test, we can form a matrix generalization of the product–moment coefficient. This leads to a generalized correlation coefficient ρ whose distribution can be estimated by the permutation method described in Section 4.2: if $\vec{X}_i$ and $\vec{X}'_i$ are uncorrelated, it should not matter which $\vec{X}_i$ goes with which $\vec{X}'_i$.

Hence, by working through a large number of random permutations of the data, and sampling many possible pairings, we can estimate the distribution of ρ.

If ρ is appreciably different from zero, we cannot use this procedure. The best we can do is to use a jackknife (Section 6.6) to estimate the standard deviation of ρ. We may then perform a test on the assumption that ρ is Normally distributed. This is a large-sample approximation. For a small sample, we could use a bootstrap.

7. A similar test can be done for **undirected lines, or axes**. Here we do not use determinants but the data are combined in similar matrices whose traces lead to definition of a test 'correlation coefficient' statistic ρ; testing of this follows a similar procedure.

   To test against the no-correlation hypothesis, we again use a permutation method, or (for large samples, $n > 25$) we compare $3n\rho$ with χ^2 with 9 degrees of freedom. If a correlation is apparent, we may again use the jackknife or bootstrap to assess significance.

8. This test leads to a quite general one, in which $\vec{X}_i$ is a vector (direction or axis) and $\vec{X}'_i$ **is a general object with** p **components**. If $p = 3 + 1$, for instance, 3 space + 1 time, this might be a problem to do with the correlation of directions with time.

9. As a last example of correlation analysis, suppose we were interested in the **coherence or serial association in a time series of directions or axes**. Thus, our data might be ordered in time and we want to know if $\vec{X}_i$ is correlated with, say, $\vec{X}_{i-1}$. A test statistic can be derived.

10. Finally, note that **regression** between unit vectors and linear variables or other unit vectors can be handled by generalizations of least squares; see Fisher *et al.* (1987) for details.

10.2 Sky representation: projection and contouring

We frequently have a sample and we want to draw a sky representation of it. It is essential to use an *equal-area projection* to preserve density of points; we know from schooldays how unsuitable the Mercator projection is in this respect. There are many such projections available, and the following three are perhaps the best known in astronomy, given right ascension α and declination δ:

(i) The Sanson–Flamsteed projection:

$$x = \alpha \cos \delta, \quad y = \delta. \tag{10.2}$$

(ii) The Aitoff projection:

$$x = 2\phi \frac{\cos\delta \sin\frac{\alpha}{2}}{\sin\phi}, \quad y = \phi\frac{\sin\delta}{\sin\phi}, \tag{10.3}$$

where $\phi = \cos^{-1}(\cos\delta\cos\frac{\alpha}{2})$.

(iii) The Hammer–Aitoff projection:

$$x = 2\phi\cos\delta\sin\frac{\alpha}{2}, \quad y = \phi\sin\delta, \tag{10.4}$$

where $\phi = \sqrt{2}/\sqrt{1 + \cos\delta\cos\frac{\alpha}{2}}$.

In both the Aitoff and the Hammer–Aitoff projections, with the exception of the equator, the lines of constant declination curl at the extremities; those of the Sanson–Flamsteed projection are straight and horizontal. The latter is also very simple arithmetically but this is offset by the shear and crowded meridians in the polar regions. Take your pick, noting that many more projections are available.

Now suppose that we have a set of points $P_1, P_2, \ldots, P_n$ on our projection and we wish to map the density of these points. Computing a weighted average is an appropriate way to do this, and a suitable weighting scheme for a given map point P is

$$W_n(P, P_i) = \frac{C_n}{4\pi n \sinh(C_n)} \exp[C_n \cos(\theta_i)], \tag{10.5}$$

where θ_i is the angular distance between P and the data P_i. The weight thus depends only on the angular distance of points from P. Smoothing is controlled by C_n and varies inversely as C_n; we should choose C_n to increase with n, as the more data we have, the less smoothing we need. Contouring from here on is a matter of choosing an appropriate grid or map P, choosing levels or log(levels), and locating a suitable contouring routine, available in most graphics packages.

Example Figure 10.1 contrasts the Sansom–Flamsteed projection with the Aitoff projection, while showing the sky coverage of the largest redshift surveys and the widest-area radio surveys. The maps are in celestial coordinates (right ascension, declination) and the Galactic equator is marked together with Galactic latitude lines at $b = \pm 10°$. The coverages of each survey are delineated by plotting positions of catalogued objects. In the upper map the darker grey (in the North at small right ascensions; in the South at large right ascensions in two streaks at declination $0°$ and $-10°$) traces the Sloan Digital Sky Survey (York *et al.*, 2000) through plotting quasar catalogue positions (Schneider *et al.*, 2005). The light grey traces the coverage of the 2dF (AAT 2-degree Field) galaxy/quasar redshift survey (Colless *et al.*, 2001). In the lower map, black dots map out sources in the NVSS (NRAO VLA Sky Survey; 1.4 GHz, Condon *et al.*, 1998) while dark grey shows the coverage of the FIRST (Faint Images of the Radio Sky at Twenty cm) survey (Becker *et al.*, 1995). Light grey at the southernmost declination maps the Sydney University Molonglo Sky Survey (SUMSS; 0.84 GHz, Bock *et al.*, 1999; Mauch *et al.*, 2003).

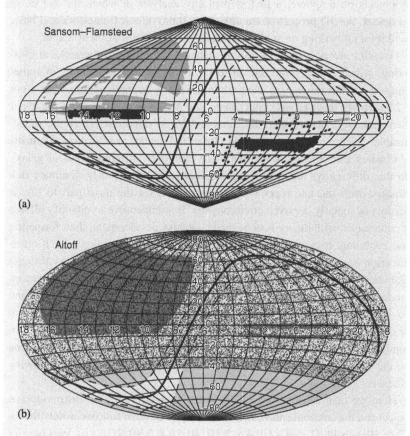

Figure 10.1   Two sky projections, Sansom–Flamsteed and Aitoff in celestial coordinates, RA and Dec, with the Galactic plane and $b \pm 10°$ shown as heavy and broken lines. (a) (Sansom–Flamsteed): coverage of the Sloan Digital Sky Survey (SDSS) in grey, 2dF survey in black. (b) (Aitoff): 1.4 GHz NVSS coverage as black dots, FIRST 1.4-GHz survey in dark grey, SUMSS 0.84-GHz survey in light grey.

10.3 The sky distribution

Much of the machinery described in the remainder of the chapter was developed
to explore the distribution of galaxies, the luminous matter in the Universe. To
a significant extent it has carried over to studies of the cosmic microwave
background, in which fluctuations were at last discovered in 1992 (Smoot
et al., 1992). This chapter concentrates on the consideration of 2D distribution
of objects on a sphere, a first step in any analysis in which the 2D or, in
particular, the 3D properties are unknown. It provides a framework and basic
toolkit for embarking on studies of the large-scale distribution of matter.

Galaxies are now recognizably distributed according to hierarchical clustering, which, when projected on to the plane of the sky, results in something of a tangle. However, there are immediately recognizable features – our
~20-member local group, Andromeda, our Milky Way galaxy and the lesser
members. Big clusters such as Coma and Virgo stand out. There are recognizable clusters of clusters, and on closer and deeper examination, filaments
of clusters and voids may be identified. Moreover, different types of galaxy
do this differently; the 'early' types (ellipticals) gregariously dominate rich
clusters, with the late types (spirals) ostracized for the most part to life as
hermits in socially deprived environments. It is imperative to quantify all this
if comparison with theory is to be made – galaxy development, their formation
and evolution, is central to modern astrophysics and cosmology. Such quantification has been recognized as vital by the pioneers, from Zwicky through
Holmberg, Abell, de Vaucouleurs, Scott and Neyman, and with most impact on
modern times, the detailed work both analytical and theoretical by Peebles and
co-workers (e.g., Peebles, 1980). The current picture – the hierarchical growth
of density perturbations in a low-density cold-dark-matter universe with substantial dark-energy density – has fed critically on studies of galaxy distribution
on the celestial sphere. The story is taken up and amplified in the following
chapter.

Here we take a fresh look at 2D statistics quantifying the distribution of
objects on the celestial sphere. Much of the presentation follows closely that in
Chris Blake's Ph.D. thesis (Blake, 2002; Blake & Wall, 2002a,b). We consider
the commonly used techniques of angular correlation functions, counts-in-cells
and power-spectrum analysis. In the course of this the relations between the
quantities are set out and some limitations of these in describing the overall
distribution are discussed. In terms of notation, angle brackets indicate expectation values, denoted elsewhere in the text as E[. . .], while barred quantities
such as $\bar{N}$ indicate averages over the survey areas in question; $\varsigma(\theta, \phi)$ denotes
object surface density.

10.4 Two-point angular correlation function

The two-point angular correlation function $w(\theta)$ is a simple and intuitive statistic to quantify clustering. Clustering increases the number of close pairs; $w(\theta)$ quantifies this increase as a function of galaxy separation θ. It is the fractional increase relative to a random distribution in the probability δP of finding objects in each of two solid angle elements $\delta\Omega_1$ and $\delta\Omega_2$ separated by angle θ:

$$\delta P = \varsigma^2 [\, 1 + w(\theta)\,]\, \delta\Omega_1\, \delta\Omega_2, \tag{10.6}$$

where ς is the object surface density.

The angular correlation function has many advantages as a clustering statistic. It is easy and quick to measure and its simplicity makes it easy to interpret (so that it can reveal systematic effects in the observational data). It directly accommodates unusually shaped survey areas with complicated boundaries and internal masked-out regions. Moreover, there is a relatively simple way to relate $w(\theta)$ to spatial clustering via the radial distribution of the objects. Hence, $w(\theta)$ is a convenient statistic to provide comparison both between data and theoretical prediction and between different observational data sets.

The angular correlation function $w(\theta)$ is not a complete description of the clustering. Phase information is lost. Two different object density fields can have identical angular correlation functions. A full field description requires a hierarchy of higher-order correlation functions that are much more difficult to measure and interpret. Moreover, $w(\theta)$ is very sensitive to shot noise, and may only be measured accurately at small angles. The error on the measurement of $w(\theta)$ is difficult to compute for small survey areas: edge effects render the simple 'Poisson error' incorrect. Furthermore, $w(\theta)$ suffers from correlated errors between adjacent $\Delta\theta$ bins, making assessment of true uncertainty in its determination notoriously difficult. Fitting a parameterized function to it is thus awkward from the point of view of minimization and error determination.

The value of $w(\theta)$ at given θ depends on density fluctuations on *all* angular scales, complicating the interpretation of the angular correlation function. In contrast, the angular power spectrum (Section 10.6) measures fluctuations on a specific angular scale. For example, a single sinusoidal density fluctuation (in one dimension) will have a δ-function angular power spectrum but a broad angular correlation function. Likewise, long-wavelength surface density gradients in the data (due to, for example, calibration problems) will offset the measured $w(\theta)$ on *all* angles (see Section 10.4.3).

Example Figure 10.2 shows two generated sky distributions, one (upper left) simulating low-contrast galaxy clusters in a regular grid on a random background, the other (upper right) simulating galaxy clusters on a background with large-scale structure in the form of a quadrupole.

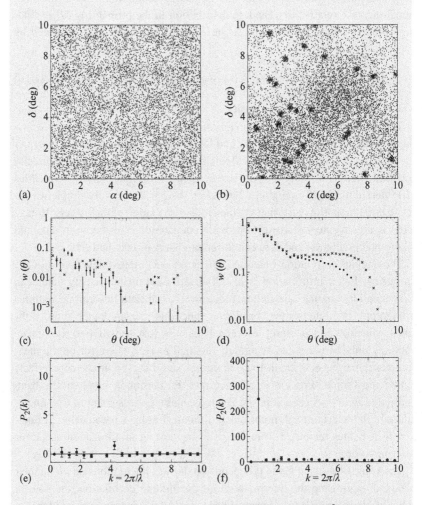

Figure 10.2 Two sky simulations; (a), (c) and (e) a uniform $2°$ grid of 25 low-surface-brightness clusters on a random background, and (b), (d) and (f) a quadrupole background with 25 randomly placed clusters. Measured $w(\theta)$ and angular power spectra appear below each. The $w(\theta)$ were evaluated with the simple estimator (crosses – Equation (10.7)) and the Landy–Szalay estimator (dots with error bars – Equation (10.10)).

The first 'sky' consists of a uniform random background of 8500 points, with a further 1500 points in 25 equal clusters of Gaussian width 0.4° placed on a uniform 2 × 2° grid across the area. The second has a background of 10 000 points generated from a power-spectrum representation of the sky (see Section 10.6) with signal in one term only – $\ell = 2$. Another 2000 points were added in 25 equal 'clusters' of Gaussian width 0.1° at random positions. Although the eye struggles to discern any features in the first sky, the two-point correlation function shows a strong signal at small θ describing the clusters themselves, and a resurgent signal at larger separations due to the 2 × 2° grid on which the clusters were placed. Both the quadrupole and the clusters are very evident in the second sky, and the correspondingly stronger $w(\theta)$ again shows two components; the signal at the smallest separations describes the galaxy clusters while the signal on degree scales is due to the quadrupole. The examples demonstrate the additive property of $w(\theta)$ as well as its ability to 'see' signal on all scales. They also demonstrate how $w(\theta)$ can mask sky information – different scale choices here could easily have resulted in a featureless $w(\theta)$. This is the case for the real sky – we see clusters, superclusters, filaments, and the result of all these scales is the well-known power-law form for $w(\theta)$ out to scales beyond tens of degrees.

The example illustrates the complementary nature of the power-spectrum analysis (Section 10.6); it is supremely sensitive to the larger structural features. For the first sky, the angular power-spectrum grabs the grid spacing unambiguously, while showing little evidence of signal on the cluster scale. Likewise with the second sky, the quadrupole signal dominates all else, although there is now some significant signal on the cluster scale.

10.4.1 Estimators and errors

In order to estimate $w(\theta)$ from a distribution of n objects, we (a) measure the angular separation θ of all galaxy pairs and (b) bin these separations to form the data-pair count $DD(\theta)$, the number of galaxy separations having lengths θ to $\theta + \mathrm{d}\theta$, and (c) calculate $RR(\theta)$, the corresponding number in each of these bins for a random sky having the average surface density of the real sky. (This latter is a simple sum: neglecting edge effects, the expected number of random pairs in the separation bin $\theta \rightarrow \theta + \delta\theta$ is $RR(\theta) = \frac{1}{2} n \, \varsigma \, 2\pi\theta \, \delta\theta$, where ς is the object surface density.) Hence an estimator for $w(\theta)$ – the fractional enhancement in pairs above random – is

$$w_0(\theta) = \frac{DD(\theta)}{RR(\theta)} - 1, \qquad (10.7)$$

and this is $w(\theta)$ in its simplest form.

However, edge effects are important when dealing with a small survey area or with weak clustering. Thus, we need to measure the average available bin area around a point ($\approx 2\pi\theta\,\delta\theta$) using Monte Carlo integration, by generating a comparison random distribution of r points over the same survey area. The result of this calculation (Blake, 2002) is an improved estimator for $w(\theta)$, the so-called 'natural' estimator:

$$w_1 = \frac{r(r-1)}{n(n-1)} \frac{DD}{RR} - 1. \qquad (10.8)$$

This estimator too has its shortcomings, the variance in this case, leading to the development of other estimators for $w(\theta)$ involving the cross-pair separation count between the sets of n data points and r random points. The following have been constructed:

$$w_2 = \frac{2r}{(n-1)} \frac{DD}{DR} - 1 \qquad (10.9)$$

$$w_3 = \frac{r(r-1)}{n(n-1)} \frac{DD}{RR} - \frac{(r-1)}{n} \frac{DR}{RR} + 1 \qquad (10.10)$$

$$w_4 = \frac{4nr}{(n-1)(r-1)} \frac{DD \times RR}{(DR)^2} - 1 \qquad (10.11)$$

with w_2, w_3 and w_4 known as the Peebles (Davis & Peebles, 1983), Landy–Szalay (Landy & Szalay, 1993) and Hamilton (Hamilton, 1993) estimators.

To reduce statistical fluctuations it is standard practice to use either a small number of densely populated random skies (of course corresponding in shape to our observed sky), or a large number of relatively low density random skies. Averaging is then carried out over the random pair counts to obtain $\overline{DR}$ and $\overline{RR}$ and the cross-products. Which is better? It is tempting to use a single high-surface-density random sky. But note that the computation time needed to measure the separations between n objects scales as n^2. For a given computation time (an important consideration for large samples), this mitigates against using a few high-density random skies and favours the use of a large number of relatively low density ones. If the number of random skies is m and the ratio of randoms in each set to the data is $k = r/n$, a reasonable guideline to adopt is $k \sim 1$, $m \gg 1$. With $m \geq 10$, the excess error is ≤ 10 per cent and may be ignored.

The best of the above estimators for $w(\theta)$ is that with the smallest bias and variance in the angular range under investigation. Excess variance comes about primarily but not exclusively through edge effects, which become more significant with increasing separation (see Exercise 10.2). Detailed analysis (Landy & Szalay, 1993; Hamilton, 1993) has quantified the non-Poisson variance and showed that for estimators w_3 and w_4, the non-Poisson terms cancel out and the error in the measurement of $w(\theta)$ is just the Poisson error. Furthermore, estimators w_2 and w_4 show small levels of bias. The Landy–Szalay estimator w_3 is thus the best bet.

It is thus possible to measure $w(\theta)$ with Poisson variance for an individual angular separation bin; but this does not mean that the errors in adjacent separation bins are uncorrelated. They *are* correlated, simply because a single object appears in many different separation bins through the numerous pairs in which it participates. Edge effects cause further correlation of errors in adjacent separation bins. If there are fewer objects on average near the boundaries, then there are systematically more close pairs in small-angle bins, and the numbers per bin are not independent. These correlations can be significant when assessing the goodness of fit of a model x_i to the N data points X_i. The correlations should be incorporated into model assessment by computing the covariance matrix (Section 4.2 and Press *et al.*, 2007).

10.4.2 Integral constraint

A point frequently not appreciated about $w(\theta)$ is that positive signal at small angles *demands* that the function becomes negative at larger separations. The total number of pairs over all bins is fixed at $\frac{1}{2}n(n-1)$; clustering shifts pairs from larger to smaller separations. This gives rise to difficulties, the first being the standard method of fitting the function with a power law. This is secondary to the main problem: if the surveyed area is sufficiently small, $w(\theta)$ appears positive for even the most distant separations sampled. The pair count *cannot* be enhanced in all separation bins while keeping the total number of pairs constant. The normalization must change, and we can formulate this in terms of an adjustment factor C as follows:

$$< DD(\theta) > = C \times \frac{1}{2}n(n-1)\,\mathrm{d}G_p\,[\,1 + w(\theta)\,], \qquad (10.12)$$

where δG_p is the equal-area fraction of the surface between θ and $\theta + \delta\theta$. (For a sphere, $\delta G_p = \frac{1}{2}\sin\theta\delta\theta$, satisfying $\int_0^\pi \mathrm{d}G_p = 1$ as it must.) It may be shown

(Exercise 10.3) that if $W = \int w(\theta)dG_p$, then $C = 1/(1 + W)$ and for all cases of interest,

$$w(\theta) \approx w(\theta)_{\text{est}} + W, \tag{10.13}$$

i.e. the estimated $w(\theta)$ is in error by a constant offset, which becomes negligible when the survey area becomes large.

10.4.3 Instrumental effects

Instrumental effects have a serious impact on $w(\theta)$. Large-scale calibration errors probably represent the most common such effect. Large-scale calibration errors in surveys produce gradients or discontinuities in object surface density. Such calibration problems may result from plate-to-plate calibration errors in a Schmidt-telescope survey or intensity-calibration changes over the area of a radio survey. Changing surface densities will spuriously enhance measured values of $w(\theta)$. This is because the number of close pairs of galaxies in any region depends on the local surface density ($DD \propto \overline{\varsigma^2}$), but the number of pairs expected over the sky by random chance depends on the global average surface density ($RR \propto (\overline{\varsigma})^2$). Systematic fluctuations mean that $\overline{\varsigma^2} > (\overline{\varsigma})^2$, increasing $w(\theta)$ by

$$\Delta w(\theta) = \frac{\overline{\varsigma^2}}{(\overline{\varsigma})^2} - 1 = \overline{\delta^2}, \tag{10.14}$$

where $\delta = (\varsigma - \overline{\varsigma})/\overline{\varsigma}$ is the surface overdensity. Equation (10.14) applies on angular scales less than those on which the surface density is typically varying; on larger scales the estimate of DD in this model is wrong. As an indication of the strength of this effect, a simple model in which a survey is divided into two equal areas with a surface-density change of 20 per cent produces an offset $\Delta w = 0.01$ (see Exercise 10.4).

Excess resolution is another source of error in $w(\theta)$. High-resolution surveys may 'break' single objects into two or more components. In deep optical/NIR surveys, faint interacting galaxies may bias $w(\theta)$. The problem is particularly acute in radio surveys in which the dominant form of the emission is double-lobed, and the lobes become catalogued as independent sources. It is evident that in such cases the amplitudes of $w(\theta)$ at the smaller spacings will be artificially inflated (see, e.g., Cress *et al.*, 1996; Magliocchetti *et al.*, 1998; Blake & Wall, 2002a). An example illustrates the issue.

Example Figure 10.3 shows the angular correlation function measured for the NVSS survey.

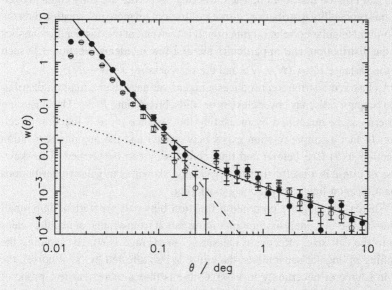

Figure 10.3   The angular correlation function $w(\theta)$ for the source catalogue of the NVSS survey, at $S_{1.4GHz} = 20$ mJy (solid circles) and 10 mJy (open circles). The best-fitting sum of two power laws for the 20-mJy data is shown as the solid line, with the two power laws shown individually, dashed due to multiple-component sources, dotted due to galaxy clustering.

Key to the interpretation of the two power laws is that the amplitude of the small-angle power law decreases with flux-density threshold as predicted by analysis (Blake & Wall, 2002a) if this signal is due to multiple source-components. The amplitude of the large-angle power law shows no such dependence on threshold, as expected to first approximation if it is due to true galaxy clustering. It shows the slope of ~ -0.8 typical of galaxy clustering.

10.5 Counts in cells

A second simple way to quantify clustering is the counts-in-cells (c-in-c) technique. This is the traditional way, in fact the way of the pioneers in clustering investigation (e.g., Shane & Wirtanen, 1954). We simply grid the sky into cells of fixed area and shape and count the objects falling in each cell. This yields the probability distribution $P(N)$ of finding N objects in a cell; for no clustering this is a Poisson distribution. The clustering properties are fully characterized by this probability distribution; c-in-c results contain more information than the two-point angular correlation function. It is convenient to consider the statistics of the distribution, and in particular the first few moments (Section 3.1) such as the variance $\mu_2 = \overline{(N - \overline{N})^2}$ and the skewness $\mu_3 = \overline{(N - \overline{N})^3}$.

A clustered distribution produces a higher variance than a random distribution because cells may cover clusters or voids, broadening $P(N)$. The clustering pattern can be quantified by measuring the variance μ_2 as a function of cell size. In fact a simple relation exists between μ_2 and the angular correlation function $w(\theta)$ (see below) and thus consistency can be verified. The skewness of c-in-c is a useful statistic physically: skewness in galaxy distributions quantifies non-linear gravitational clustering.

Whereas the angular correlation function bins pair separations into small intervals, a c-in-c analysis combines information from a range of angular scales up to the cell size, effectively measuring an average $w(\theta)$. By avoiding the binning of angular separations, the c-in-c is less affected by 'shot noise', the main source of uncertainty in $w(\theta)$. C-in-c is thus a more sensitive probe of long-range correlations than $w(\theta)$. It is, however, harder to make the connection with spatial clustering, and the values of moments of the counts distribution for different cell sizes will be highly correlated.

10.5.1 Counts-in-cells moments

Consider a non-clustered distribution of objects with surface density ς. The expectation value of objects in a cell of area S is $<N> = \varsigma S$. The expected probability distribution $P(N)$ is the Poisson distribution with mean $<N>$ and variance $<N>$. We define the following statistic to quantify the *increased variance* of a clustered distribution:

$$y = \frac{\mu_2 - \overline{N}}{(\overline{N})^2}.$$ (10.15)

Hence $<y> = 0$ for no clustering, as $<\mu_2> = <\overline{N}> = <N>$. For a given $w(\theta)$, the expected value of y (Peebles, 1980) is

$$<y> = \frac{\int_{\text{cell}} \int_{\text{cell}} w(\theta)\, dS_1\, dS_2}{S^2}$$ (10.16)

Thus $<y>$ may be calculated for an assumed form of the power-law angular correlation function, but survey resolution needs to be built in to this analysis.

Suppose in fact that the survey has angular resolution θ_{res}; then the power-law form of $w(\theta)$ can be expressed as

$$w(\theta) = \begin{cases} -1 & \theta > \theta_{res} \\ (\theta/\theta_0)^{-\alpha} & \theta < \theta_{res} \end{cases}, \tag{10.17}$$

where θ_0 is an alternative parameterization of the amplitude $A = (\theta_0)^\alpha$ of the angular correlation function. From this a general expression for $<y>$ in terms of survey resolution and $w(\theta)$ power-law parameters may be derived, general in the sense that it is for varying cell shape and size. The derivation (Blake & Wall, 2002b) is not simple and requires numerical integration; it results in an expression of the form

$$<y(L)> = a\, L^{-2} + b\, L^{-\alpha}, \tag{10.18}$$

where a and b are constants and L is the cell dimension. The detailed expression for $<y>$ shows that the angular resolution θ_{res} *reduces* the variance because the existence of an object in a cell limits the available space in which other objects can appear. This effect varies with cell size because it depends on the scale of the resolution relative to the cell size. Thus a non-clustered distribution viewed with non-zero angular resolution has a variance *less* than the Poisson value:

$$<y(L)> = -\frac{k}{2}\left(\frac{\theta_{res}}{L}\right)^2. \tag{10.19}$$

In fact if a survey has high enough angular resolution, then the first term of Equation (10.18) can be neglected (provided that $\alpha < 2$); then $<y> \propto L^{-\alpha}$.

  Skewness is of special importance. Skewness quantifies asymmetry in the non-Poisson clustering, such as a tail in the probability distribution to high cell counts. Assuming Gaussian primordial perturbations and linear growth of clustering, the skewness of c-in-c remains zero (Peebles, 1980). Measurement of a non-zero skewness therefore indicates either non-linear gravitational clustering or non-Gaussian initial conditions. As the growth of cosmic structure moves out of the linear regime, the expected skewness increases from zero. Using second-order perturbation theory, Peebles (1980) demonstrated that the density field develops a skewness $<\delta^3>/<\delta^2>^2 = 34/7$ (where δ is the overdensity, or dimensionless density contrast) assuming Gaussian initial perturbations growing purely due to gravity (see also Coles & Frenk, 1991). As fluctuations become non-linear, skewness increases because the value of δ grows large in density peaks but approaches the minimum value $\delta = -1$ in underdense regions.

Recalling that for a Poisson distribution of mean $<N>$ the expectation values for the variance μ_2 and the skewness μ_3 are both equal to $<N>$, the following statistic quantifies the increased skewness due to clustering:

$$z = \frac{\mu_3 - 3\mu_2 + 2\overline{N}}{(\overline{N})^3}. \tag{10.20}$$

Hence $<z> = 0$ for no clustering (neglecting a small bias), as $<\mu_2> = <\mu_3> = <\overline{N}> = <N>$. This statistic has the expectation value

$$<z> = \frac{\int_{\text{cell}} \int_{\text{cell}} \int_{\text{cell}} W(\theta_{12}, \theta_{13}, \theta_{23}) \, dS_1 \, dS_2 \, dS_3}{S^3}, \tag{10.21}$$

where $W(\theta_{12}, \theta_{13}, \theta_{23})$ is the three-point angular correlation function, which quantifies the excess probability (beyond two-point clustering) of finding objects in each of the solid angle elements $\delta\Omega_1, \delta\Omega_2, \delta\Omega_3$ with mutual separations $\theta_{12}, \theta_{13}, \theta_{23}$:

$$\delta P = \varsigma^3 \left[1 + w(\theta_{12}) + w(\theta_{13}) + w(\theta_{23}) + W(\theta_{12}, \theta_{13}, \theta_{23}) \right] \delta\Omega_1 \, \delta\Omega_2 \, \delta\Omega_3. \tag{10.22}$$

The statistical error on the estimator of Equation (10.20) for a grid of N_c cells is

$$\sigma_z = \sqrt{\frac{6}{N_c \, (\overline{N})^3}}. \tag{10.23}$$

10.5.2 Measuring counts-in-cells

The methodology of c-in-c measurement was revolutionized by Szapudi (1998), who showed that it was valid to throw a very large number of randomly placed cells over the sky, heavily oversampling the survey area. But of course measurement of the variance statistic y remains subject to statistical error due to averaging over a finite number of independent cells N_c. Calculating the standard error in the case of a non-clustered distribution yields

$$\sigma_y = \sqrt{\frac{2}{N_c \, (\overline{N})^2}}. \tag{10.24}$$

The probability distribution of the clustered data does not depart greatly from a Poisson distribution (i.e. $y \ll 1$) so that Equation (10.24) is a good approximation to the actual statistical error.

Surveys do not encompass the whole sky: there are boundaries and masked regions. Hence, with any form of gridding or random cell placement, some cells are partially filled, the ith cell having a fraction of useful area f_i (say). Populating the survey area with random points is an obvious way to determine f_i for each cell, the number of points falling in each cell used as a measure of cell area. It is then possible to boost the data count in the ith cell by a factor $1/f_i$, unless of course f_i turns out to be so small as to render $1/f_i$ unstable; under this thresholding circumstance, reject the cell. In the design of the Monte Carlo experiment it is essential not to add spurious variance by insufficient accuracy in determining cell areas.

When evaluating the moments of the c-in-c distribution, it is assumed that all cells are populated independently. This is not strictly true given that clustered objects have correlated positions, but the assumption should be a good approximation if the cells are large enough; a minimum cell size $L_{\min}$ must be adopted so that $\overline{N} \geq 1$.

Example A c-in-c analysis of the distribution of radio sources in the NRAO VLA Sky Survey was carried out by Blake & Wall (2002b). Figures 10.4 and 10.5 show how close the distributions are to Poissonian; how well the double power-law interpretation of $w(\theta)$ (Figure 10.3) predicts the c-in-c variance function $y(L)$; and how good the agreement is for the parameters describing the cosmological portion of $w(\theta)$ as derived from direct measurement and from c-in-c.

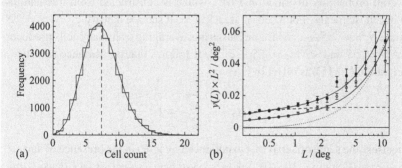

Figure 10.4 (a) Counts of NVSS radio sources with $S_{1.4GHz} > 20$ mJy in cells of diameter $1°$. Vertical dashed line – expected mean count derived from the source surface density; solid curve – the corresponding Poisson distribution. (b) The variance statistic $y(L)$ is plotted for thresholds 20 mJy (solid circles) and 10 mJy (open circles), with predictions of the double power-law $w(\theta)$ models at 20 mJy and 10 mJy (solid lines). The dashed and dotted lines show the separate contributions to $y(L)$ at 20 mJy of the steep (multiple-component) $w(\theta)$ and the shallow (cosmological) $w(\theta)$.

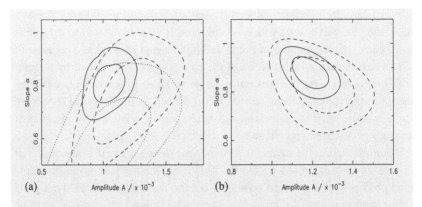

Figure 10.5   Contours of constant χ^2 are shown in the plane of the clustering parameters A and α, $w(\theta) = A\theta^{-\alpha}$; these are approximate 1σ and 2σ contours for flux-density thresholds 30 mJy (dotted), 20 mJy (dashed) and 10 mJy (solid): (a) is from fitting the correlation function (Figure 10.3) directly; (b) from fitting the c-in-c variance function shown above.

10.5.3 Instrumental effects

Systematic surface density gradients spuriously offset the c-in-c variance: a spread in the mean surface density across the cells will inevitably broaden the overall probability distribution $P(N)$, which is constructed from fluctuations about those means. For a cell of area S at local surface density ς, $<N> = \varsigma S$ and $<N^2> = \varsigma S + \varsigma^2 S^2$ for no clustering; averaging over many cells produces $<\overline{N}> = \overline{\varsigma} S$ and $<\overline{N^2}> = \overline{\varsigma} S + \overline{\varsigma^2} S^2$. It follows that the variance statistic y (Equation (10.15)) is offset by

$$<\Delta y> = \frac{\overline{\varsigma^2}}{(\overline{\varsigma})^2} - 1, \qquad (10.25)$$

precisely the same offset as that experienced by $w(\theta)$ in the presence of surface gradients (Equation (10.14)). Likewise systematic object surface density gradients offset the skewness by $<\Delta z> = \overline{\delta^3}$ (where δ is the surface overdensity).

The presence of multiple-component objects ('excess resolution') increases the c-in-c moments – the fraction of objects within a cell split into multiple components varies from cell to cell, which acts to broaden the probability distribution of c-in-c. The simplest model is to suppose that a fraction e of the objects are double (in which case two apparent objects appear as components of the same object close to the centroid position of the object), and a fraction f are triple objects. Expressions may be derived (Blake, 2002) for the expected offsets in the variance and skewness statistics. It turns out that skewness is insensitive to double objects, but very sensitive to triple objects. When triples are present, the skewness offset scales with cell dimension L as $z \propto L^{-4}$. The point is of particular importance in the analysis of radio-source catalogues; many radio sources have a triple structure with a compact nuclear component roughly centred between the extended pair of lobes.

10.6 The angular power spectrum

The angular power spectrum, denoted c_ℓ, is the third statistic we describe to quantify a surface or sky distribution. This statistic, invoked to look at cluster, galaxy and radio-source distributions by Yu & Peebles (1969), Peebles & Hauser (1974), and Webster (1976a,b), respectively, imagines that the object surface density field over the sky is expressed as a sum of angular density fluctuations of different wavelengths.

It is a Fourier analysis (Section 9.2) around the sky. The mathematical tools involved in this process are the spherical harmonic functions, the 2D analogues of sine and cosine, and the quantity c_ℓ expresses the amplitude of the ℓth multipole, which produces fluctuations on angular scales $\theta \sim 180°/\ell$.

In a theoretical sense the c_ℓ spectrum is entirely equivalent to the angular correlation function $w(\theta)$ as a description of the galaxy distribution. The two quantities are connected by the well-known relations (Peebles, 1980):

$$c_\ell = 2\pi \, \varsigma_0^2 \int_{-1}^{+1} w(\theta) \, P_\ell(\cos\theta) \, \mathrm{d}(\cos\theta), \qquad (10.26)$$

and

$$w(\theta) = \frac{1}{4\pi \, \varsigma_0^2} \sum_{\ell=1}^{\infty} (2\ell + 1) \, c_\ell \, P_\ell(\cos\theta), \qquad (10.27)$$

with ς_0 the mean object surface density and P_ℓ the Legendre polynomials. However, the angular scales on which the measured signal is highest are very different for each statistic. $w(\theta)$ can only be determined accurately at small angles, beyond which Poisson noise dominates. By contrast, c_ℓ has highest signal at small ℓ, corresponding to large angular scales $\theta \sim 180°/\ell$. The two statistics $w(\theta)$ and c_ℓ are complementary in this sense. However, note that the two statistics quantify very different properties of the galaxy distribution (Section 9.4). The value of c_ℓ quantifies the amplitude of fluctuations on the angular scale corresponding to ℓ. The value of $w(\theta)$ is the average of the product of the galaxy overdensity at any point with the overdensity at a point at angular separation θ: $w(\theta)$ depends on angular fluctuations on all scales (Equation (10.27)).

Measurement of the angular power spectrum has practical advantages in comparison with $w(\theta)$. Firstly, it is possible to produce measurements of c_ℓ at different multipoles ℓ that are uncorrelated, whereas the $w(\theta)$ statistic suffers from correlated errors between adjacent separation bins. Secondly, on small scales the evolution of structure is complicated by non-linear effects, and thus it can be advantageous to investigate larger scales where linear theory still applies. Thirdly, there is a natural relation between the c_ℓ spectrum and the spatial power spectrum $P(k)$. This latter quantity provides a very convenient means of describing structure in the Universe because its primordial form is produced by models of inflation, which prescribe the pattern of initial density fluctuations $\delta\rho/\rho$. Furthermore, in linear theory for the growth of perturbations, fluctuations described by different wavevectors k evolve independently. The angular correlation function is more naturally related to the spatial correlation function $\xi(r)$, the Fourier transform of $P(k)$.

10.6.1 Formalism for c_ℓ

A distribution of objects on the sky can be modelled in two statistical steps. Firstly, a continuous density field $\varsigma(\theta, \phi)$ is specified; this can be described in terms of its spherical harmonic coefficients $a_{\ell,m}$:

$$\varsigma(\theta, \phi) = \sum_{\ell=0}^{\infty} \sum_{m=-\ell}^{+\ell} a_{\ell,m} \, Y_{\ell,m}(\theta, \phi), \qquad (10.28)$$

where $Y_{\ell,m}$ are the spherical harmonic functions and θ $(0 \to \pi)$ and ϕ $(0 \to 2\pi)$ are spherical polar coordinates. Secondly, discrete galaxy positions are generated in a Poisson process as a realization of this density field (i.e. the probability of finding a galaxy in an element of solid angle $\delta\Omega$ at position θ, ϕ is $\delta P = \varsigma(\theta, \phi)\delta\Omega$).

The angular power spectrum c_ℓ prescribes the spherical harmonic coefficients in the first step of this model. It is defined by

$$< |a_{\ell,m}|^2 > = c_\ell, \qquad (10.29)$$

where the angled brackets imply an averaging over many realizations of the density field. The assumption of isotropy ensures that $< |a_{\ell,m}|^2 >$ is a function of ℓ alone, and not m.

The $Y_{\ell,m}$'s and $a_{\ell,m}$'s of Equation (10.28) are, in general, complex quantities. Because the density field $\varsigma(\theta, \phi)$ is real, $a_{\ell,-m} = a_{\ell,m}^*$. Thus, the independent coefficients describing the density field are $a_{\ell,0}$ (which is real) and the real and imaginary parts of $a_{\ell,m}$ for $m \geq 1$. Hence, the ℓth harmonic is described by $2\ell + 1$ independent coefficients. The assumption is usually made, motivated by inflationary models, that $\varsigma(\theta, \phi)$ is a Gaussian random field. In this model the real and imaginary parts of $a_{\ell,m}$ are drawn independently from Gaussian distributions such that the normalization satisfies Equation (10.29). Thus, $<a_{\ell,m}> = 0$ for $\ell > 0$ and $<a_{\ell,m}^* a_{\ell',m'}> = 0$ unless $\ell = \ell'$ and $m = m'$.

Consider first a fully surveyed sky. As an initial step we consider the estimation of the harmonic coefficients $a_{\ell,m}$ of the density field. The orthonormality properties of the $Y_{\ell,m}$'s mean that Equation (10.28) may be reversed:

$$a_{\ell,m} = \int \varsigma(\theta, \phi) \, Y_{\ell,m}{}^*(\theta, \phi) \, d\Omega. \tag{10.30}$$

Equation (10.30) suggests that the $a_{\ell,m}$'s may be estimated by summing over spherical harmonics at the N object positions (θ_i, ϕ_i):

$$A_{\ell,m} = \sum_{i=1}^{N} Y_{\ell,m}{}^*(\theta_i, \phi_i). \tag{10.31}$$

We denote estimators by upper-case symbols (e.g. $A_{\ell,m}$) and the underlying 'true' quantities by lower-case symbols (e.g. $a_{\ell,m}$). It can now be shown (Blake, 2002) that the expectation value of the estimator $|A_{\ell,m}|^2$ is

$$< \overline{|A_{\ell,m}|^2} > = c_\ell + < \varsigma_0 > \tag{10.32}$$

and if ς_0 is the average surface density in a given realization, the correct estimator for c_ℓ from our original distribution is

$$C_{\ell,m} = |A_{\ell,m}|^2 - \varsigma_0 \tag{10.33}$$

such that $< \overline{C_{\ell,m}} > = c_\ell$. The discreteness of the distribution causes the correction term '$-\varsigma_0$'. For a given multipole ℓ there are $\ell + 1$ different estimates of c_ℓ, corresponding to $m = 0, 1, \ldots, \ell$. The fact that the density field is real rather than complex implies that $C_{\ell,-m} = C_{\ell,m}$ and thus negative values of m provide no new information. The statistical error on the estimator of Equation (10.33) is

$$\sigma(C_{\ell,m}) = \sqrt{< \overline{C_{\ell,m}^2} > - < \overline{C_{\ell,m}} >^2} = (\varsigma_0 + c_\ell) \times \begin{cases} \sqrt{2} & m = 0 \\ 1 & m \neq 0 \end{cases}. \tag{10.34}$$

The error for the $m \neq 0$ case is reduced by a factor of $\sqrt{2}$ because there are two independent measurements built in: the real and imaginary parts of $A_{\ell,m}$.

Equation (10.34) illustrates that there are two contributions to the statistical errors:

(i) shot noise (ς_0) because the number of discrete objects is finite and does not perfectly sample the underlying density field. The magnitude of c_ℓ is proportional to ς_0^2 (Equation (10.26)); hence increasing the number of objects decreases the fractional error;

(ii) cosmic variance (c_ℓ), because we can only measure a finite number of fluctuations on a given scale around the sky.[1]

By considering $< \overline{C_{\ell,m} \, C_{\ell',m'}} >$ we can show that the estimates of Equation (10.33) are statistically independent; and we derive a better estimate of c_ℓ for a given multipole ℓ by averaging over m:

$$C_\ell = \frac{\sum_{m=0}^{\ell} C_{\ell,m}}{\ell + 1}. \tag{10.35}$$

From Equation (10.34), the resulting error on C_ℓ is

$$\sigma(C_\ell) = (\varsigma_0 + c_\ell)\frac{\sqrt{\ell + 2}}{\ell + 1}. \tag{10.36}$$

For an incomplete sky, the requisite modification to the derivation is given by Peebles (1973). A summary is as follows. Equation (10.33) becomes

$$C_{\ell,m} = \frac{|A_{\ell,m} - \varsigma_0 \, I_{\ell,m}|^2}{J_{\ell,m}} - \varsigma_0, \tag{10.37}$$

where $\varsigma_0 = N/\Delta\Omega$, $\Delta\Omega$ is the survey area and

$$I_{\ell,m} = \int_{\Delta\Omega} Y_{\ell,m}{}^* \, d\Omega, \quad J_{\ell,m} = \int_{\Delta\Omega} |Y_{\ell,m}|^2 \, d\Omega \tag{10.38}$$

with the integrals being over the survey area. Thus, the partial sky is compensated for by replacing $|A_{\ell,m}|^2$ with $|A_{\ell,m} - \varsigma_0 I_{\ell,m}|^2/J_{\ell,m}$: there is a systematic deviation in each harmonic coefficient and the overall normalization changes.

---

[1] The term *cosmic variance* is of uncertain origin and has varying interpretations. It is generally agreed that cosmic variance describes uncertainty which can never be reduced because of the limitation imposed by the (apparently) singular nature of the Universe. For example, uncertainty in the very lowest modes of the CMB angular power spectrum from all-sky surveys is inherent; we have no more sky to survey. Beyond this, some consider it to represent scatter or spread which is 'imposed' by our cosmos above that expected from Poisson statistics. For example, deep fields well separated on the sky may contain numbers of galaxies which differ significantly on the basis of $\sqrt{N}$ statistics because of clustering on very large scales. The former interpretation is favoured.

We again estimate the angular power spectrum using

$$C_\ell = \left(\sum_{m=0}^{\ell} C_{\ell,m} \right) / (\ell + 1).$$

The partial sky has some important effects on this estimate. Only for a complete sky does $< \overline{C_\ell} > = c_\ell$: for an incomplete sky there is some 'mixing' of harmonics so that the measured angular power spectrum at multipole ℓ depends on a range of $c_{\ell'}$ around $\ell' = \ell$:

$$< \overline{C_\ell} > = \sum_{\ell'=1}^{\infty} c_{\ell'} R_{\ell,\ell'}, \tag{10.39}$$

where $\sum_{\ell'} R_{\ell,\ell'} = 1$, and $R_{\ell,\ell'}$ can be determined from the geometry of the survey region (see Hauser & Peebles, 1973). In addition, for an incomplete sky the estimates $C_{\ell,m}$ are no longer statistically independent and the error of Equation (10.36) is only an approximation. The resulting measurements of c_ℓ at different ℓ are not wholly independent: the covariance matrix is no longer diagonal. MLE provides a powerful and general way to take into account the correlations induced by a partial sky area. In fact this alternative method of deriving the c_ℓ spectrum has been widely used in recent years for quantifying the CMB temperature fluctuations (e.g., Lange *et al.*, 2001). These tools can be exploited to analyse the galaxy distribution by pixellating the sky into equal-area cells (see Efstathiou & Moody, 2001; Huterer *et al.*, 2001; Tegmark *et al.*, 2002).

If the surveyed area $\Delta\Omega$ is reduced for a fixed average object density ς_0, then the signal to noise of the measurement decreases, as expected. The statistical error does not change to a first approximation, but the magnitude of the signal (Equation (10.37)) is reduced because while the denominator scales as $\Delta\Omega$, the numerator scales as $(\Delta\Omega)^2$ (depending, as it does, on the square of a sum over $N = \varsigma_0 \Delta\Omega$ objects). The estimation process takes into account the fact that the surface density ς_0 is not known in advance but is determined from the data ($\varsigma_0 = N/\Delta\Omega$); this is not a source of systematic error in the estimator.

10.6.2 Instrumental effects

Multiple-component objects from excess resolution spuriously increase the measured angular power spectrum, producing an offset that can be quantified for a given proportion of double, triple or multiply imaged objects (Blake, 2002). The presence of systematic surface gradients also distorts the angular power spectrum but in a non-straightforward way, because the harmonic coefficients need to reproduce these gradients as well as the fluctuations due to clustering. As ever it is best to fix the calibration, or to stick to thresholds at which the gradients are insignificant.

Example A radio survey maps the galaxy distribution out to very large distances $D > 10^3$ Mpc and is hence able to probe $P(k)$ on large scales $k \sim 1/D < 10^{-3}$ Mpc^{-1}, where the shape of the power spectrum is unaltered from its initial form. Determination of the c_ℓ spectrum of radio galaxies therefore has the potential to constrain the primordial pattern of density fluctuations in a manner independent of measurements of fluctuations in the cosmic microwave background (CMB) radiation. Such an analysis (Blake, 2002) was carried out for the NVSS. The results are shown in Figure 10.6 together with the predictions resulting from transforming the angular correlation function using Equation (10.26), assuming that $w(\theta) = (1 \times 10^{-3})\theta^{-0.8}$ (see Figure 10.5). The predictions turn out to be a good match to the measured c_ℓ spectrum – with the notable exception of the dipole term $\ell = 1$, 'spuriously' high due to the cosmic velocity dipole detected in this experiment (Blake & Wall, 2002c). The NVSS angular power spectrum decreases with ℓ as (roughly) a power law. The offset due to multiple-component sources is significant; analysis by Blake (2002) gives $\Delta C_\ell \approx 1 \times 10^4$ for the 10 mJy threshold.

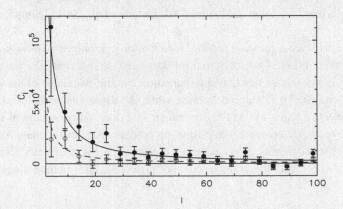

Figure 10.6   Measurement of the NVSS c_ℓ spectrum. Results are plotted for two flux-density thresholds: $S_{1.4GHz} = 10$ mJy (solid circles) and 20 mJy (open circles). The solid and dashed lines show the prediction of Equation (10.26) at these thresholds assuming that $w(\theta) = (1 \times 10^{-3})\theta^{-0.8}$. The difference in amplitude between the two measurements arises from the factor ς_0^2 in Equation (10.26).

It is initially surprising that the observed c_ℓ spectrum of Figure 10.6 agrees so well with the angular correlation function prediction of Equation (10.26): c_ℓ depends on $w(\theta)$ at all angles, but $w(\theta)$ is only measurable for $\theta < 10°$ and will deviate from a power law (and become slightly negative) at larger angles. The agreement implies that long-wavelength surface density fluctuations (generated by multipoles at low ℓ) are important in producing angular correlations at small θ. This is not a contradiction: the angular correlation function is the average of the product of the galaxy overdensity at any point with the overdensity at any other point at fixed angular separation, and positive contributions to this average are readily produced by long-wavelength fluctuations. Mathematically, agreement arises because the dominant contribution from the integrand of Equation (10.26), $dc_\ell/d\theta \propto w(\theta)\, P_\ell(\cos\theta)\, \sin\theta$, originates from small angles. As $\theta \to 0$, $dc_\ell/d\theta \to \theta^{-0.8} \times 1 \times \theta \to 0$; but as θ increases, $P_\ell(\cos\theta)$ falls off and a maximum in $dc_\ell/d\theta$ occurs at $\theta \approx$ a few degrees. At larger angles, the oscillations in $P_\ell(\cos\theta)$ ensure that subsequent contributions to c_ℓ approximately cancel out.

10.7 Galaxy distribution statistics: interpretation

Our description of some 2D statistics relevant to sky distribution is reminiscent of the first steps in the 1970s toward formalizing the exploration of galaxy distribution. Statistics to characterize 2D distribution remain relevant today in several respects: as a first look at a survey to see if (3D) clustering might be present; in examining possible associations on the sky plane; in determining if observational selection effects might be present in a survey (uneven sensitivity or gradients; shortcomings in detection algorithms; biases, maskings or resolution effects). And of course for the CMB and its fluctuations, 2D analyses are the sum total of what we get – the last scattering surface is at a redshift of $\sim$1100, so that all features and fluctuations observed are at this approximate redshift.

The early role of 2D statistics in examining galaxy clustering was played out with the use of small samples of galaxy redshifts complete over limited areas and ranges. In particular, the angular two-point correlation function was used with the *relativistic Limber's equation* (Limber, 1953; Peebles, 1980; Baugh & Efstathiou, 1993) to derive the 3D two-point correlation function, in turn offering early insight into galaxy-clustering properties. Initial evaluation of the galaxy 2D two-point correlation function showed that it followed a power law of slope ~ -0.8. An estimate of 3D clustering properties then follows:

1. Assume a power-law form for the 2D correlation function of $w(\theta) = A\theta^{(1-\gamma)}$ and likewise a power-law form for the 3D correlation function

$$\xi(r_c, z) = \left(\frac{r_c}{r_0}\right)^{-\gamma-(3+\epsilon)}, \qquad (10.40)$$

where $r_c = r(1 + z)$ is the co-moving coordinate, r_0 is the correlation length; and $\epsilon = 0$ corresponds to constant ('stable') clustering in proper coordinates, $\epsilon = \gamma - 3$ implies constant clustering in co-moving coordinates, and $\epsilon = \gamma - 1$ implies growth of clustering under linear theory (Treyer & Lahav, 1996).

2. Then using a redshift distribution $N(z)$ complete to some apparent magnitude for the galaxies involved, through Limber's equation derive values for the correlation length r_0 under the different clustering assumptions (e.g. Loan *et al.*, 1997).

Example The first 2D correlation function signals in large-area radio surveys were detected by Cress *et al.* (1996) in the 1.4-GHz FIRST survey (Becker *et al.*, 1995), and Kooiman *et al.* (1995) and Loan *et al.* (1997) in the 5-GHz 87GB (Gregory & Condon, 1991) + PMN surveys (Griffith & Wright, 1993). Using the formulation of Equation (10.40) and a best estimate at a complete $N(z)$ for radio AGN from Dunlop & Peacock (1990), Loan *et al.* deduced radio-galaxy correlation lengths for various clustering modes from $r_0 = 13/h$ to $18/h$ Mpc ($h = H_0/100$). Much improved signal to noise from careful examination of the FIRST and NVSS correlation functions reduced these high values to $r_0 \sim 6/h$ to $8/h$ (Blake & Wall, 2002a), in agreement with the Peacock (1999b) value of $r_0 = 6.5/h$, obtained directly from matching NVSS sources to the Las Campanas Redshift Survey. Magliocchetti *et al.* (1999) further showed how it is possible to use the 2D correlation function, c-in-c and estimates of the redshift distribution to derive constraints on the bias parameter describing how galaxies trace mass in the Universe (Kaiser, 1984).

It was recognized early on by many authors that two-point corelation functions, no matter how accurately correlation lengths were measured, represented a very incomplete description of galaxy distribution. This was rapidly confirmed by early studies of galaxy redshifts such as the CfA survey (Davis *et al.*, 1982).

Despite its shallow nature by modern standards ($z_{max} < 0.005$), the CfA survey demonstrated the existence of filaments and voids; galaxy topology was revealed to be sponge-like (Gott *et al.*, 1989). A complete description in terms of correlation functions of order 3 to n (Peebles & Groth, 1975) was shown to be impractical. Several other approaches to reveal the true nature of the distribution were suggested. The power-law (scale-free) form of the 2-point correlation function was indicative of a fractal nature, and *fractal analysis* was proposed and considered by Martínez *et al.* (1990). There was also the method of *minimal spanning trees* (Barrow *et al.*, 1985). *Percolation theory* had its advocates (Zeldovich *et al.*, 1982; Dekel & West, 1985; Einasto & Saar, 1987). We also need to mention *nearest-neighbour analysis* (Wagoner, 1967; Bogart & Wagoner, 1973), employed initially to assess the statistical significance of apparent associations on the sky between low-redshift galaxies and quasars (the nature of whose redshifts was still a matter of debate). Nearest-neighbour analysis, minimal spanning trees and fractal analysis can all be used in 2D surface analyses, in addition to the 2D 2-point correlation function, c-in-c and angular-spectrum analysis described in this chapter.

With the advent of huge redshift surveys and massive N-body simulations, the science of galaxy distribution has moved on, intimately bound up as it is with galaxy formation, evolution and the geometry of the Universe. We describe this progress briefly in our final chapter.

Exercises

10.1 Rayleigh's test. Why should the test statistic for Rayleigh's test be asymptotically χ^2? Compute the statistic for small numbers, say <10; see Section 3.3.3.

10.2 Variance of estimators for $w(\theta)$ (D). Generate 20 000 data points randomly in the region $0° < \alpha < 5°, 0° < \delta < 5°$. Estimate $w(\theta)$ using the Natural estimator w_1, the Peebles estimator w_2, the Landy–Szalay estimator w_3 and the Hamilton estimator w_4. (Average DR and RR over, say, 10 comparison sets each of 20 000 random points). Plot the results as a function of δ showing Poisson error bars $1/\sqrt{DD}$. Comment on the results. Which estimator is best?

10.3 Integral constraint on $w(\theta)$. (a) Show that the factor C in Equation (10.12) is $1/(1 + W)$ where $W = \int w(\theta) \, dG_p$. (b) Derive an approximate expression for W, assuming a power-law form for $w(\theta) = (\theta/\theta_0)^{-b}$.

10.4 The effect of surface density changes on $w(\theta)$ (D). (a) Estimate the magnitude of the offset in $w(\theta)$ taking a simple model in which a survey is divided into two equal areas between which there is a fractional surface density shift ϵ (Equation (10.14)). Find the expected step in $w(\theta)$ as a function of ϵ; verify that a step of 20 per cent results in $\Delta w = 0.01$. (b) Confirm this prediction with a toy-model simulation, putting, say, 100 000 random points in the region $0° < \alpha < 60°$, $-20° < \delta < +20°$ with a 20 per cent step at $\delta = 0°$. Then calculate the $w(\theta)$ using, say, the Landy–Szalay $w(\theta)_3$ estimator over the small angular-scale range $\theta < 1°$, checking that $w(\theta)$ agrees within errors with the prediction from Equation (10.14).

10.5 **The effect of surface-density changes on c-in-c (D).** (a) Use the 100 000 random points generated in the region $0° < \alpha < 60°$, $-20° < \delta < +20°$ for Exercise 10.4. Generate a set of 10 grid patterns for circular non-overlapping cells over the area, with diameter $0.03°$ to $3°$, evenly spaced in $\log \theta$. Compile $P(N)$ for each of these and show that the means and variances are as expected for Poissonian distributions. Calculate and plot the variance statistic $y(L)$ (Equation (10.15)) as a function of cell size; verify that there is no significant offset from zero. (b) Put in a 20 per cent step in surface density, dividing the field in half at $\delta = 0°$. Recalculate the $y(L)$ and verify that the apparent offset in $y(L)$ is of the expected magnitude $\Delta y = 0.01$ (Equation (10.25)).

10.6 $w(\theta)$ **and the angular power spectrum (D).** Simulate a square piece of sky $10° \times 10°$, using 10 000 points placed at random. (a) Verify that there is no significant signal either in $w(\theta)$and in the angular power spectrum. (b) Build a hierarchy of galaxy clusters and clusters of clusters using perhaps another 10 000 points, adopting Gaussian shapes for clusters, cluster-clusters, etc. (c) Show that with a few adjustments to the parameters (see Figure 10.2), it is possible to produce an approximate power law of slope ~ -1 for a single hierarchy of clustering. Relate the resultant form of the angular power spectrum and its information content to this $w(\theta)$.

In the exercises denoted by (D), data sets are provided on the book's website; or create your own.

11

Epilogue: statistics and our Universe

Extraordinary claims demand extraordinary evidence.

(Anon)

In the past it may have been true that Stephen Senn's (2003) analogy was right. Paraphrasing his words for the sake of moderate language: *scientists regarded statistics as the one-night stand: the quick fix, avoiding long-term entanglement.* This analogy is no longer apt. Statistical procedures now drive many if not most areas of current astrophysics and cosmology. In particular the currently understood nature of our Universe is a product of statistical analysis of large and combined data sets. Here we briefly describe the scene in three areas dominating definition of the current model of the Universe and its history. The three areas inextricably tie together the shape and content of the Universe and the formation of structure and galaxies, leading to life as we know it. While these sketches are not reviews, we show by cross-referencing how frequently our preceding discussions play in to current research in cosmology.

11.1 The galaxy universe

The story of galaxy formation since 1990 is based on two premises. Firstly, it was widely accepted that the matter content in the Universe is primarily cold and dark – CDM prevails. The recognition of dark matter was slow, despite Zwicky (1937) demonstrating its existence via the cosmic virial theorem. The measurements of rotation curves of spiral galaxies (e.g. Rubin *et al.*, 1980) convinced us. That the dark matter was cold emerged from the need to grow galaxies within a Hubble time, in particular through the epoch of recombination. Baryons and radiation are tightly coupled prior to recombination; as radiation perturbations decay on entering the horizon, so do baryon fluctuations. Only after decoupling can the baryons, freely released from the radiation field, fall into the waiting traps of the gravitational potentials set up by the CDM, the traps having slid through the decoupling era unscathed. Secondly, by 1990 (Kolb & Turner, 1990) it was generally recognized that we live in a universe which has gone through an inflation (positive-energy false-vacuum) phase (Guth, 1981), and the upshot was flat space (density parameter $\Omega = 1$, matching the critical density for closure) with an index for the power-law spectrum (Section 2.4.2.4) of primordial fluctuations of index $n_s = 1$, the Harrison–Zeldovich–Peebles spectrum. The assumption was that the Ω was matter, predominantly cold dark matter (CDM). Inflation solved the questions of (1) why the Universe should appear flat, isotropic and homogeneous (the latter two constituting the 'horizon problem'); and (2) why don't we see monopoles, as the vacuum phase dilutes such heavy exotic particles.

In 1990 the standard CDM model was simple: it had $\Omega_m = 1.0$ (i.e. the mean mass density of the Universe was the closure density) and a scale factor h ($= H_0/100$ with the Hubble constant H_0 in units of km s^{-1}/Mpc) of $\sim 0.5 -$ 0.8. At this point it took a severe hit. The *two-point angular correlation function* (Section 10.4) for the automatic-plate-measuring system (APM) galaxy survey (Maddox *et al.*, 1990) showed far more power on large angular scales ($> 2°$) than this model predicted. Death of CDM was in the air (e.g. Davis *et al.*, 1992). But in a brilliant and prescient paper responding to the crisis, Efstathiou *et al.* (1990) had suggested that CDM and the model ($\Omega = 1$, $n_s = 1$) could be saved with a low-mass-density universe ($\Omega_m \simeq 0.3$) and a reinstatement of Einstein's notorious cosmological constant at the level of $\Omega_\Lambda \simeq 0.7$. *In essence this is how it turned out*; but observational verification was eight or nine years away.

During the 1990s, theorists considered esoteric models of galaxy formation and/or biasing (Kaiser, 1987) to get around the problem, but the major advances came observationally, and in a different direction – cosmic evolution was established. Two things indicating epochal change in the galaxy population had been known for some time: (1) the Butcher–Oemler effect (Butcher & Oemler, 1984), more distant clusters showing an excess of blue galaxies over local clusters; and (2) a number–magnitude relation (Sections 2.4.2.4, 8.2) in the blue passband which was significantly at variance with predictions of a uniformly filled universe (Shanks *et al.*, 1984). Models of cosmic evolution of these blue galaxies were constructed using the count–redshift-distribution technique (Section 8.3.3). New data from the HST deep survey (Williams *et al.*, 1996) enabled counts of galaxies to be constructed by morphological type rather than colour (Abraham *et al.*, 1996). These showed that the excess counts in the blue of Shanks *et al.* were due to bright star-forming galaxies whose morphologies were largely unclassifiable on the Hubble tuning fork – they generally showed evidence of tidal interaction or merging, and it was suggested that they are protogalaxies, galaxies in early stages of coalescence. These objects (and not the spirals or ellipticals) show clear size evolution, being significantly smaller at large redshifts (Roche *et al.*, 1998). Further manifestation of cosmic evolution in the galaxy population came from the first version of the Lilly–Madau diagram (Lilly *et al.*, 1996; Madau *et al.*, 1996), which showed that the star-formation rate evolved massively between redshifts of 1 and the present day. (A recent compilation of star-formation rate estimates at different redshifts is given by Yüksel *et al.*, 2008.)

The framework of the Universe came back into sharp focus with two dramatic observational developments at the end of the millennium. Both relied heavily on statistical techniques. Firstly, there were the astonishing Hubble diagrams (Section 4.1) for distant supernovae type Ia (white-dwarf progenitors) (Riess *et al.*, 1998; Perlmutter *et al.*, 1999). Analyses of these revealed the modern-day Universe in a state of acceleration, strongly pointing to what was termed dark energy, either some new form of physical energy or the infamous cosmological constant. In either case the missing energy density, termed Ω_Λ rightly or wrongly, had been 'found' (but not understood). We could now have our flat Universe with $\Omega_m \sim 0.3$, $\Omega_\Lambda \sim 0.7$, and $\Omega_{tot} = 1.0$. Both SNIa papers employ a battery of statistical tests (Section 5) to verify the claims; and both rely on a marvellous correlation (Section 4) established by Phillips (1993) between SNIa luminosity (absolute magnitude) and light-curve decay time. Secondly, the first direct observations of fluctuations in the CMB (de Bernardis *et al.*, 2000) confirmed the flat nature of our Universe, the *angular power spectrum* (Section 10.6) revealing the first acoustic peak (see Section 11.3) to be in accord with $\Omega_{tot} = 1.0$. We enter the new millennium as the era of so-called precision cosmology, in a ΛCDM flat universe, recognizing that of the total mass energy of $\Omega = 1.0$, our take on this universe is via the shining baryons, constituting a mere 4 per cent of it. The baryons are a dreadfully biased tracer, but the only direct one we have got.

And so to the great galaxy surveys of the early millennium years, driven by new technology and demanding the best statistical analyses. The 2dF survey (Colless *et al.*, 2001) ran 1997 to 2002, and covered 1500 square degrees (Figure 10.1) using the 2°-diameter field at the prime focus of the AAT together with magnetically placed fibres feeding a low-resolution spectrograph. The 2dF galaxy redshift survey (2dFGRS) garnered nearly 2.4×10^5 redshifts, with objects selected from the APM galaxy survey (Maddox *et al.*, 1990). The Sloan Digital Sky Survey began in 2000 with a purpose-built 2.4-m telescope, spectrograph, and a 5-band imaging photometric system (York *et al.*, 2000).

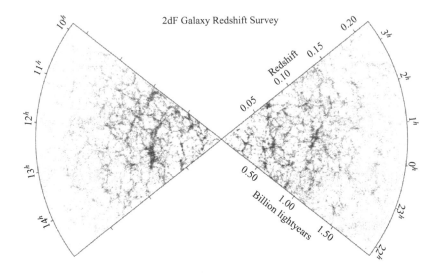

Figure 11.1   The 2dF galaxy redshift survey, with North (left) and South sections in flat projection. The distribution shows the sponge-like topology, and the known features of the distribution of nearby galaxies are clearly seen: in the North, the Shapley concentration at $z = 0.1$, $RA = 10^h$ and the Sloan Great Wall at $z = 0.08$, $RA = 11^h$ to 14^h; in the South, the Horologium–Reticulum concentration at $z = 0.07$, $RA = 03^h 30^m$ and the Pisces–Cetus concentration at $z = 0.06$, $RA = 0^h$. Reproduced by kind permission of Matthew Colless, the 2dF Galaxy Redshift Survey team and the Office of Commonwealth Copyright.

It aimed to provide a calibrated digital imaging survey of π sterad above Galactic latitude 30° (Figure 10.1) in five broad optical bands to a depth of $g' \simeq 23$ mag, together with a spectroscopic survey of the approximately 10^6 brightest galaxies and 10^5 brightest quasars found in the photometric object catalogue of the imaging survey.

The rich rewards from these surveys can be viewed on the websites msowww.anu.edu.au/2dFGRS and www.sdss.org, which, together with lists of publications, describe the surveys and how to access the public databases. We start by considering a few of the major results of the 2dF survey (Peacock, 2003; Colless, 2004), as the first galaxy redshift survey to break the 10^5 redshift mark. We have already met this survey via an example of PCA (Section 4.5); Folkes *et al.* (1999) used PCA to classify the 2dF galaxies in terms of spectral type.

This first mapping of the large-scale structure of galaxy distribution out to $600h^{-1}$ (Figure 11.1) finally yielded the unambiguous evidence that structures grow through gravitational instability. How? By the simplest statistic for clustering studies, namely the 3D *two-point correlation function* $\xi(\sigma_t, \pi)$. This

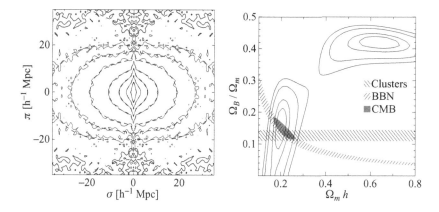

Figure 11.2 (a) The 2dFGRS 3D 2-point correlation function (Hawkins *et al.*, 2003). The axes represent transverse and radial distances, and the contours are of the two-point correlation function $\xi(\sigma_t, \pi)$; peculiar velocities in galaxy clusters produce the 'finger of God' structure at small values of transverse distances, while the overall flattening is due to gravitational infall on large scales (the Kaiser effect). (b) Cosmological parameters from the 2dFGRS power-spectrum analysis (Percival *et al.*, 2001). The likelihood contours are for the best-fit linear-regime power spectrum and were derived assuming a prior on h of 0.7 ± 10 per cent. The results were compared with estimates of the parameters from analyses of X-ray clusters, CMB anisotropies and Big-Bang nucleosynthesis as shown. Both diagrams reproduced by kind permission of the authors and John Wiley & Sons Inc.

is analogous to the two-point angular correlation function $w(\theta)$ we visited in Section 10.4; the σ_t refers to the transverse (angular) separation, while the π is the line-of-sight (radial) separation, as derived from the angular separations on the sky and the redshifts. The procedure to compile this is likewise analogous but in order to do so, the key – as it was for our 2D analysis of Section 10.4 – lies in comparing the observed count of galaxy pairs with a count estimated from a random distribution following the same selection function as the observed galaxies. This is hard; this selection function needs to mimic both the survey distribution on the sky and the redshift distribution of the observed galaxies. For this purpose a mask is needed for the sky which contains both spatial and obscuration information; and a template is needed to mimic the redshift selection effect, which amongst other things, depends also on the obscuration and on the spectral type of the galaxy. Clearly only those intimately connected with the observation process are qualified to make this measurement. The result (Peacock *et al.*, 2001; Hawkins *et al.*, 2003) is shown in Figure 11.2, a diagram which is governed by two effects. On small angular scales ($\sigma_t = 0.0 \pm 5.0h^{-1}$Mpc) the 'finger-of-God' elongation occurs because of the peculiar velocities in clusters of galaxies; on large scales there is flattening/ellipticity where we might have anticipated isotropy. The finger-of-God effect was demonstrated in the earliest galaxy redshift surveys. The ellipticity, the *Kaiser (1987) effect*, demonstrates *coherent infall on large scales*. This 2dF survey result provided the first true measurement of this ellipticity, as opposed to previous 3σ detections. This ellipticity is direct evidence of gravitational collapse on large scales, i.e. it is the evidence supporting the hypothesis that the distribution of galaxies arises from gravitational growth of small fluctuations in the initial density field of the Universe.

The flattening is governed by the parameter combination $\Omega_m^{0.6}/b$ where b is the bias parameter describing the density enhancement requirement for galaxy formation (Kaiser, 1984). To disentangle this we need a *power-spectrum analysis*. In any case the 3D clustering power spectrum (analogous to the 2D spectrum described in Section 2.4.2.4) was a key aim of the 2dFGRS. The measurement of this (Percival *et al.*, 2001) again needs the detailed (insider-only) understanding of the selection function for the galaxies in the 2dF sample, the *window function* (Sections 9.2, 9.2.2) for the survey, and consideration of the spatial sampling and the range of the power spectrum in order to ensure that the Nyquist criterion (Section 9.2) was met. Because of the convolution of the real power spectrum with the window function, the individual values of $P(k)$ have correlated errors, so that the correlation matrix (Section 6.1) must be calculated in order to assess significance of features in this spectrum. The shape of the spectrum closely resembled that of evolved linear density perturbations in a ΛCDM universe. Using a Bayesian approach (Sections 2.3, 6.5), Percival *et al.* compared the observed power spectrum with model power spectra convolved with the window function to search for the best fit, varying the appropriate parameters of hierarchical growth, in particular the combination of parameters Ω_b/Ω_m and $\Omega_m h$, the ratio of baryon to total mass and Ω_m scaled by h. Apart from the ΛCDM cosmology, the only assumption was $n_s = 1$; and it is later shown that the results are little affected by rather extreme tilting of the initial power spectrum away from $n_s = 1$.

Some results of this analysis are in Figure 11.2, from which we note the following. Firstly, here is an example of the use of a prior (Section 2.3), $h = 0.7 \pm 0.1$, well established from previous scientific investigations, in particular a Hubble Space Telescope key project (Freedman *et al.*, 2001). Secondly, the degeneracy between the two parameter combinations produces two regions of comparable maximum likelihood (Figure 11.2). The authors could easily have removed one of these areas with a prior from other scientific results;

the rightmost region, high $\Omega_m h$, had already been effectively rejected by the discovery of our low Ω_m universe as described above. Instead the authors commendably chose to show us exactly what the unbiased likelihood function looked like. They then invoked analyses of other observations, in particular Big-Bang nucleosynthesis (Burles *et al.*, 2001), X-ray clusters (Evrard, 1997), and CMB anisotropies (Netterfield *et al.*, 2002; Pryke *et al.*, 2002). These 'remove' the high Ω_m region. (It becomes slightly circular then to continue the analysis by comparing results with those from the CMB!). The analysis highlights a general issue: obtaining individual parameter estimates and their errors from marginalizing (Sections 7.6, 7.7, 7.9) is highly dependent on the *shape* of the regions mapped out by the likelihood function. Look at Figure 11.2, right – the small 'CMB region' suggests that the CMB estimates for the parameters will be much less uncertain. Not so; the CMB region is at $\sim 45°$, and the result is that projection onto either axis (the marginalization process) means that the errors along both axes are relatively large. For the $\Omega_m h$ parameter, they are virtually identical to that found from the 2dFGRS, because the 2dFGRS likelihood contours with $\Omega_m h$ along the *x*-axis are tall and narrow. The figure further highlights the enormous value added from different experiments for which the sense of degeneracy between parameters differs, working best when contours are approximately orthogonal, as we shall see. (Note that the example chosen for discussion here is 'pre-WMAP'; see Section 11.3 – much better parameter estimates are now available.)

Between the power-spectrum and the correlation-function analyses, then, the 2dF survey succeeded in confirming the gravitational instability picture, in measuring the mass fraction, the baryon fraction and the bias parameter (complicated by the further demonstration from 2dF of an unambiguous dependence of this parameter on galaxy luminosity, Norberg *et al.*, 2002). The survey had many other accomplishments including the derivation of luminosity functions (Section 8.3) for different galaxy types (Madgwick *et al.*, 2002) and an upper limit on the neutrino mass fraction (Elgarøy *et al.*, 2002).

Whatever the 2dFGRS could do, the SDSS survey could do better. Power-spectrum analyses (Tegmark *et al.*, 2006; Percival *et al.*, 2007; Padmanabhan *et al.*, 2007; Reid *et al.*, 2010) pushed the number of galaxies in the sample to 6×10^5, the range of redshift out to 0.6, and the volume sampled to $1.5h^{-3}$ Gpc^3, providing an accurate measurement of the real-space power spectrum over spatial frequencies k from 0.005 to $1h$ Mpc^{-1}. In general the analyses use samples of SDSS galaxies restricted to 'luminous red galaxies', LRGs, old red ellipticals with deep 4000 Å breaks, leading to accurate photometric redshifts and reduced luminosity effects. The important statistical procedures, some specifically developed, are described in detail: Tegmark *et al.* (2006), for instance, constructed estimators of the power spectrum with error bars minimized according to information theory, using a matrix-base pseudo-Karhunen–Loeve (Section 9.1) eigenmode method (Vogeley & Szalay, 1996). The result is a sequence of narrow window functions for the sample which ensures that the individual measurements along the power spectrum are independent; the price is in cpu time, ~ 1 yr. The achievements in providing ever tighter constraints on the parameters of our Universe were spectacular.

It is the subject of baryon acoustic oscillations (BAO) which has been the stunning revelation of SDSS LRG survey analysis. These oscillations are imprinted on the matter–photon plasma by acoustic waves at early epochs; they are frozen at the epoch of recombination (coincident approximately with the epoch of last scattering ($z = 1089$) from which we see the CMB). After recombination the photons free-stream; but the matter imprint remains. The oscillations are seen in the power spectrum of the LRGs (Figure 11.3), and from them, dramatic inferences about universe geometry and dark energy are possible (e.g. Blake & Glazebrook, 2003, and references therein). The detection of the BAO peak in the large-scale correlation function at $\sim 100h^{-1}$ Mpc (Eisenstein *et al.*, 2005) is shown in Figure 11.3. Here is yet another triumph for the two-point correlation function (Section 10.4), the feature occurring at separations so large that the function appeared to be dead and buried. The landmark paper of Eisenstein *et al.* (2005) shows the complementarity of the power spectrum and the correlation function: the acoustic oscillation series in $P(k)$ becomes a single peak in $\xi(r)$, its transform. The detection shows the following:

1. The scale and amplitude of the peak agree precisely with the prediction from the ΛCDM interpretations of the CMB power spectrum; see Section 11.3.
2. The imprint of the acoustic oscillations at low redshifts (the median redshift of the LRG sample is ~ 0.35) is a fundamental prediction of CDM cosmology in two ways. It shows that the oscillations originally occur at $z > 1000$; and that they survive to be detected at $\sim$ our cosmic epoch.
3. The small amplitude requires a dilution of the photon–baryon fluid at $z \sim 1000$ by matter that does not interact with it, namely dark matter.
4. This imprint in the late-time correlation of galaxies by baryonic physics at the epoch of recombination can be used as a cosmological standard ruler. It enables a measurement of the acoustic scale to ~ 4 per cent for the LRG sample. This length can be compared directly with the angular scale of the CMB anisotropies ($z \sim 1089$) to determine a distance ratio, a ratio which relies solely on the well-understood linear perturbation theory of the recombination epoch. This geometric measure provides an 'exquisite'

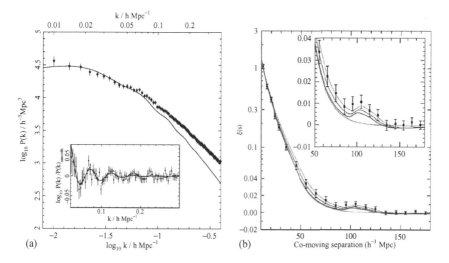

Figure 11.3   (a) The SDSS redshift-space power spectrum from the combined main galaxy and LRG samples (Percival *et al.*, 2007). The solid curve is not a best-fit model but rather, a normalized calculation from best-fit CMB parameters from WMAP 3-year data (see Section 11.3) for which $\Omega_m = 0.23$; at smaller scales the data favour $\Omega_m = 0.32$ and the discrepancy (now resolved) is at the $2-3\sigma$ level. The BAOs at low frequencies (inset) have been emphasized by dividing the region of the power spectrum by a smoothed version of the power spectrum; the curves in the inset represent oscillation models calculated from the best-fit WMAP parameters. (b) The redshift-space correlation function of the SDSS LRG sample (Eisenstein *et al.*, 2005). Error bars are from the mock-catalogue covariance matrix and the points are correlated. The inset shows an expanded view with a linear vertical axis. Of the model curves, the lowest line shows a pure CDM model, no baryons and, of course, no acoustic peak. The remaining curves are for values of $\Omega_m h^2$ from 0.12 to 0.14 with the baryon fraction $\Omega_b h^2$ set at 0.024. The bump at the $100h^{-1}$ Mpc scale, the baryon acoustic peak, disfavours the $\Omega_b = 0$ curve at the 3.4σ level. Both figures reproduced by kind permission of the authors and the AAS.

(Eisenstein *et al.*, 2005) constraint on spatial curvature $\Omega_K = -0.010 \pm 0.009$, a clear demonstration of the geometry requiring a flat and (currently) Λ-dominated universe – solid, virtually direct, geometric evidence for the existence of dark energy.

Percival *et al.* (2010) used the SDSS-7 data release to show that SDSS data alone can be redshift-sliced to track this imprinted distance scale as a function of epoch; ever tighter constraints on cosmological parameters result.

Simulations (see our primitive example of Section 2.6 et seq.) are the bedrock of all these analyses. Simulations are employed not just to check software, but (a) to determine what to assume for optimum analysis parameters and methodologies, and (b) to assess bias and errors, almost invariably correlated in both power-spectrum and correlation-function results. We know already that simulated random catalogues whose properties match the real sample are at the core of two-point correlation function analyses, with comparison done using an appropriate estimator (Section 10.4.1). Everyone therefore uses the technique of making samples from known distributions (Section 2.6), these being the galaxy samples used, trimmed and truncated from data-release catalogues requiring that selection effects are fully understood. The comparisons between model universe predictions and observed power spectra or correlation functions make use of standard model-choosing methods (Chapter 7), frequently minimum chi-square (Section 6.3), maximum likelihood (Section 6.1) or Bayesian modelling (Section 6.5). Simulations are also heavily used in experimental design for future galaxy samples targeting, for example, optimal surveys to extract maximum information from BAOs – see, e.g., Blake *et al.* (2006); Seo & Eisenstein (2007). (This latter reference has five mentions of the Fisher matrix (Section 6.1) in the abstract alone.) Following design, new surveys aimed at exploiting the potential of BAO are underway. BOSS (Baryon Oscillation Spectroscopic Survey; Schlegel *et al.*, 2009) is part of SDSS-III, and has taken 2×10^5 spectra by July 2010 (Padmanabhan *et al.*, 2010). The WiggleZ survey on the Anglo-Australian Telescope (www.astronomy. swin.edu.au/wigglez) has its first results published (Drinkwater *et al.*, 2010; Blake *et al.*, 2010). The Joint Dark Energy Mission (JDEM, now known as WFIRST) will in time lead to 2×10^8 galaxy redshifts in the range 0.7 to 2.0, sampling a volume of 200 Gpc^3 and redefining the terms 'exquisite' and 'precision cosmology' (Eisenstein *et al.*, 2009).

With regard to simulations (Section 2.6), all the foregoing does not make galaxies as we know them – it just grows gravitational perturbations in accordance with observed structure in the Universe. 'Semi-analytic galaxy formation' is a major simulation industry to understand galaxy formation in the context of ever-refined parameters of universe structure, galaxy distribution, and physics of gas collapse and star formation. The N-body simulations follow as many as 10^{10} 'galaxies' over cosmic time, using days of cpu on the most powerful parallel-processing computer systems available (Springel *et al.*, 2005; Benson & Bower, 2010; see www.mpa-garching.mpg. de/galform/virgo/millennium/). The latter reference is particularly illustrative in listing the vast data sets with which comparison is made, and in making extensive use of PCA (Section 4.5) to examine the huge parameter space involved. One mystery that the process may have solved is 'cosmic downsizing' (Cowie *et al.*, 1996), the observation that the nearby universe is dominated by galaxies that are small, old, red and spent in terms of star formation, rather than luminous, blue and ever-accreting giant galaxies. This latter is what hierarchical galaxy formation might suggest. The semi-analytic modellers came up with an AGN solution (Croton *et al.*, 2006; Bower *et al.*, 2006) – radio galaxies represent a short-lived 'fireworks' phase lasting perhaps just 10^7 years, but this phase may be vital in the lives of galaxies in expelling the accreting star-forming gas and halting their otherwise inexorable growth over cosmic timescales.

11.2 The weak lensing universe

Weak gravitational lensing offers unique opportunities to map the total gravitating matter, to constrain cosmological parameters and to probe the properties of dark energy. It is *entirely* a statistical procedure; no results are apparent without statistical analysis, by definition.

Weak lensing developed through our acquaintance with strong gravitational lensing, lensing in which the observed effects are evident in individual cases. This lensing provided the first observational proof of general relativity. During the solar eclipse of 1919, Eddington and his team measured the deflection of starlight by the gravitational field of the Sun, and the results (Dyson *et al.*, 1920) were in accord with Einstein's prediction. Sixty years later came the next advance: the first gravitational lens system (the 'double quasar') was discovered in 1979 (the quasar 0957+561; Walsh *et al.*, 1979), and many more such systems are now known. Subsequently, arcs were noted in the images of nearby large clusters Abell 370, CL 2244-02 and Abell 2218 (Soucail *et al.*, 1987a; Lynds & Petrosian, 1988a). The blue arc in Abell 370, the most distant Abell cluster, was initially interpreted as a star-forming region (Soucail *et al.*, 1987a), but it was rapidly recognized along with the other arcs as the distorted image of a blue background galaxy, strongly lensed by the gravitational field of the cluster (Soucail *et al.*, 1987b; Lynds & Petrosian, 1988b). Zwicky predicted such lensing in 1937 (Zwicky, 1937a,b), while Newton and Cavendish had raised the possibility of gravitational lensing in the eighteenth century.

By the early 1990s it was recognized that strong lensing implied that there were vast numbers of weakly lensed galaxies – lensed so weakly that detection could only be made in a statistical sense. These could provide a powerful cosmological probe to examine two basic issues:

1. The real gravitational mass – including, therefore, the dark matter – along the line of sight, and in particular a mapping of the total mass in foreground clusters (e.g., Tyson *et al.*, 1990; Miralda-Escude, 1991a), and

2. large-scale structure, including cosmological parameters (e.g. Blandford
et al., 1991; Miralda-Escude, 1991b), as it is one of the few astronomical observations measuring the power spectrum of *total* gravitating mass,
while at the same time exploring the dependence of this power spectrum on
cosmological epoch. Doing so enables a mapping of the cosmic expansion
rate at redshifts <1, the region in which the effects of the dark energy are
strongest; the equation of state may therefore be constrained, offering the
possibility of determining the elusive nature of dark energy.

Weak lensing has no formal definition except what it is not – its distortion
effects are not directly visible in an image; it is not strong lensing. *The technique,
now known to be one of the most powerful cosmological tools on offer, is entirely
statistical in nature.* The subject is fortunate to have two fine reviews available
to date: the initial comprehensive review article by Bartelmann & Schneider
(2001) is regarded by weak-lensers as *the* fundamental reference, while progress
over the subsequent years is described by Munshi *et al.* (2008).

All analyses start from the thin-lens equation, which may be written in terms
of a 2D lensing potential ψ, the 2D analogue of the Newtonian gravitational
potential, with the deflection angle θ the gradient of ψ. Deflection angles
can be calculated in terms of surface density of mass Σ, which is the total
mass along the line of sight. This surface density is often expressed in terms
of κ, the *convergence*, given by $\kappa(\theta) = \Sigma(\theta)/\Sigma_c$, where the *critical surface
density* Σ_c is simply a function of the geometry of the system, namely object-
lens-observer distances. If $\kappa > 1$ multiple images can be formed; thus $\kappa < 1$,
$\Sigma < \Sigma_c$, represents a division between weak and strong lensing. However, the
weak-lensing regime is usually regarded as having $\kappa \ll 1$.

With the lens equation holding for all points on the source plane, an extended
source, treated as a bundle of point sources, will have these points shifted differently; the image will be distorted. If $A(\vec{\theta})$ represents the distorted image it may
be shown (Bartelmann & Schneider, 2001) that the *distortion or amplification
matrix* is characterized by the convergence κ and the complex variable γ. The
variable $g(\theta) = |\gamma(\theta)|/[1 - \kappa(\theta)]$ is known as the *reduced shear*. Convergence
κ produces focusing and magnification; it does not change shape. The *com-
plex shear* γ produces image stretching via anisotropic mapping, with $|\gamma|$ the
amount of stretch. The matrix will produce mapping of a circular image to an
ellipse with ratio of the semi-axes to the original radius of $1 - \kappa \pm |\gamma|$.

It is the recognition that this shear produces distortions over many galaxies,
ellipticities in the first instance, which is the basis for using weak lensing to
map the gravitational field, and hence matter density along the line of sight.
As weak lensing occurs over the entire light path, the thin-lens approximation

fails; but 'effective convergence' by integrating along the line of sight allows use of the distortion matrix to determine both size and shape. The distortions are of order 1 per cent.

In the standard analysis, namely the method of second-order surface-brightness moments (Bartelmann & Schneider, 2001), there arrives an intuitive result: $e_o = e + g \approx e + |\gamma|$, i.e. the *observed ellipticity* e_0 under shear is the galaxy *intrinsic ellipticity* (e) plus an extra component due to shape distortion along the line of sight. If the average intrinsic ellipticity $<e>$ is zero and there is no preferred universal orientation for the galaxies (not necessarily the case!), then $<e_o> = <g>$ so that an estimator (Chapter 3) for the shear is $\gamma = <g> = <e_0>$. (The ensemble averages here ignore the vanishing cross-terms, and represent local estimates in the region to be mass-mapped.) Because there is a limited range and sample, a window function (Sections 9.2, 9.2.2) is always involved and must be taken into account. The result holds generally for a population of lensed galaxies with an observed distribution in redshift (Bartelmann & Schneider, 2001), such as a sample selected via a deep lensing survey would have. Weak-lensing effects grow with the redshift of the sources as the line of sight extends; but distant galaxies are fainter and more difficult to measure so that current weak-lensing surveys mainly probe redshifts $\lesssim 1$. This is ideal for exploring the properties of dark energy via the expansion it has produced.

κ is the average of the local density contrast along the line of sight and if we could measure it, we would get the projected density field on the sky. However, the quantities κ and γ are uniquely related (in the weak-lensing regime and under a number of assumptions) through the *reduced shear g* so that the κ map of projected mass may be reconstructed directly from the shear field. The procedure, then, is simple in essence: map the shear field with measurements of the ellipticities of the chosen galaxy sample, with corrections to these for systematic effects produced by seeing and instrumental point-spread functions (PSFs). (These 'corrections' are an industry in their own right, frequently overwhelming in magnitude the intrinsic or observed ellipticities.) These ellipticities (with corrections) become the 'shear map'; and this may be transformed via inversion of the distortion/amplification matrix into a mass distribution.

It may be shown that the two-point correlation function of the complex shear is the Fourier transform of the 2D power spectrum of the mass (Bartelmann & Schneider, 2001; Munshi *et al.*, 2008). There is clear analogy with our discussion of surface density in terms of the two-point correlation function (Section 10.4) as the Fourier transform of the 2D (angular) power spectrum (Section 10.6).

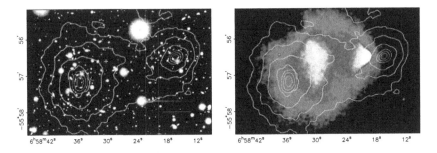

Figure 11.4   A direct demonstration of dark matter (Clowe *et al.*, 2006). (a) Magellan telescope image of the merging cluster 1E 0657-558 (z = 0.296), with the contours tracing the gravitating mass (κ reconstruction via weak lensing). (b) 500-ksec Chandra X-ray image of the cluster, tracing the bulk of the cluster baryons, which is in the form of the hot intergalactic plasma. This gas is centralized and shocked in the collision between the two sub-clusters, while the dissipationless stars and dark matter retain the initial shape of their gravitational potentials. Reproduced by kind permission of the authors and the AAS.

Much care is required in the choice of galaxy sampling (to obtain enough to beat down sample variance), redshift estimation, estimates of PSFs, and reconstructions (see below) with smoothing functions to obtain variances which do not fly off to infinity. However, recent results are spectacular: see Figure 11.4, a colliding pair of clusters in which the bulk of the baryons and the matter (mostly dark) do *not* coincide (Clowe *et al.*, 2006). Here the selection of lensing galaxies is made in a redshift slice which contains the cluster and the result of the shear mapping is a 2D surface potential for the cluster plane.

This is an example of 2D weak lensing. The drive is to obtain 3D mapping of matter in the Universe, or at least the power spectrum of this matter and its redshift dependence. Following Munshi *et al.* (2008), there are several techniques in use and under development.

(1) *Inversion* The potential may be determined directly through an inversion. As the shear pattern is derivable from the lensing potential, the inverse problem to construct the lensing potential from the shear-map information may be attempted. It may be the obvious way to go; but it is hard (Taylor, 2001; Bacon & Taylor, 2003). The intrinsic ellipticity of objects is a particularly serious source of noise, as it is for other techniques. This can be overcome to some extent by Wiener filtering (Section 9.3.1).

(2) *Tomography* With distance information available for the individual sources (generally through photo-redshifts), the method divides the survey into consecutive redshift slices, with the shear pattern determined for each slice. In addition to the cross-correlation (Section 9.4) in each slice, the

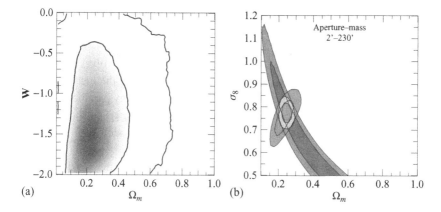

Figure 11.5   (a) Constraints on Ω_m and w from 3D weak-lensing analysis, HST COSMOS field, assuming a flat ΛCDM cosmology (Schrabback *et al.*, 2010). The contours indicate the 68 and 95 per cent credibility regions, with marginalization (Section 6.5) over parameters which are not shown. The grey scale indicates the highest-density region of the posterior distribution. (b) Comparison (Fu *et al.*, 2008) between WMAP 3-year results (1σ and 2σ contours at approximately $60°$ angle) and 3D weak-lensing tomography results from the CFHT Legacy Survey (long open contours, 1σ and 2σ). The small central region represents the combined likelihood contours. Both figures reproduced by kind permission ©ESO.

cross-correlations between the slices are measured. The result is a matter power spectrum (Section 10.6) for each slice as well as a z-dependence of this power spectrum. Tomography does not add greatly to the accuracy of determination of the amplitude of the power spectrum (σ_8), but because it is so sensitive to low-redshift changes in the power spectrum, it can contribute significantly to elucidating dark-energy properties (characterized by w). Recent demonstrations of the power of the technique come from weak-lensing analysis of the HST COSMOS field by Schrabback *et al.* (2010), and weak-lensing analysis of some of the CFHT Legacy Survey (CFHTLS) fields by Fu *et al.* (2008); see Figure 11.5. The technique is critically dependent on the accuracy of redshift estimates of the objects involved, the photo-redshifts.

(3) *Bayesian modelling: full 3D analysis* With redshift estimates, the result is a noisy and imperfectly sampled 3D field, although the noise and the sampling window function need to be estimated. But with a Bayesian modelling (Sections 2.3, 6.5) of the matter power spectrum and its redshift dependence, the binning involved in tomography is avoided and smaller statistical errors should result. The approach is the method of choice for predicting how future experiments will perform in constraining universal parameters (Heavens *et al.*,

2006, 2007). The latter paper considers the Fisher matrix (Section 6.1) – that vital tool in experimental design which calculates how well a given experiment will obtain model parameters. In practice, Heavens *et al.* (2007) generalize the Fisher matrix approach from a Bayesian standpoint, which involves computing the Bayesian evidence (Section 7.8). The authors illustrate the method with a study of how well future experiments will derive dark-energy parameters. They find that proposed large-scale weak-lensing surveys from space should readily be able to distinguish general relativity dark-energy models from modified gravity models.

With regard to weak-lensing experiments, two sky surveys hold sway at present; the COSMOS field of the HST, and the CFHT Legacy Survey. The former is deepest and has the narrower PSF, but covers a mere 1.64 square degrees; the latter has the smallest PSF available from the ground and covers about 170 square degrees. The COSMOS field has 17 bands of photometry compared to only 5 for the CFHTLS, providing much more accurate photo-redshifts. The future lies in space, where PSFs are tiny and stable, and dedicated missions can map large areas while obtaining simultaneous and accurate photo-redshifts. The Large Synoptic Space Telescope (LSST, www.lsst.org/lsst), the Joint Dark Energy Mission (JDEM, now known as WFIRST, jdem.gsfc.nasa.gov/) and the EUCLID mission (sci.esa.int/euclid) represent the future.

In the meantime, weak lensers have demonstrated the power of the technique: they have produced stark 'pictures' of dark matter, and they have provided powerful constraints to cosmological parameters highly complementary to those from BAO (Section 11.1), CMB (Section 11.3) and SNIa results. They continue to wrestle with issues concerning the PSF, which include variable seeing and changes in the instrumentation point-spread function across the field, in which PCA methods (Jarvis & Jain, 2004; Section 4.5) are to the fore. International challenges have been set up to determine the optimum techniques to tease out mass maps from known simulations (the STEP collaboration – STEP 1: Heymans *et al.*, 2006); to determine the best filters and algorithms for photo-redshifts (Ma *et al.*, 2006); and to characterize galaxy shapes (the GREAT'08 challenge: Bridle *et al.*, 2009). Additional and powerful techniques are under development for image classification, more sophisticated ways of describing galaxy shapes than simple ellipses, including 'shapelets' (Refregier, 2003; Refregier & Bacon, 2003), analogous to wavelets (Section 9.6), which provide a minimum-variance estimator for the shear, and 'flexions', analysis of higher-order shape distortions (Schneider & Er, 2008: 'Weak lensing goes bananas').

11.3 The cosmic microwave background universe

Excess isotropic radiation at a brightness temperature of about 3K was discovered serendipitously by Penzias & Wilson (1965) and identified as cosmic microwave background (CMB) radiation in a companion paper by Dicke *et al.* (1965). It seems a definitive starting point. The background to the background however, is long and complex, detailed in a fascinating compilation of contributions from those involved by Peebles *et al.* (2009).

Many years of measuring the isotropy of the radiation and its spectrum ensued; it was discovered to be isotropic on large scales except for a dipole moment, implying a velocity of the Sun through space of $\sim$400 km s^{-1} (Smoot *et al.*, 1977). The hunt for deviations from a black-body spectrum, together with attempts to detect small-scale fluctuations evidencing the seeds of galaxy formation, occupied much ground-based telescope time from 1970 to 1990 to no great effect. The radio telescopes in our arsenal were too low in altitude and in operating frequency, and too large in extent. 3K radiation peaks at $\sim$160 GHz. Its exploration requires relatively low resolution at altitudes above the water vapour layer. Thus, the CMB radiation is really the province of high-flying balloon and dedicated satellite missions, although substantial advances are now being made from the ground with high-frequency experiments at high altitudes.

The first CMB satellite mission COBE (Cosmic Background Explorer) reaped the rewards (including Nobel prizes for principal investigators Mather and Smoot). The CMB spectrum was shown to follow pure black-body radiation at a temperature of 2.728 ± 0.004 K (Mather *et al.*, 1990), the spectral purity alone ruling out many competing cosmological models. The dipole was accurately measured; the Earth's velocity through space was found to be 371 ± 1 km s^{-1} towards $(\ell, b) = (264°, 48°)$. Removing the rotation of the Galaxy and the peculiar motion of our Galaxy in the Local Group yielded the motion of our Local Group of galaxies to be 600 km s^{-1} towards $(\ell, b) = (270°, 30°)$, not far off the direction to the Virgo cluster. COBE found the long-sought fluctuations (Smoot *et al.*, 1992), albeit only in a statistical sense; the background on isotropic black-body emission was shown to have fluctuations at a level $\Delta T / T \approx 10^{-5}$. The fluctuations were consistent with an inflationary universe with approximately the Harrison–Zeldovich–Peebles spectrum, a power law (Section 2.4.2.4) of index $-n_s$ with $n_s = 1$. What is more, the amplitude of these fluctuations produced the 'COBE normalization', the size of perturbations at a fixed length scale; and this turned out to be close to the prediction of simple assumptions for the gravitational instability picture in an inflationary universe.

See Peacock (1999a, Chapter 18) for the technical story; see Partridge (1995) for an excellent descriptive account. There was much relief amongst theorists. One commented that 'cosmology can start now'; the 5-page paper received nearly 2000 citations; the 'blobby' images in it received worldwide attention including a front-page spread 24 April 1992 in the *Independent*, a premier UK newspaper, under the banner headline 'How the Universe began'.

The results in 2000 from two balloon missions from Antarctica, BOOMERanG and MAXIMA, provided the next milestone in CMB research and indeed in cosmology, with direct detections of the fluctuations. Together the two missions were designed to reveal the primary anisotropies of the CMB, those which can be calculated using linear perturbation theory. The achievement of these experiments (subsequent to calibration and removal of instrumental effects) was a first view of the highly prized angular power spectrum (Section 10.6) of the temperature fluctuations.

The angular power spectrum is a mainstay of CMB investigation since the pioneering papers of Bond & Efstathiou (1984, 1987); these and subsequent papers by these authors showed how this power spectrum of CMB fluctuations can be used to infer most of the parameters describing the cosmology of our Universe. (Calculations of anisotropies in the CMB for different universes go back to Sachs & Wolfe (1967); Peebles & Yu (1970); Sunyaev & Zeldovich (1970) and beyond; see the references in the Bond & Efstathiou papers.) There are degeneracies amongst these parameters from CMB analysis alone, and these were examined with simulations (Section 2.6) by Efstathiou & Bond (1999), in which extensive use was made of the Fisher matrix (Section 6.1), PCA (Section 4.5), and the likelihood function (Sections 2.3, 6.1). The analyses have underpinned and to a large extent driven the missions to explore the CMB (cf. Bond & Efstathiou, 1987: 'We hope that the parameters for the CDM models presented ... will serve as a challenge for prospective experimenters').

The balloon experiments were designed to be complementary, BOOMERanG covering 1800 square degrees, MAXIMA only 124, but to a much fainter level. Between the two of them they cover the angular scales from 10 arcmin to 5°. The power-spectrum analysis of the BOOMERanG fluctuations (de Bernardis *et al.*, 2000) defined the *first acoustic peak*, subsequently refined by the MAXIMA results (Hanany *et al.*, 2000). The position of this peak is dominated by the total mass density parameter Ω_{tot}. If the Universe is of flat geometry, then this parameter is unity and the first acoustic peak should be at $\ell \approx 200$ – de Bernardis *et al.* (2000) and Hanany *et al.* (2000) showed that it was. Bond & the MaxiBoom collaboration (2000) put the results of the two missions together and showed that the data clearly defined both the first and second peaks in the spectrum (Figure 11.6).

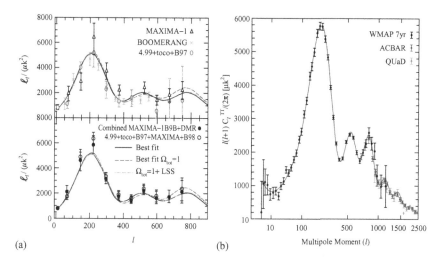

Figure 11.6   The CMB angular power spectrum. (a) Status in 2000, as determined from the BOOMERanG and MAXIMA balloon missions (Bond *et al.*, 2000); the upper panel shows the individual data from the two missions, the lower panel the combined data. Reproduced by kind permission of the authors. (b) Status in 2010 from the 7-year WMAP data (Larson *et al.*, 2010), image courtesy of the WMAP team and NASA GSFC public archive. The curve is computed for the best-fit 7-parameter ΛCDM model. Data at high ℓ values are from the ACBAR (Reichardt *et al.*, 2009) and QUaD (Brown *et al.*, 2009) experiments. The ACBAR and QUaD data are shown only at $\ell > 690$, where the errors in the WMAP power spectrum are dominated by noise. Features are discussed in the text. Reproduced by kind permission of the authors and the AAS.

In each region of the temperature power spectrum, a different physical process dominates. The regions are as follows.

a. Angular scales larger than the horizon size at decoupling as observed today. These correspond to $\theta \gtrsim 2°$ or, equivalently $\ell \lesssim 90°$. The low ℓ portion is termed the Sachs–Wolfe plateau (Sachs & Wolfe, 1967). In this region one observes the relatively unprocessed primordial fluctuation spectrum because patches of sky with larger separations could not have been in causal contact at decoupling. The error in defining this region is generally dominated by cosmic variance (Section 10.6.1).

b. The acoustic peak region, $0.2° \lesssim \theta \lesssim 2°$ or $90 \lesssim \ell \lesssim 900$, which is described by the physics of a 3000 K plasma of number density $n_e = 300$ cm^{-3} responding to fluctuations in the gravitational potential produced by the dark matter. The first peak represents a compression phase of the acoustic wave; the second, a rarefaction phase; the third, a compressional phase at the second harmonic of the first peak, and the fourth, a rarefaction at the third harmonic of the first peak – see Hu & Dodelson (2002). If there are second and higher peaks (and there are!) then the photon–baryon fluid underwent acoustic oscillations.

c. The Silk damping tail (Silk, 1968), $\theta \lesssim 0.2°$ or $\ell \gtrsim 900$, which is produced by diffusion of the photons from the potential fluctuations and the washing out of the net observed fluctuations by the large number of hot and cold regions along the line of sight.

It is over to the triumphant Wilkinson Microwave Anisotropy Probe (WMAP) mission to carry out this detailed exploration of the temperature angular power spectrum, and much else. Sent out to the Earth–Sun L2 point (Bennett *et al.*, 2003b) and beginning observations in August 2001, it was designed to run for two years, but instead ran *for 9 near-flawless years.* WMAP repeatedly mapped the entire sky at five frequencies, 23, 33, 41, 61 and 94 GHz, with corresponding beam sizes of 53, 40, 31, 21 and 13 arcmin, in Stokes parameters I, Q and U, using 10 horn pairs with beam separation 140° feeding 10 differencing receivers. The mission description and results, both popular and technical versions, bibliography, maps, images and tables of best-fit parameters resulting for our Universe are all available courtesy of the WMAP team and NASA on the website `lambda.gsfc.nasa.gov`.

There is a long and arduous road, paved with statistical techniques, leading from the complex all-sky scan pattern producing the time-ordered data from 10 radiometers to the angular power spectrum and then to cosmology. As a first step (Bennett *et al.*, 2003b) the data stream was 'prewhitened' with a high-pass Wiener filter (Section 9.3.1) to remove inherent $1/f$ noise (Section 9.7). Then two aspects required detailed attention before there is a map from which cosmology may be deduced: the characteristics of the instrument and data set, and the contaminants, the Galactic and point-source foregrounds.

With regard to the instrumentation, there are 10 beams, two differencing beams at each of the five frequencies. The beam smoothing tapers off the sensitivity to high spatial frequencies, acting as a low-pass filter as in Section 9.3.1. The transform of the beam profile in harmonic space is the *beam transfer function*. A raw CMB power spectrum is divided by the *square* of the beam transfer function, namely the window function (Sections 9.2, 9.2.2), in order to invert the filtering done by finite-width beams and hence to produce the beam-corrected power spectra to be used for cosmological investigations. Thus, the beam patterns must be established as precisely as possible. The techniques to calibrate the signal amplitude and to measure beam properties with precision, together with the year-on-year improvements, are described by Page *et al.* (2003); Jarosik *et al.* (2007); Hill *et al.* (2009); Weiland *et al.* (2011). Observations of the planets Mars and Saturn provided primary flux calibration, with Jupiter, effectively a point source but variable in flux, used to confirm primary beam shapes. First approximations to beam shapes were calculated from physical optics. The final beams (Hill *et al.*, 2009) are the result of heavy-duty processing by parallel processors, and include the use of statistical techniques such as additional Fourier-mode (Section 9.2) construction optimized by a modified conjugate gradient method to descend into the main minimum χ^2 valley (Section 5.3.1), and the augmenting of primary-mirror modes using Gram–Schmidt orthogonal polynomials. Many simulations (Section 2.6) were run to establish errors introduced into the angular power spectrum by errors in the beam measurements, and to search for systematic errors.

With beams and total response well defined, the real impediment to cosmological interpretation of any CMB measurement is our Galaxy, whose radio emission at the lower WMAP frequencies dwarfs the CMB signal by a factor $>10^3$. The main purpose for WMAP to have five frequency bands is to measure and remove this foreground. The rigorous efforts to do this are documented in the papers by Bennett *et al.* (2003a); Barnes *et al.* (2003); Hinshaw *et al.* (2007); Kogut *et al.* (2007); Gold *et al.* (2009); Dunkley *et al.* (2009b); Gold *et al.* (2011). The separation of the CMB and foreground signal components relies on their differing spectral and spatial distributions. The Galactic emission was modelled primarily from three components: free-free emission (from electron–ion scattering), synchrotron emission (from the acceleration of cosmic-ray electrons in magnetic fields, occurring primarily in Type Ib and Type II supernova remnants), and thermal dust emission. At least four techniques were used in 'galaxy removal' with progressive data releases. In each, a mask was applied over ∼15 per cent of the sky nearest the Galactic plane, a region deemed impossible to use for CMB analysis. For the rest of the sky, the various approaches then used different fitting techniques incorporating the three components of Galactic emission as well as the old 408 MHz map of Haslam *et al.* (1981). The techniques include Bayesian modelling (Section 6.5) with the 408 MHz map as a prior (Section 2.3), maximum-entropy methods (MEM; Section 7.10) to explore the detailed properties of each component, and for the five-year maps, an MCMC (Section 7.7) technique in which each pixel was modelled independently by four components, the three Galactic components plus the CMB itself, with marginalization over all other pixels. The resulting variance of ∼100 μK^2 is far below the CMB power.

Of course a sky mask creates a window function (Sections 9.2, 9.2.2), coupling multipole modes (Section 10.6) so that the power-spectrum covariance matrices (Section 6.1) are no longer diagonal. All techniques made extensive use of Monte Carlo simulations (Section 2.6) – inputting known components of Galactic emission onto a modelled CMB, for example – to examine bias and noise properties.

Foreground radio AGN were found by searching for 5σ deviations above a smoothed CMB, with Gaussians (Section 2.4.2.3) plus baselines fitted at such positions to the real CMB data.

This saga of Galactic foreground analysis by at least four techniques in the one, three, five and seven-year data acquisition cycles contains much repetition. It is not the only case of repetition: in polarization analysis, the removal of polarized foregrounds is discussed in different papers via three different techniques; and several papers, two or more in each series, consider the details of the power spectrum itself and the cosmological implications. However, this is not merely the sociology of a large collaboration in which several first authors from different institutes are required. Firstly, the repetition in analyses for each of the four data releases is important in demonstrating both consistency of results and the steadily improving parameter constraints, providing clear indication that error assessment is on the money. Secondly, the apparent repetition *represents the thorough examination of the data set through many different eyes.* The lesson is salutary. With a great data set, examine it from as many directions as possible; simulate it; subject it to everyone's best efforts, individually and collectively; play with it; gain total familiarity – and reap the rewards: here, consistent results of such significance that cosmology will never be the same.

To return to the WMAP story, we now have a calibrated CMB, Galactic and AGN foregrounds removed, and well-defined beams and response structures. We can thus complete the tale with the climax – those cosmological results. From the 7-year data, the temperature spectrum is now defined up to $\ell \approx 1200$; the third peak is seen at high-level significance; the temperature spectrum has a signal to noise exceeding unity for all $\ell < 919$; and it is limited by cosmic variance (Section 10.6.1) for all $\ell < 548$. All that is needed is to fit the power spectra as best as possible. The process is made simpler with the ready availability of online programmes (e.g. CAMB, Lewis *et al.*, 2000, see camb.info) to compute instantaneously the power spectra in different ΛCDM universes. The repeated likelihood analyses (e.g. Verde *et al.*, 2003; Spergel *et al.*, 2007; Dunkley *et al.*, 2009a; Komatsu *et al.*, 2009, 2011; Larson *et al.*, 2011) employ with increasing levels of sophistication the tools we have met already – Bayesian analyses (Section 6.5), χ^2 or least squares as the goodness-of-fit parameter (Sections 6.2, 6.3), MCMC (Section 7.7), Fisher matrix (i.e. the inverse covariance, or 'curvature' matrix, Section 6.1), Monte Carlo simulations (Section 2.6) to test and calibrate the likelihood code so that the χ^2 values may be used for goodness-of-fit tests as well as model comparison; and simulations again to test for bias in the derived parameter set. The analyses use the MCMC approach to examine the likelihood function in a classic application; as Verde *et al.* (2003) point out, exploring just a six-parameter model via a grid search would require perhaps 1200 days of cpu time. In fact this initial description of parameter estimation methodology by Verde *et al.* (2003) is very clear, and merits close reading; it provides a particularly succinct description of how the fluctuations appearing in the CMB angular power spectrum (Figure 11.6) transform into the non-linear regime to become the galaxy power spectrum (Figure 11.3). Also clearly described is how the galaxy power spectrum results are used with CMB WMAP results to provide tightened constraints on the ΛCDM universe parameters. The Spergel *et al.* (2007) paper has received over 4500 citations, the Komatsu *et al.* (2009) paper 2400.

The simple result from all the cosmological analyses is that a six-parameter ΛCDM universe not only fits the data – it accounts for it all. This flat universe, starting out with near-scale-invariant[1] adiabatic perturbations falls out not just from WMAP, but, as Komatsu *et al.* (2011) point out, from large-scale structure data, supernova Type Ia data, cluster measurements, distance measurements, and strong and weak lensing data. The WMAP data have been examined carefully for evidence of non-Gaussianity and anomalies and none has been found to be significant (Komatsu *et al.*, 2003, 2011; Benson & Bower, 2010). The result is not just a powerful endorsement of an inflationary universe; it signifies that *the angular power spectra of both total and polarized emission fully encode all information in the CMB maps.*

Despite the chorus of acclaim for the simple ΛCDM cosmology and ever-refined values for its six parameters, it is important to test if the data support a model with more parameters, and this is done in more than one of the investigations, e.g. Page *et al.* (2007), Spergel *et al.* (2007) and Komatsu *et al.* (2011). The tests for justification of increased model complexity (see Sections 7.8, 7.2) conclude against it. The six parameters of our flat ΛCDM universe are h (the Hubble constant); τ (reionization optical depth); σ_8 (linear-theory amplitude of matter fluctuations on $8h^{-1}$ Mpc scales); and two out of three of the mass fractions Ω_b (baryon density), Ω_c (CDM density) and Ω_Λ (dark-energy density), as the sum of these three is unity. Dunkley *et al.* (2009a) show how the WMAP data alone constrain these parameters (Figure 11.7), while Komatsu *et al.* (2011)

---

[1] It is customary to use $n_s \equiv 1$ and scale-invariance interchangeably to describe the Harrison–Zeldovich–Peebles primordial spectrum. Of course any power law is scale invariant (Section 2.4.2.4) or at least scale free or scale independent, but the sense here is a restricted one – namely the power per logarithmic interval is constant; and there are physical explanations as well.

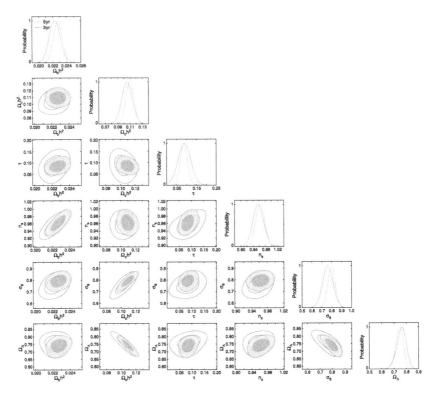

Figure 11.7   Constraints from the five-year WMAP data on the 6 ΛCDM parameters (shaded regions), showing marginalized 1D distributions and 2D 68 per cent
and 95 per cent limits (from Dunkley *et al.*, 2009, and the WMAP archive). The
six parameters, the x-axes in columns (1) through (6), the y-axes running down
from row (1) to row (6), are $\Omega_b h^2$, the baryon density; $\Omega_c h^2$, the matter density;
τ, the optical depth to reionization; n_s, the spectral index ('tilt') of the power-law
spectrum of initial perturbations; σ_8, the scale of unity density enhancement at
the present day; and Ω_Λ, the dark-energy density. Five-year parameters (shaded
contours) are consistent with the three-year limits (open contours) from Spergel
et al. (2007), and are all more tightly constrained. Reproduced by kind permission
of the authors and the AAS.

show how these constraints are dramatically tightened with the incorporation
of BAO and Hubble-constant data from studies described in Section 11.1. The
parameters are all consistent with previous – even historical – estimates and
prejudices, with the possible exception of $n_s = 0.968 \pm 0.012$. The Harrison–
Zeldovich–Peebles spectrum does not hold for our kind of inflation, at least
to the 99.5 per cent level of confidence (although the tilt to a 'red' spectrum
retains consistency with the simplest inflationary scenarios). The simulations

used to test for parameter bias take on particular importance when a result of this significance is at stake.

The detection of the E-mode polarization (Kogut *et al.*, 2003) is critical in establishing the WMAP ΛCDM picture of the Universe with inflation. The detection of superhorizon fluctuations is a distinctive signature of this early epoch of acceleration. Acoustic peaks in the temperature angular power spectrum do not prove that the fluctuations are superhorizon, as causal sources acting purely through gravity can mimic the observed peak pattern (Turok, 1996b,a). However, the large-angle ($50 < \ell < 150$) temperature–polarization anti-correlation detected by WMAP (Kogut *et al.*, 2003) is a distinctive signature of superhorizon adiabatic fluctuations (Spergel & Zaldarriaga, 1997). In addition Kogut *et al.* (2003) show how the peaks detected in the E-mode power spectrum imply a reionization redshift of about 20 ± 10, larger than that suggested by the Gunn–Peterson trough in $z \geq 6$ QSOs, and implying a complex and lengthy ionization history for our Universe.

Although the six-parameter model describes the data adequately, strong constraints may nevertheless be derived on additional cosmological and physical aspects, as explored by Komatsu *et al.* (2011). Information on additional aspects of our Universe includes new limits on (the lack of) curvature, independent confirmation of pre-stellar helium, an estimate of its mass fraction, estimates of the total mass of neutrinos and the effective number of neutrino species, limitations on the dark-energy equation of state, limits on parity violation, limits on gravitational waves, and details of the Sunyaev–Zeldovich (SZ) effect (with clear differences evident between the effect for cooling-flow and non-cooling-flow clusters); and the beginnings of quantitative tests on inflationary models. Moreover the WMAP results contain a wealth of new information on both Galactic emission and extragalactic radio sources; WMAP has produced all-sky surveys and catalogues for the latter at each of its five observing frequencies.

The WMAP mission has forged an apparently unassailable ΛCDM picture of the Universe, while showing that ever-improving quality of data leads to scrutiny of fundamentally new areas of physics and cosmology as outlined by Komatsu *et al.* (2009, 2011). WMAP has thrown down the gauntlet for new-era cosmology and physics. It is for Planck (returning data as this is written) and subsequent missions to pick it up. Those with full knowledge of the instrumentation *and statistical techniques* will be richly rewarded.

Appendix A
The literature

Of the vast literature, we point to some works which we have found useful, enlightening or just plain entertaining. We bin these into six types (somewhat arbitrarily as there is much overlap): popular, the basic text, the rigorous text, the data analysis manual, the texts considering statistical packages, and the statistics treatments of specialist interest to astronomers.

1. The classic *popular books* have legendary titles: *How to Lie with Statistics* (Huff, 1973), *Facts from Figures* (Moroney, 1965), *Statistics in Action* (Sprent, 1977) and *Statistics without Tears* (Rowntree, 1981). They are all fun. To this list we can now add *The Lady Tasting Tea: How Statistics Revolutionized Science in the Twentieth Century* (Salsburg, 2002), an entertaining exposition of the development of modern statistics; *Struck by Lightning: the Curious World of Probabilities* (Rosenthal, 2006); *Making Sense of Statistics: A Non-mathematical Approach* (Wood, 2003), and *Dicing with Death: Chance, Risk and Health* (Senn, 2003). This latter is a devastatingly blunt, funny and erudite exposition of the importance and application of statistics in decision processes which may affect the lives of millions. As a popular book it is heavy-going in parts; but for scientists, budding or mature, it is a rewarding read.

2. *Textbooks* come in types (a) and (b), both of which cover similar material for the first two-thirds of each book. They start with descriptive or summarizing statistics (mean, standard deviation), the distributions of these statistics, and move to the concept of probability and hence statistical inference and hypothesis testing, including correlation of two variables. They then diverge, choosing from a menu including analysis of variance (ANOVA), regression analysis, non-parametric statistics, etc. Modern versions come in bright colours and dramatic layouts, perhaps to help presentation to undergraduates of a subject with which excitement is not always associated. The value of

many such books is exceptional because of the sales they generate. They are complete with tables, ready summaries of tests and formulae inside covers or in coloured insets, and frequently contain CD-ROMs including test data sets. Those of type (a) are essentially devoid of calculus but with much arithmetic in the form of worked examples, and are statistics primers for undergraduates in non-scientific disciplines. Type (b) has basic mathematics which may run as far as simple calculus. A wonderfully readable example of the former is *Statistics* by Freedman *et al.* (2007), in which a non-conventional approach is adopted, very successfully. Another which gets substantially further, for example to ANOVA and to non-parametric tests, is *Introductory Statistics* by Weiss (2007), entertaining through inclusion of short biographical sketches of the founding fathers of statistical science. A further very readable member of set (a) – with a twist in the title – is *Seeing through Statistics* (Utts, 2004), which entertains and serves masterfully as a statistics primer. Then, if you can, have a serious look at *The Cartoon Guide to Statistics* (Gonick & Smith, 2005) – of course it entertains, but it presents a very solid base for frequentist statistics and includes a short (but illuminating) glimpse of Bayesian methodology. Of type (b), more appropriate in the present context but not necessarily so entertaining, an outstanding example is *Mathematical Statistics and Data Analysis* by Rice (2006), basic but erudite and thorough; it goes so far as to discuss covariance matrices, Bayesian inference, moment-generating functions, multiple linear regression, and computer-intensive methods such as the bootstrap; and it includes a CD-ROM with examples. Unfortunately, non-parametric tests do not get a mention. They do in other basic texts of type (b), such as that by Hogg & Tanis (2009): *Probability and Statistical Inference*, although the presentation here is not to everyone's taste. *Weighing the Odds* (Williams, 2001) is a textbook providing a fine marriage of probability and statistics, frequentist and Bayesian methodologies, basic to advanced, theory and practicality. In the Bayesian framework, we can highly recommend *Bayesian Statistics: An Introduction* by Lee (2004).

3. The *more scholarly volumes* which go beyond the undergraduate level include *Statistics: Concepts and Applications* by Frank & Altheon (1994), a thorough and well-set-out description of classical statistics (although it may now be out of print); and *Statistical Inference* by Casella & Berger (2002), in which the theory is presented in a highly accessible manner, although the Bayesian approach and non-parametric statistics hardly get a mention. A classic reference which we have consulted frequently is the *Introduction to the Theory of Statistics* (Mood *et al.*, 1974). Speaking of classics, *Kendall's Advanced Theory of Statistics*, the three volumes being *Distribution Theory*

(Stuart & Ord, 1994), *Classical Inference and Relationship* (Stuart & Ord, 1991), and *Bayesian Inference* (O'Hagan, 1994), is a complete reference; not easy going, though. Another very useful standard work is *Probability, Random Variables and Stochastic Processes* by Papoulis & Unnikrishna Pillai (2002), strong mathematically and biased towards classical real-time signal processing issues. *Probability and Random Processes* (Grimmett & Stirzaker, 2001) presents a rigorous introduction to probability theory and, as the title implies, is strong on random process, Monte Carlo and MCMC techniques. For strictly Bayesian analysis, note the excellent and comprehensive *Bayesian Logical Data Analysis for the Physical Sciences* (Gregory, 2004). *Principles of Statistical Inference* by Cox (2006) emphasizes the modern rapprochement between frequentist and Bayes positions: it is now recognized (see our Chapter 7) that melding of Bayes and frequentist methods can prove extremely powerful.

Various works by Jaynes (Jaynes, 2003, 1986, 1983, 1976, 1968) are indispensable reading on the concepts of probability. There is an archive of his writings at `bayes.wustl.edu`.

4. The *data analysis texts* are led by the highly practical Bevington & Robinson (2002), *Data Reduction and Error Analysis for the Physical Sciences*. *A Guide to the Use of Statistical Methods* by Barlow (1989) is excellent. A useful little monograph is *A Practical Guide to Data Analysis for Physical Science Students* by Lyons (1991). Lyons (1986) has also written *Statistics for Nuclear and Particle Physicists*, highly practical and strong on parameter-fitting, hypothesis testing and Monte Carlo methods. A valuable monograph for the Bayesian framework is *Data Analysis: A Bayesian Tutorial* by Sivia (2006), extremely practical. Carlin & Louis (2000) *Bayes and Empirical Bayes Methods for Data Analysis* gives a very thorough treatment of Bayesian techniques in data reduction and is excellent on Bayesian (MCMC) integration problems. Gelman *et al.* (2004) *Bayesian Data Analysis* is a comprehensive advanced textbook on Bayesian methods, strong on hierarchical modelling, advanced computational methods and MCMC.

With regard to applying non-parametric statistical tests, the books by Conover (1999) *Practical Nonparametric Statistics*, and Siegel & Castellan (1988) *Nonparametric Statistics for the Behavioural Sciences* are very straightforward, the latter particularly recommended.

The dominant force in physical analysis books remains *Numerical Recipes: The Art of Scientific Computing* (Press *et al.*, 2007), unparalled in breadth and containing much common sense, and sub-routines to do everything described. These invariably work! No scientist should be without access to this superb book. There is an economical alternative (without coded sub-routines, however): the outstanding paperback *Numerical Methods for Scientists and Engineers* (Hamming, 1987).

Finally note the three books by Tufte, *The Visual Display of Quantitative Information, Envisaging Information* and *Visual Explanations: Images and Quantities, Evidence and Narrative* (Tufte 2001, 1990, 1997), magnificent in presentation and essential browsing for anybody wishing to present data in graphical form effectively.

5. As for *statistical packages*, the manuals for the now highly developed packages, e.g. MINITAB, SPSS, GENSTAT, S-PLUS and R, contain much practical advice. Note for examples *Statistics Using SPSS* (Weinberg & Abramowitz, 2008), and *Modern Applied Statistics with S* (Venables & Ripley, 2002). The R package has a considerable recent literature, e.g. *A First Course in Statistical Inference with R* (Braun & Murdoch, 2007); *Statistics: An Introduction Using R* (Crawley, 2005); *Bayesian Computation with R* (Albert, 2009); *An Introduction to R: Revised and Updated* (Venables & Smith, 2002); and at least 50 more. General-purpose mathematical packages, such as Matlab, Maple and MATHEMATICA, offer substantial support for statistical analysis, integrated with their other numerous helpful features.

Using a statistics/data analysis package of course saves time in that someone has done the coding for you, but it is important to consult the manuals and/or these or similar references before embarking on any serious science. Black boxes are not necessarily conducive to best understanding your data, its strengths and its limitations.

6. Finally to the *astronomy specialist* end of the spectrum, the high-octane and exciting statistical and analytical techniques which been developed recently as essential in the handling of the huge data sets now available or eagerly anticipated. The growth of interest by astronomers in statistical methods, driven by the data explosion, is demonstrated by a series of conferences which have resulted in the collection of much useful information. The first of these, *Statistical Methods in Astronomy* (Rolfe, 1983), contains useful background bibliographies in time-series analysis and in non-parametric statistics. Two later conferences, *Errors, Bias and Uncertainties in Astronomy* (Jaschek & Murtagh, 1990) and *Statistical Challenges in Modern Astronomy* (Feigelson & Babu, 1992) reflect the dramatic change in what we consider to be the important data sets over a 15-year period, and are instructive reading for this alone. Following the success of the 1992 meeting, Jogesh Babu and Eric Feigelson instituted 'Statistical Challenges in Modern Astronomy' conferences at Pennsylvania State University at five-year intervals. The subsequent conference proceedings *Statistical Challenges in*

Modern Astronomy II, III, IV (Babu & Feigelson, 1997; Feigelson & Babu, 2003; Babu & Feigelson, 2007) reflect the impressive growth of statistical application to astronomy and cosmology, and describe in detail many of these applications. Scott Dodelson's excellent *Modern Cosmology* has its last chapter entitled *Analysis*, and contains detail on the likelihood function and its evaluation, the signal covariance matrix, the Fisher matrix, map-making and inversion, and systematics. Dodelson's descriptions and examples of these topics go well beyond the scope of our treatment. A number of useful references are included. We also need to mention *Bayesian Methods in Cosmology* by Hobson *et al.* (2010).

It will not have escaped today's reader that the www is an amazing resource – if used with discretion. For instance, most of the books above are available through large online booksellers, and as such, most have customer reviews. *With few exceptions* it is obvious that the bad reviews are from customers who have picked the wrong book for their particular application, course or mathematical background.

Appendix B

Statistical tables

Table B.1 *Area under Normal (Gaussian) distribution*

					$\int_0^z \exp[-\frac{1}{2}z^2]\,dz$, with $z = (x - \mu)/\sigma$					
$z +$	.00	.01	.02	.03	.04	.05	.06	.07	.08	.09
0.0	.0000	.0040	.0080	.0120	.0160	.0199	.0239	.0279	.0319	.0359
0.1	.0398	.0438	.0478	.0517	.0557	.0596	.0636	.0675	.0714	.0754
0.2	.0793	.0832	.0871	.0910	.0948	.0987	.1026	.1064	.1103	.1141
0.3	.1179	.1217	.1255	.1293	.1331	.1368	.1406	.1443	.1480	.1517
0.4	.1554	.1591	.1628	.1664	.1700	.1736	.1772	.1808	.1844	.1879
0.5	.1915	.1950	.1985	.2019	.2054	.2088	.2123	.2157	.2190	.2224
0.6	.2258	.2291	.2324	.2357	.2389	.2422	.2454	.2486	.2518	.2549
0.7	.2580	.2612	.2642	.2673	.2704	.2734	.2764	.2794	.2823	.2852
0.8	.2881	.2910	.2939	.2967	.2996	.3023	.3051	.3079	.3106	.3133
0.9	.3159	.3186	.3212	.3238	.3264	.3289	.3315	.3340	.3365	.3389
1.0	.3413	.3438	.3461	.3485	.3508	.3531	.3554	.3577	.3599	.3621
1.1	.3643	.3665	.3686	.3708	.3729	.3749	.3770	.3790	.3810	.3830
1.2	.3849	.3869	.3888	.3907	.3925	.3944	.3962	.3980	.3997	.4015
1.3	.4032	.4049	.4066	.4082	.4099	.4115	.4131	.4147	.4162	.4177
1.4	.4192	.4207	.4222	.4236	.4251	.4265	.4279	.4292	.4306	.4319
1.5	.4332	.4345	.4357	.4370	.4382	.4394	.4406	.4418	.4430	.4441
1.6	.4452	.4463	.4474	.4485	.4495	.4505	.4515	.4525	.4535	.4545
1.7	.4554	.4564	.4573	.4582	.4591	.4599	.4608	.4616	.4625	.4633
1.8	.4641	.4649	.4656	.4664	.4671	.4678	.4686	.4693	.4700	.4706
1.9	.4713	.4719	.4726	.4732	.4738	.4744	.4750	.4756	.4762	.4767
2.0	.4773	.4778	.4783	.4788	.4793	.4798	.4803	.4808	.4812	.4817
2.1	.4821	.4826	.4830	.4834	.4838	.4842	.4846	.4850	.4854	.4857
2.2	.4861	.4865	.4868	.4871	.4875	.4878	.4881	.4884	.4887	.4890
2.3	.4893	.4896	.4898	.4901	.4904	.4906	.4909	.4911	.4913	.4916
2.4	.4918	.4920	.4922	.4925	.4927	.4929	.4931	.4932	.4934	.4936

(cont.)

Table B.1   (*cont.*)

	\multicolumn{10}{c}{$\int_0^z \exp[-\frac{1}{2}z^2]\,dz$, with $z = (x - \mu)/\sigma$}									
$z +$	.00	.01	.02	.03	.04	.05	.06	.07	.08	.09
2.5	.4938	.4940	.4941	.4943	.4945	.4946	.4948	.4949	.4951	.4952
2.6	.4953	.4955	.4956	.4957	.4959	.4960	.4961	.4962	.4963	.4964
2.7	.4965	.4966	.4967	.4968	.4969	.4970	.4971	.4972	.4973	.4974
2.8	.4974	.4975	.4976	.4977	.4977	.4978	.4979	.4980	.4980	.4981
2.9	.4981	.4982	.4983	.4983	.4984	.4984	.4985	.4985	.4986	.4986
3.0	.4987	.4987	.4987	.4988	.4988	.4989	.4989	.4989	.4990	.4990
3.1	.4990	.4991	.4991	.4991	.4992	.4992	.4992	.4992	.4993	.4993
3.2	.4993	.4993	.4994	.4994	.4994	.4994	.4994	.4995	.4995	.4995
3.3	.4995	.4995	.4996	.4996	.4996	.4996	.4996	.4996	.4996	.4997
3.4	.4997	.4997	.4997	.4997	.4997	.4997	.4997	.4997	.4998	.4998

Table B.2   *The tails of the Gaussian distribution*

m	\multicolumn{3}{c}{Percentage area under the Gaussian curve in region:}		
	$>m\sigma$ (one tail)	$<-m\sigma, >m\sigma$ (both tails)	$-m\sigma < m\sigma$ (between tails)
0.0	50.0	100.00	0.00
0.5	30.85	61.71	38.29
1.0	15.87	31.73	68.27
1.5	6.681	13.36	86.64
2.0	2.275	4.550	95.45
2.5	0.621	1.24	98.76
3.0	0.135	0.270	99.73
3.5	0.0233	0.0465	99.954
4.0	0.00317	0.00633	99.9937
4.5	0.000340	0.000680	99.99932
5.0	0.0000287	0.0000573	99.999943

Table B.3   *Critical values of Student's t distribution*

	Level of significance for one-tailed test					
	0.100	0.050	0.025	0.010	0.005	0.0005
	Level of significance for two-tailed test					
	0.200	0.100	0.050	0.020	0.010	0.001
$\nu = 1$	3.078	6.314	12.706	31.821	63.657	636.619
2	1.886	2.920	4.303	6.965	9.925	31.598
3	1.638	2.353	3.182	4.541	5.841	12.941
4	1.533	2.132	2.776	3.747	4.604	8.610
5	1.476	2.015	2.571	3.365	4.032	6.859
6	1.440	1.943	2.447	3.143	3.707	5.959
7	1.415	1.895	2.365	2.998	3.499	5.405
8	1.397	1.860	2.306	2.896	3.355	5.041
9	1.383	1.833	2.262	2.821	3.250	4.781
10	1.372	1.812	2.228	2.764	3.169	4.587
11	1.363	1.796	2.201	2.718	3.106	4.437
12	1.356	1.782	2.179	2.681	3.055	4.318
13	1.350	1.771	2.160	2.650	3.012	4.221
14	1.345	1.761	2.145	2.624	2.977	4.140
15	1.341	1.753	2.131	2.602	2.947	4.073
16	1.337	1.746	2.120	2.583	2.921	4.015
17	1.333	1.740	2.110	2.567	2.898	3.965
18	1.330	1.734	2.101	2.552	2.878	3.922
19	1.328	1.729	2.093	2.539	2.861	3.883
20	1.325	1.725	2.086	2.528	2.845	3.850
21	1.323	1.721	2.080	2.518	2.831	3.819
22	1.321	1.717	2.074	2.508	2.819	3.792
23	1.319	1.714	2.069	2.500	2.807	3.767
24	1.318	1.711	2.064	2.492	2.797	3.745
25	1.316	1.708	2.060	2.485	2.787	3.725
26	1.315	1.706	2.056	2.479	2.779	3.707
27	1.314	1.703	2.052	2.473	2.771	3.690
28	1.313	1.701	2.048	2.467	2.763	3.674
29	1.311	1.699	2.045	2.462	2.756	3.659
30	1.310	1.697	2.042	2.457	2.750	3.646
40	1.303	1.684	2.021	2.423	2.704	3.551
60	1.296	1.671	2.000	2.390	2.660	3.460
120	1.289	1.658	1.980	2.358	2.617	3.373
∞	1.282	1.645	1.960	2.326	2.576	3.291

Appendix B

Table B.4 *Critical values of F distribution*

Level of significance = 0.90									
n	*m* = 5	10	15	20	25	30	35	40	45	50
5	3.45	2.52	2.27	2.16	2.09	2.05	2.02	2.00	1.98	1.97
10	3.30	2.32	2.06	1.94	1.87	1.82	1.79	1.76	1.74	1.73
15	3.24	2.24	1.97	1.84	1.77	1.72	1.69	1.66	1.64	1.63
20	3.21	2.20	1.92	1.79	1.72	1.67	1.63	1.61	1.58	1.57
25	3.19	2.17	1.89	1.76	1.68	1.63	1.60	1.57	1.55	1.53
30	3.17	2.16	1.87	1.74	1.66	1.61	1.57	1.54	1.52	1.50
35	3.16	2.14	1.86	1.72	1.64	1.59	1.55	1.52	1.50	1.48
40	3.16	2.13	1.85	1.71	1.63	1.57	1.53	1.51	1.48	1.46
45	3.15	2.12	1.84	1.70	1.62	1.56	1.52	1.49	1.47	1.45
50	3.15	2.12	1.83	1.69	1.61	1.55	1.51	1.48	1.46	1.44

Level of significance = 0.95									
n	*m* = 5	10	15	20	25	30	35	40	45	50
5	5.05	3.33	2.90	2.71	2.60	2.53	2.49	2.45	2.42	2.40
10	4.74	2.98	2.54	2.35	2.24	2.16	2.11	2.08	2.05	2.03
15	4.62	2.85	2.40	2.20	2.09	2.01	1.96	1.92	1.89	1.87
20	4.56	2.77	2.33	2.12	2.01	1.93	1.88	1.84	1.81	1.78
25	4.52	2.73	2.28	2.07	1.96	1.88	1.82	1.78	1.75	1.73
30	4.50	2.70	2.25	2.04	1.92	1.84	1.79	1.74	1.71	1.69
35	4.48	2.68	2.22	2.01	1.89	1.81	1.76	1.72	1.68	1.66
40	4.46	2.66	2.20	1.99	1.87	1.79	1.74	1.69	1.66	1.63
45	4.45	2.65	2.19	1.98	1.86	1.77	1.72	1.67	1.64	1.61
50	4.44	2.64	2.18	1.97	1.84	1.76	1.70	1.66	1.63	1.60

Level of significance = 0.99									
n	*m* = 5	10	15	20	25	30	35	40	45	50
5	10.97	5.64	4.56	4.10	3.85	3.70	3.59	3.51	3.45	3.41
10	10.05	4.85	3.80	3.37	3.13	2.98	2.88	2.80	2.74	2.70
15	9.72	4.56	3.52	3.09	2.85	2.70	2.60	2.52	2.46	2.42
20	9.55	4.41	3.37	2.94	2.70	2.55	2.44	2.37	2.31	2.27
25	9.45	4.31	3.28	2.84	2.60	2.45	2.35	2.27	2.21	2.17
30	9.38	4.25	3.21	2.78	2.54	2.39	2.28	2.20	2.14	2.10
35	9.33	4.20	3.17	2.73	2.49	2.34	2.23	2.15	2.09	2.05
40	9.29	4.17	3.13	2.69	2.45	2.30	2.19	2.11	2.05	2.01
45	9.26	4.14	3.10	2.67	2.42	2.27	2.16	2.08	2.02	1.97
50	9.24	4.12	3.08	2.64	2.40	2.25	2.14	2.06	2.00	1.95

Table B.5   *Critical values of r_s, Spearman rank correlation coefficient*

	Level of significance for one-tailed test								
	0.250	0.100	0.050	0.025	0.010	0.005	0.0025	0.0010	0.0005
	Level of significance for two-tailed test								
	0.500	0.200	0.100	0.050	0.020	0.010	0.005	0.002	0.001
$N = 4$	0.600	1.000	1.000	–	–	–	–	–	–
5	0.500	0.800	0.900	1.000	1.000	–	–	–	–
6	0.371	0.657	0.829	0.886	0.943	1.000	1.000	–	–
7	0.321	0.571	0.714	0.786	0.893	0.929	0.964	1.000	1.000
8	0.310	0.524	0.643	0.738	0.833	0.881	0.905	0.952	0.976
9	0.267	0.483	0.600	0.700	0.783	0.833	0.867	0.917	0.933
10	0.248	0.455	0.564	0.648	0.745	0.794	0.830	0.879	0.903
11	0.236	0.427	0.536	0.618	0.709	0.755	0.800	0.845	0.873
12	0.224	0.406	0.503	0.587	0.671	0.727	0.776	0.825	0.860
13	0.209	0.385	0.484	0.560	0.648	0.703	0.747	0.802	0.835
14	0.200	0.367	0.464	0.538	0.622	0.675	0.723	0.776	0.811
15	0.189	0.354	0.443	0.521	0.604	0.654	0.700	0.754	0.786
16	0.182	0.341	0.429	0.503	0.582	0.635	0.679	0.732	0.765
17	0.176	0.328	0.414	0.485	0.566	0.615	0.662	0.713	0.748
18	0.170	0.317	0.401	0.472	0.550	0.600	0.643	0.695	0.728
19	0.165	0.309	0.391	0.460	0.535	0.584	0.628	0.677	0.712
20	0.161	0.299	0.380	0.447	0.520	0.570	0.612	0.662	0.696
21	0.156	0.292	0.370	0.435	0.508	0.556	0.599	0.648	0.681
22	0.152	0.284	0.361	0.425	0.496	0.544	0.586	0.634	0.667
23	0.148	0.278	0.353	0.415	0.486	0.532	0.573	0.622	0.654
24	0.144	0.271	0.344	0.406	0.476	0.521	0.562	0.610	0.642
25	0.142	0.265	0.337	0.398	0.466	0.511	0.551	0.598	0.630
26	0.138	0.259	0.331	0.390	0.457	0.501	0.541	0.587	0.619
27	0.136	0.255	0.324	0.382	0.448	0.491	0.531	0.577	0.608
28	0.133	0.250	0.317	0.375	0.440	0.483	0.522	0.567	0.598
29	0.130	0.245	0.312	0.368	0.433	0.475	0.513	0.558	0.589
30	0.128	0.240	0.306	0.362	0.425	0.467	0.504	0.549	0.580
31	0.126	0.236	0.301	0.356	0.418	0.459	0.496	0.541	0.571
32	0.124	0.232	0.296	0.350	0.412	0.452	0.489	0.533	0.563
33	0.121	0.229	0.291	0.345	0.405	0.446	0.482	0.525	0.554
34	0.120	0.225	0.287	0.340	0.399	0.439	0.475	0.517	0.547
35	0.118	0.222	0.283	0.335	0.394	0.433	0.468	0.510	0.539
36	0.116	0.219	0.279	0.330	0.388	0.427	0.462	0.504	0.533
37	0.114	0.216	0.275	0.325	0.383	0.421	0.456	0.497	0.526
38	0.113	0.212	0.271	0.321	0.378	0.415	0.450	0.491	0.519
39	0.111	0.210	0.267	0.317	0.373	0.410	0.444	0.485	0.513
40	0.110	0.207	0.264	0.313	0.368	0.405	0.439	0.479	0.507
41	0.108	0.204	0.261	0.309	0.364	0.400	0.433	0.473	0.501
42	0.107	0.202	0.257	0.305	0.359	0.395	0.428	0.468	0.495
43	0.105	0.199	0.254	0.301	0.355	0.391	0.423	0.463	0.490
44	0.104	0.197	0.251	0.298	0.351	0.386	0.419	0.458	0.484
45	0.103	0.194	0.248	0.294	0.347	0.382	0.414	0.453	0.479
46	0.102	0.192	0.246	0.291	0.343	0.378	0.410	0.448	0.474
47	0.101	0.190	0.243	0.288	0.340	0.374	0.405	0.443	0.469
48	0.100	0.188	0.240	0.285	0.336	0.370	0.401	0.439	0.465
49	0.098	0.186	0.238	0.282	0.333	0.366	0.397	0.434	0.460
50	0.097	0.184	0.235	0.279	0.329	0.363	0.393	0.430	0.456

Table B.6 *Critical values of the chi-square distribution for v degrees of freedom*

Probability under H_0 that χ^2 exceeds listed value

v	0.99	0.98	0.95	0.90	0.80	0.50	0.30	0.20	0.10	0.05	0.02	0.01	0.001
$v=1$	0.00016	0.00063	0.0039	0.016	0.15	0.46	1.07	1.64	2.71	3.84	5.41	6.64	10.83
2	0.02	0.04	0.10	0.21	0.71	1.39	2.41	3.22	4.60	5.99	7.82	9.21	13.82
3	0.12	0.18	0.35	0.58	1.42	2.37	3.66	4.64	6.25	7.82	9.84	11.34	16.27
4	0.30	0.43	0.71	1.06	2.20	3.36	4.88	5.99	7.78	9.49	11.67	13.28	18.46
5	0.55	0.75	1.14	1.61	3.00	4.35	6.06	7.29	9.24	11.07	13.39	15.09	20.52
6	0.87	1.13	1.64	2.20	3.83	5.35	7.23	8.56	10.64	12.59	15.03	16.81	22.46
7	1.24	1.56	2.17	2.83	4.67	6.35	8.38	9.80	12.02	14.07	16.62	18.48	24.32
8	1.65	2.03	2.73	3.49	5.53	7.34	9.52	11.03	13.36	15.51	18.17	20.09	26.12
9	2.09	2.53	3.32	4.17	6.39	8.34	10.66	12.24	14.68	16.92	19.68	21.67	27.88
10	2.56	3.06	3.94	4.86	7.27	9.34	11.78	13.44	15.99	18.31	21.16	23.21	29.59
11	3.05	3.61	4.58	5.85	8.15	10.34	12.90	14.63	17.28	19.68	22.62	24.72	31.26
12	3.57	4.18	5.23	6.30	9.03	11.34	14.01	15.81	18.55	21.03	24.05	26.22	32.91
13	4.11	4.76	5.89	7.04	9.93	12.34	15.12	16.98	19.81	22.36	25.47	27.69	34.53
14	4.66	5.37	6.57	7.79	10.82	13.34	16.22	18.15	21.06	23.68	26.87	29.14	36.12
15	5.23	5.98	7.26	8.55	11.72	14.34	17.32	19.30	22.31	25.00	28.26	30.58	37.70
16	5.81	6.61	7.96	9.31	12.62	15.34	18.42	20.46	23.54	26.30	29.63	32.00	39.29
17	6.41	7.26	8.67	10.08	13.53	16.34	19.51	21.62	24.77	27.59	31.00	33.41	40.75
18	7.02	7.91	9.39	10.86	14.44	17.34	20.60	22.76	25.99	28.87	32.35	34.80	42.31
19	7.63	8.57	10.12	11.65	15.35	18.34	21.69	23.90	27.20	30.14	33.69	36.19	43.82
20	8.26	9.24	10.85	12.44	16.27	19.34	22.78	25.04	28.41	31.41	35.02	37.57	45.32
21	8.90	9.92	11.59	13.24	17.18	20.34	23.86	26.17	29.62	32.67	36.34	38.93	46.80
22	9.54	10.60	12.34	14.04	18.10	21.24	24.94	27.30	30.81	33.92	37.66	40.29	48.27
23	10.20	11.29	13.09	14.85	19.02	22.34	26.02	28.43	32.01	35.17	38.97	41.64	49.73
24	10.86	11.99	13.85	15.66	19.94	23.34	27.10	29.55	33.20	36.42	40.27	42.98	51.18
25	11.52	12.70	14.61	16.47	20.87	24.34	28.17	30.68	34.38	37.65	41.57	44.31	52.62
26	12.20	15.38	15.38	17.29	21.79	25.34	29.25	31.80	35.56	38.88	42.86	45.64	54.05
27	12.88	16.15	16.15	18.11	22.72	26.34	30.32	32.91	36.74	40.11	44.14	46.96	55.48
28	13.56	16.93	16.93	18.94	23.65	27.34	31.39	34.03	37.92	41.34	45.42	48.28	56.89
29	14.26	17.71	17.71	19.77	24.58	28.34	32.46	35.14	39.09	42.56	46.69	49.59	58.30
30	14.95	18.49	18.49	20.60	25.51	29.34	33.53	36.25	40.26	43.77	47.96	50.89	59.70

Table B.7   *Critical values of D, Kolmogorov–Smirnov one-sample test*

| | Level of significance for $D = \max[F_0(X) - S_N(X)]$ | | | | |
	0.200	0.150	0.100	0.050	0.010
$N = 1$	0.900	0.925	0.950	0.975	0.995
2	0.684	0.726	0.776	0.842	0.929
3	0.565	0.597	0.642	0.708	0.828
4	0.494	0.525	0.564	0.624	0.733
5	0.446	0.474	0.510	0.565	0.669
6	0.410	0.436	0.470	0.521	0.618
7	0.381	0.405	0.498	0.466	0.577
8	0.358	0.381	0.411	0.457	0.543
9	0.339	0.360	0.388	0.432	0.514
10	0.322	0.342	0.368	0.410	0.490
11	0.307	0.326	0.352	0.391	0.468
12	0.295	0.313	0.338	0.375	0.450
13	0.284	0.302	0.325	0.361	0.433
14	0.274	0.292	0.314	0.349	0.418
15	0.266	0.283	0.304	0.338	0.404
16	0.258	0.274	0.295	0.328	0.392
17	0.250	0.266	0.286	0.318	0.381
18	0.244	0.259	0.278	0.309	0.371
19	0.237	0.252	0.272	0.301	0.363
20	0.231	0.246	0.264	0.294	0.356
25	0.210	0.220	0.240	0.270	0.320
30	0.190	0.200	0.220	0.240	0.290
35	0.180	0.190	0.210	0.230	0.270
>35	$1.07/\sqrt{N}$	$1.14/\sqrt{N}$	$1.22/\sqrt{N}$	$1.36/\sqrt{N}$	$1.63/\sqrt{N}$

Table B.8 *Critical values of r in the one-sample runs test*

$r \leq$ (smaller value) or $\geq$ (larger value) indicates significance at $\alpha = 0.05$

	n = 2	3	4	5	6	7	8	9	10	11	12	13	14	15	16	17	18	19	20
m = 2	–	–	–	–	–	–	–	–	–	–	2	2	2	2	2	2	2	2	2
	–	–	–	–	–	–	–	–	–	–	–	–	–	–	–	–	–	–	–
3	–	–	–	–	2	2	2	2	2	2	2	2	2	3	3	3	3	3	3
	–	–	–	–	–	–	–	–	–	–	–	–	–	–	–	–	–	–	–
4	–	–	–	2	2	2	3	3	3	3	3	3	3	3	4	4	4	4	4
	–	–	–	9	9	–	–	–	–	–	–	–	–	–	–	–	–	–	–
5	–	–	2	2	3	3	3	3	3	4	4	4	4	4	4	4	5	5	5
	–	–	9	10	10	11	11	–	–	–	–	–	–	–	–	–	–	–	–
6	–	2	2	3	3	3	3	4	4	4	4	5	5	5	5	5	5	6	6
	–	–	9	10	11	12	12	13	13	13	13	–	–	–	–	–	–	–	–
7	–	2	2	3	3	3	4	4	5	5	5	5	5	6	6	6	6	6	6
	–	–	–	11	12	13	13	14	14	14	14	15	15	15	–	–	–	–	–
8	–	2	3	3	3	4	4	5	5	5	6	6	6	6	6	7	7	7	7
	–	–	–	11	12	13	14	14	15	15	16	16	16	16	17	17	17	17	17
9	–	2	3	3	4	4	5	5	5	6	6	6	7	7	7	7	8	8	8
	–	–	–	–	13	14	14	15	16	16	16	17	17	18	18	18	18	18	18
10	–	2	3	3	4	5	5	5	6	6	7	7	7	7	8	8	8	8	9
	–	–	–	–	13	14	15	16	16	17	17	18	18	18	19	19	19	20	20
11	–	2	3	4	4	5	5	6	6	7	7	7	8	8	8	9	9	9	9
	–	–	–	–	13	14	15	16	17	17	18	19	19	19	20	20	20	21	21
12	2	2	3	4	4	5	6	6	7	7	7	8	8	8	9	9	9	10	10
	–	–	–	–	13	14	16	16	17	18	19	19	20	20	21	21	21	22	22
13	2	2	3	4	5	5	6	6	7	7	8	8	9	9	9	10	10	10	10
	–	–	–	–	–	15	16	17	18	19	19	20	20	21	21	22	22	23	23
14	2	2	3	4	5	5	6	7	7	8	8	9	9	9	10	10	10	11	11
	–	–	–	–	–	15	16	17	18	19	19	20	20	21	21	22	22	23	23
15	2	3	3	4	5	6	6	7	7	8	8	9	9	10	10	11	11	11	12
	–	–	–	–	–	15	16	18	18	19	20	21	22	22	23	23	24	24	25
16	2	3	4	4	5	6	6	7	8	8	9	9	10	10	11	11	11	12	12
	–	–	–	–	–	–	17	18	19	20	21	21	22	23	23	24	25	25	25
17	2	3	4	4	5	6	7	7	8	9	9	10	10	11	11	11	12	12	13
	–	–	–	–	–	–	17	18	19	20	21	22	23	23	24	25	25	26	26
18	2	3	4	5	5	6	7	8	8	9	9	10	10	11	11	12	12	13	13
	–	–	–	–	–	–	17	18	19	20	21	22	23	24	25	25	26	26	27
19	2	3	4	5	6	6	7	8	8	9	10	10	11	11	12	12	13	13	13
	–	–	–	–	–	–	17	18	20	21	22	23	23	24	25	25	26	26	27
20	2	3	4	5	6	6	7	8	9	9	10	10	11	12	12	13	13	13	14
	–	–	–	–	–	–	17	18	20	21	22	23	24	25	25	26	27	27	28

Table B.9 *Lower- and upper-tail probabilities for U, the Wilcoxon–Mann–Whitney rank-sum statistic*

Entries are $P(U < c_l)$ and $P(U > c_u)$. U is the rank-sum for the smaller group.

$m = 3$

c_l	$n=3$	c_u	$n=4$	c_u	$n=5$	C_u	$n=6$	c_u	$n=7$	c_u	$n=8$	c_u	$n=9$	c_u	$n=10$	c_u
6	0.0500	15	0.0286	18	0.0179	21	0.0119	24	0.0083	27	0.0061	30	0.0045	33	0.0035	36
7	0.1000	14	0.0571	17	0.0357	20	0.0238	23	0.0167	26	0.0121	29	0.0091	32	0.0070	35
8	0.2000	13	0.1143	16	0.0714	19	0.0476	22	0.0333	25	0.0242	28	0.0182	31	0.0140	34
9	0.3500	12	0.2000	15	0.1250	18	0.0833	21	0.0583	24	0.0424	27	0.0318	30	0.0245	33
10	0.5000	11	0.3143	14	0.1964	17	0.1310	20	0.0917	23	0.0667	26	0.0500	29	0.0385	32
11	0.6500	10	0.4286	13	0.2857	16	0.1905	19	0.1333	22	0.0970	25	0.0727	28	0.0559	31
12	0.8000	9	0.5714	12	0.3929	15	0.2738	18	0.1917	21	0.1394	24	0.1045	27	0.0804	30
13	0.9000	8	0.6857	11	0.5000	14	0.3571	17	0.2583	20	0.1879	23	0.1409	26	0.1084	29
14	0.9500	7	0.8000	10	0.6071	13	0.4524	16	0.3333	19	0.2485	22	0.1864	25	0.1434	28
15	1.0000	6	0.8857	9	0.7143	12	0.5476	15	0.4167	18	0.3152	21	0.2409	24	0.1853	27
16	—	—	0.9429	8	0.8036	11	0.6429	14	0.5000	17	0.3879	20	0.3000	23	0.2343	26
17	—	—	0.9714	7	0.8750	10	0.7262	13	0.5833	16	0.4606	19	0.3636	22	0.2867	25
18	—	—	1.0000	6	0.9286	9	0.8095	12	0.6667	15	0.5394	18	0.4318	21	0.3462	24
19	—	—	—	—	0.9643	8	0.8690	11	0.7417	14	0.6121	17	0.5000	20	0.4056	23
20	—	—	—	—	0.9821	7	0.9167	10	0.8083	13	0.6848	16	0.8682	19	0.4685	22
21	—	—	—	—	1.0000	6	0.9524	9	0.8667	12	01515	15	0.6364	18	0.5315	21
22	—	—	—	—	—	—	0.9762	8	0.9083	11	0.8121	14	0.7000	17	0.5944	20
23	—	—	—	—	—	—	0.9881	7	0.9417	10	0.8606	13	0.7591	16	0.6538	19
24	—	—	—	—	—	—	1.0000	6	0.9667	9	0.9030	12	0.8136	15	0.7133	18

(cont.)

Table B.9   (cont.)

Entries are $P(U < c_l)$ and $P(U > c_u)$. U is the rank-sum for the smaller group.

$m = 4$

c_l	$n = 4$	c_u	$n = 5$	C_u	$n = 6$	c_u	$n = 7$	c_u	$n = 8$	c_u	$n = 9$	c_u	$n = 10$	c_u
10	0.0143	26	0.0079	30	0.0048	34	0.0030	38	0.0020	42	0.0014	46	0.0010	50
11	0.0286	25	0.0159	29	0.0095	33	0.0061	37	0.0040	41	0.0028	45	0.0020	49
12	0.0571	24	0.0317	28	0.0190	32	0.0121	36	0.0081	40	0.0056	44	0.0040	48
13	0.1000	23	0.0556	27	0.0333	31	0.0212	35	0.0141	39	0.0098	43	0.0070	47
14	0.1714	22	0.0952	26	0.0571	30	0.0364	34	0.0242	38	0.0168	42	0.0120	46
15	0.2429	21	0.1429	25	0.0857	29	0.0545	33	0.0364	37	0.0252	41	0.0180	45
16	0.3429	20	0.2063	24	0.1286	28	0.0818	32	0.0545	36	0.0378	40	0.0270	44
17	0.4429	19	0.2778	23	0.1762	27	0.1152	31	0.0768	35	0.0531	39	0.0380	43
18	0.5571	18	0.3651	22	0.2381	26	0.1576	30	0.1071	34	0.0741	38	0.0529	42
19	0.6571	17	0.4524	21	0.3048	25	0.2061	29	0.1414	33	0.0993	37	0.0709	41
20	0.7571	16	0.5476	20	0.3810	24	0.2636	28	0.1838	32	0.1301	36	0.0939	40
21	0.8286	15	0.6349	19	0.4571	23	0.3242	27	0.2303	31	0.1650	35	0.1199	39
22	0.9000	14	0.7222	18	0.5429	22	0.3939	26	0.2848	30	0.2070	34	0.1518	38
23	0.9429	13	0.7937	17	0.6190	21	0.4636	25	0.3414	29	0.2517	33	0.1868	37
24	0.9714	12	0.8571	16	0.6952	20	0.5364	24	0.4040	28	0.3021	32	0.2268	36
25	0.9857	11	0.9048	15	0.7619	19	0.6061	23	0.4667	27	0.3552	31	0.2697	35

Table B.10   *Kolmogorov–Smirnov two-sample test*

Critical values for one-tailed rejection region for $mn\,D_{m,n} \geq c$.
The upper, middle and lower values are $c_{0.10}$, $c_{0.05}$ and $c_{0.01}$ for each (m, n)

$m =$	3	4	5	6	7	8	9	10	11	12	13	14	15	16	17	18	19	20
$n = 3$	9	10	11	15	15	16	21	19	22	24	25	26	30	30	32	36	36	37
	9	10	13	15	16	19	21	22	25	27	28	31	33	34	35	39	40	41
	**	**	**	**	19	22	27	28	31	33	34	37	42	43	43	48	49	52
4	10	16	13	16	18	24	21	24	26	32	29	32	34	40	37	40	41	48
	10	16	16	18	21	24	25	28	29	36	33	38	38	44	44	46	49	52
	**	**	17	22	25	32	29	34	37	40	41	46	46	52	53	56	57	64
5	11	13	20	19	21	23	26	30	30	32	35	37	45	41	44	46	47	55
	13	16	20	21	24	26	28	35	35	36	40	42	50	46	49	51	56	60
	**	17	25	26	29	33	36	40	41	46	48	51	60	56	61	63	67	75
6	15	16	19	24	24	26	30	32	33	42	37	42	45	48	49	54	54	56
	15	18	21	30	25	30	33	36	38	48	43	48	51	54	56	66	61	66
	**	22	26	36	31	38	42	44	49	54	54	60	63	66	68	78	77	80
7	15	18	21	24	35	28	32	34	38	40	44	49	48	51	54	56	59	61
	16	21	24	25	35	34	36	40	43	45	50	56	56	58	61	64	68	72
	19	25	29	31	42	42	46	50	53	57	59	70	70	71	75	81	85	87
8	16	24	23	26	28	40	33	40	41	48	47	50	52	64	57	62	64	72
	19	24	26	30	34	40	40	44	48	52	53	58	60	72	65	72	73	80
	22	32	33	38	42	48	49	56	59	64	66	72	75	88	81	88	91	100
9	21	21	26	30	32	33	45	43	45	51	51	54	60	61	65	72	70	73
	21	25	28	33	36	40	54	46	51	57	57	63	69	68	74	81	80	83
	27	29	36	42	46	49	63	61	62	69	73	77	84	86	92	99	99	103
10	19	24	30	32	34	40	43	50	48	52	55	60	65	66	69	72	74	90
	22	28	35	36	40	44	46	60	57	60	62	68	75	76	77	82	85	100
	28	34	40	44	50	56	61	70	69	74	78	84	90	94	97	104	104	120
11	22	26	30	33	38	41	45	48	66	54	59	63	66	69	72	76	79	84
	25	29	35	38	43	48	51	57	66	64	67	72	76	80	83	87	92	95
	31	37	41	49	53	59	62	69	88	77	85	89	95	100	104	108	114	117
12	24	32	32	42	40	48	51	52	54	72	61	68	72	76	77	84	85	92
	27	36	36	48	45	52	57	60	64	72	71	78	84	88	89	96	98	104
	33	40	46	54	57	64	69	74	77	96	92	94	102	108	111	120	121	128
13	25	29	35	37	44	47	51	55	59	61	78	72	75	79	81	87	89	95
	28	33	40	43	50	53	57	62	67	71	91	78	86	90	94	98	102	108
	34	41	48	54	59	66	73	78	85	92	104	102	106	112	118	121	127	135
14	26	32	37	42	49	50	54	60	63	68	72	84	80	84	87	92	94	100
	31	38	42	48	56	58	63	68	72	78	78	98	92	96	99	104	108	114
	37	46	51	60	70	72	77	84	89	94	102	112	111	120	124	130	135	142
15	30	34	45	45	48	52	60	65	66	72	75	80	90	87	91	99	100	110
	33	38	50	51	56	60	69	75	76	84	86	92	105	101	105	111	113	125
	42	46	60	63	70	75	84	90	95	102	106	111	135	120	130	138	142	150

** Statistic cannot achieve this significance level.

Table B.11   *Kolmogorov–Smirnov two-sample test*

Critical values for two-tailed rejection region for $mn\,D_{m,n} \geq c$.
The upper, middle and lower values are $c_{0.10}$, $c_{0.05}$ and $c_{0.01}$ for each (m, n)

m =	2	3	4	5	6	7	8	9	10	11	12	13	14	15	16	17	18	19	20
n = 1	–	–	–	–	–	–	–	–	–	–	–	–	–	–	–	–	–	19	20
2	–	–	–	10	12	14	16	18	18	20	22	24	24	26	28	30	32	32	34
	–	–	–	–	–	–	16	18	20	22	24	26	26	28	30	32	34	36	38
	–	–	–	–	–	–	–	–	–	–	–	–	–	–	–	–	–	38	40
3	–	9	12	15	15	18	21	21	24	27	27	30	33	33	36	36	39	42	42
	–	–	–	15	18	21	21	24	27	30	30	33	36	36	39	42	45	45	48
	–	–	–	–	–	–	–	27	30	33	36	39	42	42	45	48	51	54	57
4	–	12	16	16	18	21	24	27	28	29	36	35	38	40	44	44	46	49	52
	–	–	16	20	20	24	28	28	30	33	36	39	42	44	48	48	50	53	60
	–	–	–	–	24	28	32	36	36	40	44	48	48	52	56	60	60	64	68
5	10	15	16	20	24	25	27	30	35	35	36	40	42	50	48	50	52	56	60
	–	15	20	25	24	28	30	35	40	39	43	45	46	55	54	55	60	61	65
	–	–	–	25	30	35	35	40	45	45	50	52	56	60	64	68	70	71	80
6	12	15	18	24	30	28	30	33	36	38	48	46	48	51	54	56	66	64	66
	–	18	20	24	30	30	34	39	40	43	48	52	54	57	60	62	72	70	72
	–	–	24	30	36	36	40	45	48	54	60	60	64	69	70	73	84	83	88
7	14	18	21	25	28	38	34	36	40	44	46	50	56	56	59	61	65	69	72
	–	21	24	28	30	42	40	42	46	48	53	56	63	62	64	68	70	76	79
	–	–	28	35	36	42	48	49	53	59	60	65	77	75	77	84	87	91	93
8	16	21	24	27	30	34	40	40	44	48	52	54	58	60	72	68	72	74	80
	16	21	28	30	34	40	48	46	48	53	60	62	64	67	80	77	80	82	88
	–	–	32	35	40	48	56	55	60	64	68	72	76	81	88	88	94	98	104
9	18	21	27	30	33	36	40	54	50	52	57	59	63	69	69	74	81	80	84
	18	24	28	35	39	42	46	54	53	59	63	65	70	75	78	82	92	89	93
	–	27	36	40	45	49	55	63	63	70	75	78	84	90	94	99	108	107	111
10	18	24	28	35	36	40	44	50	60	57	60	64	68	75	76	79	82	85	100
	20	27	30	40	40	46	48	53	70	60	66	70	74	80	84	89	92	94	110
	–	30	36	45	48	53	60	63	80	77	80	84	90	100	100	106	108	113	130
11	20	27	29	35	38	44	48	52	57	66	64	67	73	76	80	85	88	92	96
	22	30	33	39	43	48	53	59	60	77	72	75	8	84	89	93	97	102	107
	–	33	40	45	54	59	64	70	77	88	86	91	96	102	106	110	118	122	127
12	22	27	36	36	48	46	52	57	60	64	72	71	78	84	88	90	96	99	104
	24	30	36	43	48	53	60	63	66	72	84	81	86	93	96	100	108	108	116
	–	36	44	50	60	60	68	75	80	86	96	95	104	108	116	119	126	130	140
13	24	30	35	40	46	50	54	59	64	67	71	91	78	87	91	96	99	104	108
	26	33	39	45	52	56	62	65	70	75	81	91	89	96	101	105	110	114	120
	–	39	48	52	60	65	72	78	84	91	95	117	104	115	121	127	131	138	143
14	24	33	38	42	48	56	58	63	68	73	78	78	98	92	96	100	104	110	114
	26	36	42	46	54	63	64	70	74	82	86	89	112	98	106	111	116	121	126
	–	47	48	56	64	77	76	84	90	96	104	104	126	123	126	134	140	148	152
15	26	33	40	50	51	56	60	69	75	76	84	87	92	105	101	105	111	114	125
	28	36	44	55	57	62	67	75	80	84	93	96	98	120	114	116	123	127	135
	–	42	52	60	69	75	81	90	100	102	108	115	123	135	133	142	147	152	160

Table B.12   *Critical values of D for Kolmogorov–*
Smirnov two-sample test: large samples, two-tailed

Level of significance	Value of D so large as to require rejection of H_0 at indicated significance, where $D = \max[S_m(X) - S_n(X)]$
0.100	$1.22\sqrt{\frac{m+n}{mn}}$
0.050	$1.36\sqrt{\frac{m+n}{mn}}$
0.025	$1.48\sqrt{\frac{m+n}{mn}}$
0.010	$1.63\sqrt{\frac{m+n}{mn}}$
0.005	$1.73\sqrt{\frac{m+n}{mn}}$
0.001	$1.95\sqrt{\frac{m+n}{mn}}$

Table B.13   *Critical values of R in the Rayleigh test*

Percentiles of the resultant length R in samples of size n from the uniform distribution on the sphere; H_0 is rejected at significance level α for values tabulated.

n	$\alpha = 10\%$	5%	2%	1%
4	2.85	3.10	3.35	3.49
5	3.19	3.50	3.83	4.02
6	3.50	3.85	4.24	4.48
7	3.78	4.18	4.61	4.89
8	4.05	4.48	4.96	5.26
9	4.30	4.76	5.28	5.61
10	4.54	5.03	5.58	5.94
11	4.76	5.28	5.87	6.25
12	4.97	5.52	6.14	6.55
13	5.18	5.75	6.40	6.83
14	5.38	5.98	6.65	7.10
15	5.57	6.19	6.90	7.37

(*cont.*)

Table B.13   (*cont.*)

16	5.75	6.40	7.13	7.62
17	5.93	6.60	7.36	7.86
18	6.10	6.79	7.58	8.10
19	6.27	6.98	7.79	8.33
20	6.44	7.17	8.00	8.55
21	6.60	7.35	8.20	8.77
22	6.75	7.52	8.40	8.99
23	6.90	7.69	8.59	9.19
24	7.05	7.86	8.78	9.40
25	7.20	8.02	8.96	9.60

References

Abraham, R. G., van den Bergh, S., Glazebrook, K., Ellis, R. S., Santiago, B. X., Surma, P. & Griffiths, R. E., 1996, *Astrophys. J. Suppl.*, **107**, 1.

Akritas, M. J. & Siebert, J., 1996, *Mon. Not. R. Astr. Soc.*, **278**, 919.

Albert, J., 2009, *Bayesian Computation with R* (Springer).

Andrews, L. C., 1985, *Special Functions for Engineers and Applied Mathematicians* (MacMillan Publishers Ltd).

Anscombe, F. J., 1973, *Am. Stat.*, **27**, 17.

Avni, Y., 1976, *Astrophys. J.*, **210**, 642.

Avni, Y., 1978, *Astron. Astrophys.*, **66**, 307.

Avni, Y., Soltan, A., Tananbaum, H. & Zamorani, G., 1980, *Astrophys. J.*, **238**, 800.

Babu, G. J. & Feigelson, E. D., 1992, *Comm. Stat. Comp. Simul.*, **22**, 533.

Babu, G. J. & Feigelson, E. D., eds., 1997, *Statistical Challenges in Modern Astronomy II* (Springer).

Babu, G. J. & Feigelson, E. D., eds., 2007, *Statistical Challenges in Modern Astronomy IV*, ASP Conf. Ser. 371.

Bacon, D. J. & Taylor, A. N., 2003, *Mon. Not. R. Astr. Soc.*, **344**, 1307.

Ball, P., 2004, *Critical Mass: How One Thing Leads to Another* (William Heinemann).

Barcons, X., Raymont, G. B., Warwick, R. S., Fabian, A. C., Mason, K. O., McHardy, I. & Rowan-Robinson, M., 1994, *Mon. Not. R. Astr. Soc.*, **268**, 833.

Barlow, R. J., 1989, *A Guide to the Use of Statistical Methods* (John Wiley & Sons).

Barnes, C., *et al.*, 2003, *Astrophys. J. Suppl.*, **148**, 51.

Barrow, J. D., Bhavsar, S. P. & Sonada, D. H., 1985, *Mon. Not. R. Astr. Soc.*, **216**, 17.

Bartelmann, M. & Schneider, P., 2001, *Phys. Rep.*, **340**, 291.

Baugh, C. M. & Efstathiou, G., 1993, *Mon. Not. R. Astron. Soc.*, **265**, 145.

Becker, R. H., White, R. L. & Helfand, D. J., 1995, *Astrophys. J.*, **450**, 559.

Bendat, J. S. & Piersol, A. G., 1971, *Measurement and Analysis of Random Data* (John Wiley & Sons).

Bennett, C. L., *et al.*, 2003a, *Astrophys. J. Suppl.*, **148**, 97.

Bennett, C. L., *et al.*, 2003b, *Astrophys. J. Suppl.*, **148**, 1.

Benson, A. J. & Bower, R., 2010, *Mon. Not. R. Astr. Soc.*, **405**, 1573.

Berger, J. O. & Sellke, T., 1987, *J. Am. Stat. Assoc.*, **82**, 112.

Bevington, P. R. & Robinson, D. K., 2002, *Data Reduction and Error Analysis for the Physical Sciences*, 3rd Ed. (McGraw-Hill).

Bhavsar, S., 1990, in *Errors, Bias and Uncertainties in Astronomy*, ed. C. Jaschek & F. Murtagh (Cambridge University Press), 107.

Birnbaum, Z. W. & Tingey, F. H., 1951, *Ann. Math. Stat.*, **22**, 592.

Blake, C., 2002, Ph.D. Thesis, University of Oxford.

Blake, C. & Glazebrook, K., 2003, *Astrophys. J.*, **594**, 665.

Blake, C. & Wall, J., 2002a, *Mon. Not. R. Astr. Soc.*, **329**, L37.

Blake, C. & Wall, J., 2002b, *Mon. Not. R. Astr. Soc.*, **337**, 993.

Blake, C. & Wall, J., 2002c, *Nature*, **416**, 150.

Blake, C., Parkinson, D., Bassett, B., Glazebrook, K., Kunz, M. & Nichol, R. C., 2006, *Mon. Not. R. Astr. Soc.*, **365**, 255.

Blake, C., *et al.*, 2010, *Mon. Not. R. Astr. Soc.*, **406**, 803.

Blandford, R. D., Saust, A. B., Brainerd, T. G. & Villumsen, J. V., 1991, *Mon. Not. R. Astr. Soc.*, **251**, 600.

Blinnikov, S. & Moessner, R., 1998, *Astron. Astrophys. Suppl.*, **130**, 193.

Bock, D., Large, M. I. & Sadler, E. M., 1999, *Astron. J.*, **117**, 1578.

Bogart, R. S. & Wagoner, R. V., 1973, *Astrophys. J.*, **181**, 609.

Bond, J. R. & Efstathiou, G., 1984, *Astrophys. J. Lett.*, **285**, L45.

Bond, J. R. & Efstathiou, G., 1987, *Mon. Not. R. Astr. Soc.*, **226**, 655.

Bond, J. R. & the MaxiBoom collaboration, 2000, *arXiv eprint astro-ph/0011378*.

Bower, R. G., Benson, A. J., Malbon, R., Helly, J. C., Frenk, C. S., Baugh, C. M., Cole, S. & Lacey, C. G., 2006, *Mon. Not. R. Astr. Soc.*, **370**, 645.

Bracewell, R. N., 1999, *The Fourier Transform and its Applications*, 3rd Ed. (McGraw-Hill).

Braun, W. J. & Murdoch, D. J., 2007, *A First Course in Statistical Inference with R* (Cambridge University Press).

Bridle, S., *et al.*, 2009, *Ann. Appl. Stat.*, **3**, 6.

Brown, M. L., *et al.*, 2009, *Astrophys. J.*, **705**, 978.

Bruce, A., Donoho, D. & Gao, H.-Y., 1996, *IEEE Spectrum*, October, 26.

Burles, S., Nollett, K. M. & Turner, M. S., 2001, *Astrophys. J. Lett.*, **552**, L1.

Butcher, H. & Oemler, Jr., A., 1984, *Astrophys. J.*, **285**, 426.

Carlin, B. P. & Louis, T. A., 2000, *Bayes and Empirical Bayes Methods for Data Analysis* (Chapman & Hall).

Casella, G. & Berger, R. L., 2002, *Statistical Inference*, 2nd Ed. (Duxbury Press).

Chib, S. & Greenberg, E., 1995, *Am. Stat.*, **49**, 327.

Chołoniewski, J., 1987, *Mon. Not. R. Astr. Soc.*, **226**, 273.

Clowe, D., Bradač, M., Gonzalez, A. H., Markevitch, M., Randall, S. W., Jones, C. & Zaritsky, D., 2006, *Astrophys. J. Lett.*, **648**, L109.

Cochran, W. G., 1952, *Ann. Math. Stat.*, **23**, 315.

Coles, P. & Frenk, C. S., 1991, *Mon. Not. R. Astr. Soc.*, **253**, 727.

Colless, M., 2004, in *Measuring and Modeling the Universe*, ed. W. L. Freeman (Cambridge University Press), 196.

Colless, M., *et al.*, 2001, *Mon. Not. R. Astr. Soc.*, **328**, 1039.

Condon, J. J., 1974, *Astrophys. J.*, **188**, 279.

Condon, J. J., Cotton, W. D., Greisen, E. W., Yin, Q. F., Perley, R. A., Taylor, G. B. & Broderick, J. J., 1998, *Astron. J.*, **115**, 1693.

Conover, W. J., 1999, *Practical Nonparametric Statistics*, 3rd Ed. (John Wiley & Sons).

Cooley, O. W. & Tukey, J. W., 1965, *Math. Comput.*, **19**, 297.

Coppin, K., *et al.*, 2006, *Mon. Not. R. Astr. Soc.*, **372**, 1621.

Cowie, L. L., Songaila, A., Hu, E. M. & Cohen, J. G., 1996, *Astron. J.*, **112**, 839.

Cox, D. R., 2006, *Principles of Statistical Inference* (Cambridge University Press).

Cox, R. T., 1946, *Am. J. Phys.*, **14**, 1.

Crawley, M. J., 2005, *Statistics: An Introduction Using R* (John Wiley & Sons).

Cress, C. M., Helfand, D. J., Becker, R. H., Gregg, M. D. & White, R. L., 1996, *Astrophys. J.*, **473**, 7.

Croton, D. J., *et al.*, 2006, *Mon. Not. R. Astr. Soc.*, **365**, 11.

Daubechies, I., 1992, *Ten Lectures on Wavelets* (SIAM Press).

Davis, M. & Peebles, P. J. E., 1983, *Astrophys. J.*, **267**, 465.

Davis, M., Efstathiou, G., Frenk, C. S. & White, S. D. M., 1992, *Nature*, **356**, 489.

Davis, M., Huchra, J., Latham, D. W. & Tonry, J., 1982, *Astrophys. J.*, **253**, 423.

Davison, A. C. & Hinkley, D. V., 1997, *Bootstrap Methods and their Applications* (Cambridge University Press).

de Bernardis, P., *et al.*, 2000, *Nature*, **404**, 955.

De Jager, O. C., Swanepoel, J. W. H. & Raubenheimer, B. C., 1989, *Astron. Astrophys.*, **221**, 180.

Dekel, A. & West, M. J., 1985, *Astrophys. J.*, **288**, 11.

Diaconis, P. & Efron, B., 1983, *Sci. Am.*, **100**, 96.

Dicke, R. H., Peebles, P. J. E., Roll, P. G. & Wilkinson, D. T., 1965, *Astrophys. J.*, **142**, 414.

Disney, M. J., Sparks, W. B. & Wall, J. V., 1984, *Mon. Not. R. Astr. Soc.*, **206**, 899.

Dixon, R. S. & Kraus, J. D., 1968, *Astron. J.*, **73**, 381.

Drinkwater, M. J., *et al.*, 2010, *Mon. Not. R. Astr. Soc.*, **401**, 1429.

Dunkley, J., Bucher, M., Ferreira, P. G., Moodley, K. & Skordis, C., 2005, *Mon. Not. R. Astr. Soc.*, **356**, 925.

Dunkley, J., *et al.*, 2009a, *Astrophys. J. Suppl.*, **180**, 306.

Dunkley, J., *et al.*, 2009b, *Astrophys. J.*, **701**, 1804.

Dunlop, J. S. & Peacock, J., 1990, *Mon. Not. R. Astr. Soc.*, **247**, 19.

Dyson, F. W., Eddington, A. S. & Davidson, C., 1920, *Roy. Soc. Lond. Phil. Trans. Ser. A*, **220**, 291.

Eddington, A. S., 1913, *Mon. Not. R. Astr. Soc.*, **73**, 359.

Edmunds, M. G. & George, G. H., 1985, *Mon. Not. R. Astr. Soc.*, **213**, 905.

Efron, B., 1979, *Ann. Stat.*, **7**, 1.

Efron, B. & Tibshirani, R., 1986, *Stat. Sci.*, **1**, 54.

Efron, B. & Tibshirani, R. J., 1993, *An Introduction to the Bootstrap* (Chapman & Hall).

Efstathiou, G. & Bond, J. R., 1999, *Mon. Not. R. Astr. Soc.*, **304**, 75.

Efstathiou, G. & Moody, S. J., 2001, *Mon. Not. R. Astr. Soc.*, **325**, 1603.

Efstathiou, G., Sutherland, W. J. & Maddox, S. J., 1990, *Nature*, **348**, 705.

Einasto, J. & Saar, E., 1987, in *Observational Cosmology*, Proc. IAU Symp. 124 (Reidel: Dordrecht), 349.

Eisenstein, D., *et al.*, 2009, in *Astro2010: The Astronomy and Astrophysics Decadal Survey*, Science White Papers No. 70, http://adsabs.harvard.edu/abs/2009astro2010S..70E.

Eisenstein, D. J., *et al.*, 2005, *Astrophys. J.*, **633**, 560.

Elgarøy, Ø., *et al.*, 2002, *Phys. Rev. Lett.*, **89**, 061301.

Evans, M. & Swartz, T. B., 1995, *Statistical Sci.*, **10**, 254.

Evans, M. & Swartz, T. B., 1996, in *Computing Science and Statistics 27*, ed. M. M. Meyer & J. L. Rosenberger (Interface Foundation of North America, Inc.), 456.

Evrard, A. E., 1997, *Mon. Not. R. Astr. Soc.*, **292**, 289.

Feigelson, E. D. & Babu, G. J., 1992, *Astrophys. J.*, **397**, 55.

Feigelson, E. D. & Babu, G. J., eds., 1992, *Statistical Challenges in Modern Astronomy* (Springer-Verlag).

Feigelson, E. D. & Babu, G. J., eds., 2003, *Statistical Challenges in Modern Astronomy III* (Springer).

Feigelson, E. D. & Nelson, P. I., 1985, *Astrophys. J.*, **293**, 192.

Fisher, N. I., Lewis, T. & Embleton, B. J. J., 1987, *Statistical Analysis of Spherical Data* (Cambridge University Press).

Fisher, R. A., 1935, *J. Roy. Stat. Soc.*, **98**, 39.

Fisher, R. A., 1944, *Statistical Methods for Research Workers* (Oliver & Boyd).

Folkes, S., *et al.*, 1999, *Mon. Not. R. Astr. Soc.*, **308**, 459.

Francis, P. J. & Wills, B. J., 1999, in ASP Conf. Ser. 162: *Quasars and Cosmology*, ed. G. Ferland & J. Baldwin, 363.

Francis, P. J., Hewett, P. C., Foltz, C. B. & Chaffee, F. H., 1992, *Astrophys. J.*, **398**, 476.

Frank, H. & Altheon, S. C., 1994, *Statistics: Concepts and Applications* (Cambridge University Press).

Freedman, D., Pisani, R. & Purves, R., 2007, *Statistics*, 4th Ed. (W. W. Norton & Co.).

Freedman, W. L., *et al.*, 2001, *Astrophys. J.*, **553**, 47.

Fu, L., *et al.*, 2008, *Astron. Astrophys.*, **479**, 9.

Galton, F., 1889, *Natural Inheritance* (MacMillan & Co.).

Gaskill, J. D., 1978, *Linear Systems, Fourier Transforms and Optics* (John Wiley & Sons).

Gelman, A. & Rubin, D. B., 1992, *Stat. Sci.*, **7**, 457.

Gelman, A., Carlin, J. B., Stern, H. S. & Rubin, D. B., 2004, *Bayesian Data Analysis*, 2nd Ed. (Chapman & Hall).

Gigerenzer, G., 2004, *J. Socio-Econ.*, **33**, 587.

Glenn, J., *et al.*, 2010, *Mon. Not. R. Astr. Soc.*, **409**, 109.

Gold, B., *et al.*, 2009, *Astrophys. J. Suppl.*, **180**, 265.

Gold, B., *et al.*, 2011, *Astrophys. J. Suppl.*, **192**, 15.

Gonick, L. & Smith, W., 2005, *The Cartoon Guide to Statistics* (Collins Reference).

Goodman, L. A., 1954, *Psychol. Bull.*, **51**, 160.

Gordon, I. & Sorkin, S., 1959, *The Armchair Science Reader* (Simon & Schuster).

Gott, III, J. R., *et al.*, 1989, *Astrophys. J.*, **340**, 625.

Gott, III, J. R., Vogeley, M. S., Podariu, S. & Ratra, B., 2001, *Astrophys. J.*, **549**, 1.

Gregory, P. C., 2004, *Bayesian Logical Data Analysis for the Physical Sciences* (Cambridge University Press).

Gregory, P. C. & Condon, J. J., 1991, *Astrophys. J. Suppl.*, **75**, 1011.

Griffith, M. R. & Wright, A. E., 1993, *Astron. J.*, **105**, 1666.

Grimmett, G. & Stirzaker, D., 2001, *Probability and Random Processes*, 3rd Ed. (Oxford University Press).

Gull, S. F., 1989, in *Maximum Entropy and Bayesian Methods in Applied Statistics*, ed. J. Skilling (Kluwer Academic Publishers).

Gull, S. F. & Fielden, J., 1986, in *Maximum Entropy and Bayesian Methods in Applied Statistics*, ed. J. H. Justice (Cambridge University Press), 85.

Guth, A. H., 1981, *Phys. Rev. D*, **23**, 347.

Haigh, J., 1999, *Taking Chances* (Oxford University Press).

Hald, A., 1990, *A History of Probability and Statistics and their Applications before 1750* (John Wiley & Sons).

Hald, A., 1998, *A History of Mathematical Statistics from 1750 to 1930* (John Wiley & Sons).

Haller, H. & Krauss, S., 2002, *Methods Psychol. Res.*, **7**, 1.

Hamilton, A. J. S., 1993, *Astrophys. J.*, **417**, 19.

Hamming, R., 1987, *Numerical Methods for Scientists and Engineers* (Dover Publications).

Hanany, S., *et al.*, 2000, *Astrophys. J. Lett.*, **545**, L5.

Haslam, C. G. T., Klein, U., Salter, C. J., Stoffel, H., Wilson, W. E., Cleary, M. N., Cooke, D. J. & Thomasson, P., 1981, *Astron. Astrophys.*, **100**, 209.

Hastings, W. K., 1970, *Biometrika*, **57**, 97.

Hauser, M. G. & Peebles, P. J. E., 1973, *Astrophys. J.*, **185**, 757.

Hawkins, E., *et al.*, 2003, *Mon. Not. R. Astr. Soc.*, **346**, 78.

Hawking, S. W., 1988, *A Brief History of Time* (Bantam Books).

Heavens, A. F., Kitching, T. D. & Taylor, A. N., 2006, *Mon. Not. R. Astr. Soc.*, **373**, 105.

Heavens, A. F., Kitching, T. D. & Verde, L., 2007, *Mon. Not. R. Astr. Soc.*, **380**, 1029.

Hewish, A., Bell, S. J., Pilkington, J. D. H., Scott, P. F. & Collins, R. A., 1968, *Nature*, **217**, 709.

Heymans, C., *et al.*, 2006, *Mon. Not. R. Astr. Soc.*, **368**, 1323.

Hill, R. S., *et al.*, 2009, *Astrophys. J. Suppl.*, **180**, 246.

Hinshaw, G., *et al.*, 2007, *Astrophys. J. Suppl.*, **170**, 288.

Hobson, M. P. & McLachlan, C., 2003, *Mon. Not. R. Astr. Soc.*, **338**, 765.

Hobson, M. P., Bridle, S. L. & Lahav, O., 2002, *Mon. Not. R. Astr. Soc.*, **335**, 377.

Hobson, M. P., Jaffe, A. H. & Liddle, A. R., 2010, *Bayesian Methods in Cosmology* (Cambridge University Press).

Hogg, R. V. & Tanis, E. A., 2009, *Probability and Statistical Inference*, 8th Ed. (Prentice Hall).

Horne, J. H. & Baliunas, S. L., 1986, *Astrophys. J.*, **302**, 757.

Hoyle, F., 1958, *The Black Cloud* (Heinemann).

Hu, W. & Dodelson, S., 2002, *Ann. Rev. Astron. Astrophys.*, **40**, 171.

Hubble, E., 1936, *The Realm of the Nebulae* (Yale University Press).

Huff, D., 1973, *How to Lie with Statistics* (Penguin).

Hughes, D. H., *et al.*, 1998, *Nature*, **394**, 241.

Huterer, D., Knox, L. & Nichol, R., 2001, *Astrophys. J.*, **555**, 547.

Isobe, T., Feigelson, E. D. & Nelson, P. I., 1986, *Astrophys. J.*, **306**, 490.

Isobe, T., Feigelson, E. D., Akritas, M. J. & Babu, G. J., 1990, *Astrophys. J.*, **364**, 104.

James, J. F., 1995, *A Student's Guide to Fourier Transforms* (Cambridge University Press).

Jarosik, N., *et al.*, 2007, *Astrophys. J. Suppl.*, **170**, 263.

Jarvis, M. & Jain, B., 2004, *arXiv:astro-ph/0412234.*

Jaschek, C. & Murtagh, F., 1990, *Errors, Bias and Uncertainties in Astronomy* (Cambridge University Press).

Jauncey, D. L., 1967, *Nature*, **216**, 877.

Jaynes, E. T., 1968, *IEEE Trans. System Sci. Cybern.*, **4**, 227.

Jaynes, E. T., 1976, in *Foundations of Probability Theory, Statistical Inference, and Statistical Theories of Science*, ed. W. L. Harper & C. A. Hooker (Reidel: Dordrecht), 175.

Jaynes, E. T., 1983, *Papers on Probability Theory, Statistics and Statistical Physics*, ed. R. D. Rosenkrantz (ESO Garching).

Jaynes, E. T., 1986, in *Maximum Entropy and Bayesian Methods in Applied Statistics*, ed. J. H. Justice (Cambridge University Press), 1.

Jaynes, E. T., 2003, *Probability Theory: The Logic of Science* (Cambridge University Press).

Jeffreys, H., 1961, *Theory of Probability* (Clarendon Press).

Jenkins, C. R., 1987, *Mon. Not. R. Astr. Soc.*, **226**, 341.

Jenkins, C. R., 1989, *Observatory*, **109**, 69.

Jenkins, C. R. & Reid, I. N., 1991, *Astron. J.*, **101**, 1595.

Jensen, J. B., Tonry, J. L. & Luppino, G. A., 1998, *Astrophys. J.*, **505**, 111.

Joliffe, I. T., 2002, *Principal Component Analysis*, 2nd Ed. (Springer-Verlag).

Kaiser, N., 1984, *Astrophys. J. Lett.*, **284**, L9.

Kaiser, N., 1987, *Mon. Not. R. Astr. Soc.*, **227**, 1.

Kalbfleish, J. D. & Prentice, R. L., 2002, *The Statistical Analysis of Failure Time Data*, 2nd Ed. (John Wiley & Sons).

Kass, R. & Raftery, A., 1995, *J. Am. Stat. Soc.*, **90**, 773.

Kendall, M. G., 1980, *Multivariate Analysis*, 2nd Ed. (Charles Griffin & Co.).

Kennicut, R., 1992, *Astrophys. J. Suppl.*, **79**, 255.

Knight, F. H., 1921, *Risk, Uncertainty and Profit* (Houghton Mifflin).

Kogut, A., *et al.*, 2003, *Astrophys. J. Suppl.*, **148**, 161.

Kogut, A., *et al.*, 2007, *Astrophys. J.*, **665**, 355.

Kolb, E. W. & Turner, M. S., 1990, *Frontiers Phy.*, 69.

Komatsu, E., *et al.*, 2003, *Astrophys. J. Suppl.*, **148**, 119.

Komatsu, E., *et al.*, 2009, *Astrophys. J. Suppl.*, **180**, 330.

Komatsu, E., *et al.*, 2011, *Astrophys. J. Suppl.*, **192**, 18.

Kooiman, B. L., Burns, J. O. & Klypin, A. A., 1995, *Astrophys. J.*, **448**, 500.

Koornwinder, T. H., 1993, *Wavelets: An Elementary Treatment in Theory and Applications* (World Scientific).

Laing, R. A., Jenkins, C. R., Wall, J. V. & Unger, S. W., 1994, in ASP Conf. Ser. 54: *The First Stromlo Symposium: The Physics of Active Galaxies*, ed. G. V. Bicknell & M. A. Dopita, 201.

Landy, S. D. & Szalay, A. S., 1993, *Astrophys. J.*, **412**, 64.

Lange, A. E., *et al.*, 2001, *Phys. Rev. D.*, **63**, 042001.

Larson, D., *et al.*, 2011, *Astrophys. J. Suppl.*, **192**, 16.

Lee, P. M., 2004, *Bayesian Statistics: An Introduction*, 3rd Ed. (John Wiley & Sons).

LePage, R. & Billiard, L., 1993, *Exploring the Limits of the Bootstrap* (John Wiley & Sons).

Lewis, A., Challinor, A. & Lasenby, A., 2000, *Astrophys. J.*, **538**, 473.

Liddle, A. R., Mukherjee, P. & Parkinson, D., 2006, *Astron. & Geophys.*, **47**, 4, 30.

Lilly, S. J., Le Fevre, O., Hammer, F. & Crampton, D., 1996, *Astrophys. J. Lett.*, **460**, L1.

Limber, D. N., 1953, *Astrophys. J.*, **117**, 134.

Limber, D. N., 1954, *Astrophys. J.*, **119**, 655.

Linnik, Y. V., 1961, *Methods of Least-squares and Principles of the Theory of Observations* (Pergamon Press).

Loan, A. J., Wall, J. V. & Lahav, O., 1997, *Mon. Not. R. Astr. Soc.*, **286**, 994.

Lomb, N. R., 1976, *Astrophys. Space Sci.*, **39**, 447.

Lynden-Bell, D., 1971, *Mon. Not. R. Astr. Soc.*, **155**, 95.

Lynds, C. R. & Petrosian, V., 1988a, in IAU Symposium, Vol. 130, *Large Scale Structures of the Universe*, ed. J. Audouze, M.-C. Pelletan & S. Szalay, 467.

Lynds, C. R. & Petrosian, V., 1988b, in *Bull. Am. Astron. Soc.*, **20**, 644.

Lyons, L., 1986, *Statistics for Nuclear and Particle Physicists* (Cambridge University Press).

Lyons, L., 1991, *A Practical Guide to Data Analysis for Physical Science Students* (Cambridge University Press).

Lyons, R. G., 1997, *Understanding Digital Signal Processing* (Addison-Wesley Publishing Co.).

Ma, Z., Hu, W. & Huterer, D., 2006, *Astrophys. J.*, **636**, 21.

MacKay, D. J. C., 2003, *Information Theory, Inference and Learning Algorithms* (Cambridge University Press).

Macklin, J. T., 1982, *Mon. Not. R. Astr. Soc.*, **199**, 1119.

Madau, P., Ferguson, H. C., Dickinson, M. E., Giavalisco, M., Steidel, C. C. & Fruchter, A., 1996, *Mon. Not. R. Astr. Soc.*, **283**, 1388.

Maddox, S. J., Efstathiou, G., Sutherland, W. J. & Loveday, J., 1990, *Mon. Not. R. Astr. Soc.*, **242**, 43P.

Madgwick, D. S., *et al.*, 2002, *Mon. Not. R. Astr. Soc.*, **333**, 133.

Magliocchetti, M., Maddox, S. J., Lahav, O. & Wall, J. V., 1998, *Mon. Not. R. Astr. Soc.*, **300**, 257.

Magliocchetti, M., Maddox, S. J., Lahav, O. & Wall, J. V., 1999, *Mon. Not. R. Astr. Soc.*, **306**, 943.

Manly, B. J. F., 1994, *Multivariate Statistical Methods: A Primer*, 2nd Ed. (Chapman & Hall).

Marshall, H. L., Avni, Y., Tananbaum, H. & Zamorani, G., 1983, *Astrophys. J.*, **269**, 35.

Martin, B. R., 1971, *Statistics for Physicists* (Academic Press).

Martínez, V. J., Jones, B. J. T., Dominguez-Tenreiro, R. & van de Weygaert, R., 1990, *Astrophys. J.*, **357**, 50.

Masson, C. R. & Wall, J. V., 1977, *Mon. Not. R. Astr. Soc.*, **180**, 193.

Mather, J. C., *et al.*, 1990, *Astrophys. J. Lett.*, **354**, L37.

Mauch, T., Murphy, T., Buttery, H. J., Curran, J., Hunstead, R. W., Piestrzynski, B., Robertson, J. G. & Sadler, E. M., 2003, *Mon. Not. R. Astr. Soc.*, **342**, 1117.

Maxted, P. F. L., Hill, G. & Hilditch, R. W., 1994, *Astron. Astrophys.*, **282**, 821.

Metropolis, N., Rosenbluth, A. W., Rosenbluth, M. N., Teller, A. H. & Teller, E., 1953, *J. Chem. Phys.*, **21**, 1087.

Mihalas, D. & Binney, J., 1981, *Galactic Astronomy: Structure and Kinematics*, 2nd Ed. (W. H. Freeman & Co.).

Miralda-Escude, J., 1991a, *Astrophys. J.*, **370**, 1.

Miralda-Escude, J., 1991b, *Astrophys. J.*, **380**, 1.

Mittaz, J. P. D., Penston, M. V. & Snijders, M. A. J., 1990, *Mon. Not. R. Astr. Soc.*, **242**, 370.

Montgomery, D. C. & Peck, E. A., 1992, *Introduction to Linear Regression Analysis*, 2nd Ed. (John Wiley & Sons).

Mood, A. M., Graybill, F. A. & Boes, D. B., 1974, *Introduction to the Theory of Statistics*, 3rd Ed. (McGraw-Hill).

Moroney, M. J., 1965, *Facts from Figures* (Penguin).

Munshi, D., Valageas, P., van Waerbeke, L. & Heavens, A., 2008, *Phys. Rep.*, **462**, 67.

Murray, C. A., 1983, *Vectorial Astrometry* (Adam Hilger).

Netterfield, C. B., *et al.*, 2002, *Astrophys. J.*, **571**, 604.

Newman, W. I., Haynes, M. P. & Terzian, Y., 1992, in *Statistical Challenges in Modern Astronomy*, eds. E. D. Feigelson & G. J. Babu (Springer-Verlag), 137.

Neyman, J., Scott, E. L. & Shane, C. D., 1953, *Astrophys. J.*, **117**, 92.

Norberg, P., *et al.*, 2002, *Mon. Not. R. Astr. Soc.*, **332**, 827.

Oakes, M., 1986, *Statistical Inference: A Commentary for the Social and Behavioural Sciences* (John Wiley & Sons).

O'Hagan, A., 1994, *Kendall's Advanced Theory of Statistics, Vol. 2b: Bayesian Inference* (Edward Arnold).

O'Ruanaidh, J. J. K. & Fitzgerald, W. J., 1996, *Numerical Bayesian Methods Applied to Signal Processing* (Springer Verlag).

Padmanabhan, N., *et al.*, 2007, *Mon. Not. R. Astr. Soc.*, **378**, 852.

Padmanabhan, N., *et al.*, 2010, in *Bull. Am. Astron. Soc.*, **41**, 518.

Page, L., *et al.*, 2003, *Astrophys. J. Suppl.*, **148**, 39.

Page, L., *et al.*, 2007, *Astrophys. J. Suppl.*, **170**, 335.

Papoulis, A. & Unnikrishna Pillai, S., 2002, *Probability, Random Variables and Stochastic Processes*, 4th Ed. (McGraw-Hill).

Partridge, R. B., 1995, *3K: The Cosmic Microwave Background Radiation* (Cambridge University Press).

Peacock, J. A., 1983, *Mon. Not. R. Astr. Soc.*, **202**, 615.

Peacock, J. A., 1985, *Mon. Not. R. Astr. Soc.*, **217**, 601.

Peacock, J. A., 1999a, *Cosmological Physics* (Cambridge University Press).

Peacock, J. A., 1999b, in *The Most Distant Radio Galaxies*, ed. H. J. A. Röttgering, P. N. Best & M. D. Lehnert (Royal Netherlands Academy of Arts and Sciences), 377.

Peacock, J. A., 2003, in AIP Conf. Ser. 666: *The Emergence of Cosmic Structure*, ed. S. H. Holt & C. S. Reynolds, 275.

Peacock, J. A., *et al.*, 2001, *Nature*, **410**, 169.

Pearson, K., 1900, *Phil. Mag. Series 5*, **50**, 157.

Peebles, P. J. E., 1973, *Astrophys. J.*, **185**, 413.

Peebles, P. J. E., 1980, *The Large-Scale Structure of the Universe* (Princeton University Press).

Peebles, P. J. E. & Groth, E. J., 1975, *Astrophys. J.*, **196**, 1.

Peebles, P. J. E. & Hauser, M. G., 1974, *Astrophys. J. Suppl.*, **28**, 19.

Peebles, P. J. E. & Yu, J. T., 1970, *Astrophys. J.*, **162**, 815.

Peebles, P. J. E., Page., L. A. & Partridge, R. B., 2009, *Finding the Big Bang* (Cambridge University Press).

Penzias, A. A. & Wilson, R. W., 1965, *Astrophys. J.*, **142**, 419.

Percival, W. J., *et al.*, 2001, *Mon. Not. R. Astr. Soc.*, **327**, 1297.

Percival, W. J., *et al.*, 2007, *Astrophys. J.*, **657**, 645.

Percival, W. J., *et al.*, 2010, *Mon. Not. R. Astr. Soc.*, **401**, 2148.

Perlmutter, S., *et al.*, 1999, *Astrophys. J.*, **517**, 565.

Phillips, M. M., 1993, *Astrophys. J. Lett.*, **413**, L105.

Phillips, M. M., Jenkins, C. R., Dopita, M. A., Sadler, E. M. & Binette, L., 1986, *Astron. J.*, **91**, 1062.

Press, W. H., 1978, *Comments Astrophys.*, **7**, 103.

Press, W. H., 1997, in *Unsolved Problems in Astrophysics*, ed. J. N. Bahcall & J. P. Ostriker (Princeton University Press), 49.

Press, W. H., Teukolsky, S. A., Vetterling, W. T. & Flannery, B. P., 2007, *Numerical Recipes: The Art of Scientific Computing*, 3rd Ed. (Cambridge University Press).

Pryke, C., Halverson, N. W., Leitch, E. M., Kovac, J., Carlstrom, J. E., Holzapfel, W. L. & Dragovan, M., 2002, *Astrophys. J.*, **568**, 46.

Refregier, A., 2003, *Mon. Not. R. Astr. Soc.*, **338**, 35.

Refregier, A. & Bacon, D., 2003, *Mon. Not. R. Astr. Soc.*, **338**, 48.

Reichard, C. L., *et al.*, 2009, *Astrophys. J.*, **694**, 1200.

Reid, B. A., *et al.*, 2010, *Mon. Not. R. Astr. Soc.*, **404**, 60.

Reinking, J. T., 2002, *Mathematica J.*, **8**, 473.

Rice, J. A., 2006, *Mathematical Statistics and Data Analysis*, 3rd Ed. (Duxbury Press).

Riess, A. G., *et al.*, 1998, *Astron. J.*, **116**, 1009.

Roche, N., Ratnatunga, K., Griffiths, R. E., Im, M. & Naim, A., 1998, *Mon. Not. R. Astr. Soc.*, **293**, 157.

Rolfe, E., 1983, *Statistical Methods in Astronomy* (ESA Scientific and Technical Publications).

Rosenthal, J. S., 2006, *Struck by Lightning: The Curious World of Probabilities* (Joseph Henry Press).

Rowan-Robinson, M., 1968, *Mon. Not. R. Astr. Soc.*, **138**, 445.

Rowntree, D., 1981, *Statistics Without Tears* (Penguin).

Rubin, V. C., Ford, W. K. J. & Thonnard, N., 1980, *Astrophys. J.*, **238**, 471.

Sachs, R. K. & Wolfe, A. M., 1967, *Astrophys. J.*, **147**, 73.

Sadler, E. M., Jenkins, C. R. & Kotanyi, C. G., 1989, *Mon. Not. R. Astr. Soc.*, **240**, 591.

Saha, P., 1995, *Astron. J.*, **110**, 916.

Salsburg, D., 2002, *The Lady Tasting Tea: How Statistics Revolutionized Science in the Twentieth Century* (Holt Paperbacks).

Sandage, A., 1972, *Astrophys. J.*, **178**, 25.

Sargent, W. L. W., Schechter, P. L., Boksenberg, A. & Shortridge, K., 1977, *Astrophys. J.*, **212**, 326.

Scargle, J. D., 1982, *Astrophys. J.*, **263**, 835.

Scheuer, P. A. G., 1957, *Proc. Cambridge Phil. Soc.*, **53**, 764.

Scheuer, P. A. G., 1974, *Mon. Not. R. Astr. Soc.*, **166**, 329.

Scheuer, P. A. G., 1991, in *Modern Cosmology in Retrospect*, ed. B. Bertotti & *et al.* (Cambridge University Press), 331.

Schlegel, D., White, M. & Eisenstein, D., 2009, *ArXiV e-print 0902.4680*.

Schmidt, M., 1968, *Astrophys. J.*, **151**, 393.

Schmitt, J. H. M. M., 1985, *Astrophys. J.*, **293**, 178.

Schneider, D. P., *et al.*, 2005, *Astron. J.*, **130**, 367.

Schneider, P. & Er, X., 2008, *Astron. Astrophys.*, **485**, 363.

Schrabback, T., *et al.*, 2010, *Astron. Astrophys.*, **516**, A63.

Sellke, T., Bayarri, M. J. & Berger, J. O., 2001, *Am. Stat.*, **55**, 62.

Senn, S., 2003, *Dicing with Death: Chance, Risk and Health* (Cambridge University Press).

Seo, H. & Eisenstein, D. J., 2007, *Astrophys. J.*, **665**, 14.

Shane, C. D. & Wirtanen, C. A., 1954, *Astron. J.*, **59**, 285.

Shanks, T., Stevenson, P. R. F., Fong, R. & MacGillivray, H. T., 1984, *Mon. Not. R. Astr. Soc.*, **206**, 767.

Shannon, C. E., 1949, *Proc. IRE*, **37**, 10.

Siegel, S. & Castellan, N. J., 1988, *Nonparametric Statistics for the Behavioural Sciences* (McGraw-Hill).

Silk, J., 1968, *Astrophys. J.*, **151**, 459.

Sivia, D. S., 2006, *Data Analysis: A Bayesian Tutorial*, 2nd Ed. (Oxford University Press).

Sivia, D. S. & Carlile, C. J., 1992, *J. Chem. Phys.*, **96**, 170.

Skilling, J., 2007, in *Bayesian Statistics* 8. Proceedings of the 8th Valencia International Meeting, 2–6 June 2006, ed. J. M. Bernardo *et al.* (Oxford University Press), 491.

Smoot, G. F., Gorenstein, M. V. & Muller, R. A., 1977, *Phys. Rev. Lett.*, **39**, 898.

Smoot, G. F., *et al.*, 1992, *Astrophys. J. Lett.*, **396**, L1.

Soucail, G., Fort, B., Mellier, Y. & Picat, J. P., 1987a, *Astron. Astrophys.*, **172**, L14.

Soucail, G., Mellier, Y., Fort, B., Mathez, G. & Hammer, F., 1987b, *Astron. Astrophys.*, **184**, L7.

Spergel, D. N. & Zaldarriaga, M., 1997, *Phys. Rev. Lett.*, **79**, 2180.

Spergel, D. N., *et al.*, 2007, *Astrophys. J. Suppl.*, **170**, 377.

Sprent, P., 1977, *Statistics in Action* (Penguin).

Springel, V., *et al.*, 2005, *Nature*, **435**, 629.

Strang, G., 1994, *Am. Scientist*, **82**, 250.

Stuart, A. & Ord, J. K., 1991, *Kendall's Advanced Theory of Statistics, Volume 2a: Classical Inference and Relationship*, 5th Ed. (Edward Arnold).

Stuart, A. & Ord, J. K., 1994, *Kendall's Advanced Theory of Statistics, Volume 1: Distribution Theory*, 6th Ed. (Edward Arnold).

Sunyaev, R. A. & Zeldovich, Y. B., 1970, *Astrophys. Space Sci.*, **7**, 3.

Suzuki, N., *et al.*, 2011, *ArXiv e-print 1105.3470*.

Szapudi, I., 1998, *Astrophys. J.*, **497**, 16.

Taleb N. N., 2010, *The Black Swan: The Impact of the Highly Improbable* (Random House).

Taylor, A. N., 2001, *ArXiv:astro-ph/0111605*.

Tegmark, M., Taylor, A. N. & Heavens, A. F., 1997, *Astrophys. J.*, **480**, 22.

Tegmark, M., *et al.*, 2002, *Astrophys. J.*, **571**, 191.

Tegmark, M., *et al.*, 2006, *Phys. Rev. D*, **74**, 123507.

Thompson, A. R., Moran, J. M. & Swenson, Jr., G. W., 2001, *Interferometry and Synthesis in Radio Astronomy* (John Wiley & Sons).

Tonry, J. & Davis, M., 1979, *Astron. J.*, **84**, 1511.

Treyer, M. A. & Lahav, O., 1996, *Mon. Not. R. Astr. Soc.*, **280**, 469.

Tufte, E. R., 1990, *Envisaging Information* (Graphics Press).

Tufte, E. R., 1997, *Visual Explanations: Images and Quantities, Evidence and Narrative* (Graphics Press).

Tufte, E. R., 2001, *The Visual Display of Quantitative Information*, 2nd Ed. (Graphics Press).

Turler, M. & Courvoisier, T. J.-L., 1998, *Astron. Astrophys.*, **329**, 863.

Turok, N., 1996a, *Phys. Rev. Lett.*, **77**, 4138.

Turok, N., 1996b, *Phys. Rev. D*, **54**, 3686.

Tyson, J. A., Wenk, R. A. & Valdes, F., 1990, *Astrophys. J. Lett.*, **349**, L1.

Utts, J. M., 2004, *Seeing through Statistics*, 3rd Ed. (Duxbury Press).

Venables, W. N. & Ripley, B. D., 2002, *Modern Applied Statistics with S*, 4th Ed. (Springer).

Venables, W. N. & Smith, D. M., 2002, *An Introduction to R: Revised and Updated* (Network Theory Ltd., UK).

Verde, L., *et al.*, 2003, *Astrophys. J. Suppl.*, **148**, 195.

Vogeley, M. S. & Szalay, A. S., 1996, *Astrophys. J.*, **465**, 34.

Wagoner, R. V., 1967, *Nature*, **214**, 766.

Walker, J. S., 1999, *A Primer on Wavelets and their Scientific Applications* (Chapman & Hall).

Wall, J. V., 1997, *Astron. Astrophys.*, **122**, 371.

Wall, J. V. & Cooke, D. J., 1975, *Mon. Not. R. Astr. Soc.*, **171**, 9.

Wall, J. V., Pearson, T. J. & Longair, M. S., 1980, *Mon. Not. R. Astr. Soc.*, **193**, 683.

Wall, J. V., Scheuer, P. A. G., Pauliny-Toth, I. I. K. & Witzel, A., 1982, *Mon. Not. R. Astr. Soc.*, **198**, 221.

Wall, J. V., Rixon, G. T. & Benn, C. R., 1993, in ASP Conf. Ser. 51: *Observational Cosmology*, ed. G. L. Chincarini, A. Iovino, T. Maccacaro & D. Maccagni, 576.

Wall, J. V., Jackson, C. A., Shaver, P. A., Hook, I. M. & Kellermann, K. I., 2005, *Astron. Astrophys.*, **434**, 133.

Wall, J. V., Pope, A. & Scott, D., 2008, *Mon. Not. R. Astr. Soc.*, **383**, 435.

Walsh, D., Carswell, R. F. & Weymann, R. J., 1979, *Nature*, **279**, 381.

Webb, J. K., Barcons, X., Carswell, R. F. & Parnell, H. C., 1992, *Mon. Not. R. Astr. Soc.*, **255**, 319.

Webster, A. S., 1976a, *Mon. Not. R. Astr. Soc.*, **175**, 61.

Webster, A. S., 1976b, *Mon. Not. R. Astr. Soc.*, **175**, 71.

Weiland, J. L., *et al.*, 2011, *Astrophys. J. Suppl.*, **192**, 19.

Weinberg, S. L. & Abramowitz, S. K., 2008, *Statistics Using SPSS*, 2nd Ed. (Cambridge University Press).

Weiss, N. A., 2007, *Introductory Statistics*, 8th Ed. (Addison-Wesley).

Wilkinson, J. H., 1978, in *Numerical Software: Needs and Availability*, ed. D. A. H. Jacobs (Academic Press).

Williams, D., 2001, *Weighing the Odds* (Cambridge University Press).

Williams, E. J., 1959, *Regression Analysis* (John Wiley & Sons).

Williams, R. E., *et al.*, 1996, *Astron. J.*, **112**, 1335.

Willmer, C. N. A., 1997, *Astron. J.*, **114**, 898.

Wills, B. J., *et al.*, 1997, in ASP Conf. Ser. 113: *Emission Lines in Active Galaxies: New Methods and Techniques*, Proc. IAU Colloq. 159, 104.

Windhorst, R. A., Fomalont, E. B., Partridge, R. B. & Lowenthal, J. D., 1993, *Astrophys. J.*, **405**, 498.

Wittaker, E. T., 1915, *Proc. Roy. Soc. Edinburgh A.*, 35.

Wood, M., 2003, *Making Sense of Statistics: A Non-mathematical Approach* (Palgrave/Macmillan).

York, D. G., *et al.*, 2000, *Astron. J.*, **120**, 1579.

Yu, J. T. & Peebles, P. J. E., 1969, *Astrophys. J.*, **158**, 103.

Yüksel, H., Kistler, M. D., Beacom, J. F. & Hopkins, A. M., 2008, *Astrophys. J. Lett.*, **683**, L5.

Zeldovich, Y. B., Einasto, J. & Shandarin, S. F., 1982, *Nature*, **300**, 407.

Zellner, A., 1987, *An Introduction to Bayesian Inference in Econometrics* (John Wiley & Sons).

Zwicky, F., 1937, *Astrophys. J.*, **86**, 217.

Zwicky, F., 1937a, *Phys. Rev.*, **51**, 290.

Zwicky, F., 1937b, *Phys. Rev.*, **51**, 679.

Index

Printed in the United States
By Bookmasters